Wildlife Viewing

T0338158

Wildlife Viewing

A Management Handbook

edited by
Michael J. Manfredo

Oregon State University Press
Corvallis

The paper in this book meets the guidelines for permanence and durability of the Committee on Production Guidelines for Book Longevity of the Council on Library Resources and the minimum requirements of the American National Standard for Permanence of Paper for Printed Library Materials Z39.48-1984.

Library of Congress Cataloging-in-Publication Data
Manfredo, Michael J.
  Wildlife viewing : a management handbook / edited by Michael J. Manfredo.— 1st ed.
      p. cm.
Includes bibliographical references (p.   ).
  ISBN 0-87071-548-8 (alk. paper)
  1. Wildlife viewing sites—North America—Management. I. Manfredo, Michael J. II. Title.
  QL60 .M36 2002
  590'.0723—dc21

                                        2002006486
                                        Rev.

Oregon State University Press
101 Waldo Hall Corvallis
OR 97331-6407
541-737-3166
fax 541-737-3170
http://oregonstate.edu/dept/press

OREGON STATE
UNIVERSITY

Contents

Wildlife Viewing

A Management Handbook

Planning and Managing for Wildlife-viewing Recreation: An Introduction

Michael J. Manfredo

Purpose

EARLY IN THE FALL OF 1999, my wife, Brenda, and I wandered the wooded mountainside of the Southern Alps for hours in pursuit of a glimpse of the native wildlife. The guide book we carried (Jepsen 1994) proclaimed the Eastern Dolomites "one of the most surreal landscapes on the planet" (p. 38) and nothing I experienced would lead me to disagree. We were exploring the Italian National Park Fanes-Sennes-Braies, an area comprising close-packed massifs with jutting saw-toothed ridges, steep chasms, tumbling mountain streams, crystal-clear water, and high meadows -sprinkled with blooming wildflowers. The guide book indicated that, although hunting is permitted in this area, views of deer, chamois, and hare are possible.

Just one week prior to this visit, I had traveled through Yellowstone National Park. Like all travelers to Yellowstone, I was treated to an unparalleled wildlife spectacle—elk, moose, deer, and buffalo, readily available to the casual viewer. By the end of my trip through Yellowstone, I endured yet another traffic slow up—this one created by a small herd of elk lounging by the roadside, totally unaffected by the proximity of visitors. A crowd of onlookers gathered slowly to observe the animals. Children bounded from parked cars talking excitedly, amateur photographers fumbled anxiously to assemble their equipment, curious observers inched closer, and a park official patiently observed the event while directing traffic.

I thought of that experience as we were exiting this beautiful area in the Alps. Our trek had resulted in a dearth of wildlife sightings with a corroborating lack of sign. Allowing for the host of possibilities that can lead to such a day, I must confess that the experience had somewhat changed my initial impression of this spectacular place. Though it was as picturesque as any place on earth, I now viewed it as more lifeless and tame.

Like a painting from the Romantic period, it made me wonder, "What was it like in earlier times?" But more importantly, the contrast of these two trips sharpened my awareness of the effects that wildlife has on the human experience.

1

A spectacular view of the Eastern Dolomites in Italy. (Photo by Mike Manfredo)

Humans are fascinated with wildlife because it is linked to their most basic needs. For more than 99 percent of human existence, we were organized in hunter and gatherer bands (Lee and Devore 1968). During that time, human survival was dependent on an ability to understand, pursue, and harvest wildlife. Our physical and psychological make-up evolved as we adapted to the challenges and opportunities posed by wildlife. Certainly, in today's post-industrialized society, we do not rely upon wildlife as we once did. But while not tied to our basic protection and survival needs, wildlife is undoubtedly linked to higher-order needs. Our association with wildlife is, for example, believed to serve our needs for nurturing and caring, learning, competition and challenge, and affiliation and kinship, as well as our spiritual needs. These needs all give direction to our behavior toward wildlife.

I recently heard a radio advertisement for a popular outdoor magazine that put the wildlife experience into an interesting perspective. Referring to the many recreationists who travel in pursuit of wildlife, the announcer proclaimed, "They say they come to find deer, turkey, elk, and even bear. But we know why they come. They come to find themselves." This offers an important implication for wildlife professionals: "We focus most of our attention on enhancing wildlife populations and their habitat, developing facilities and sites, and providing educational brochures. But the reason we do that is to provide unique human experiences and to extend benefits to people, communities, and ecosystems." The purpose of this book is, quite simply, to urge managers to attain and implement that perspective.

Philosophical Orientations Toward Recreation Management

In a recent planning meeting, a high-ranking state wildlife official asked, "What is there to giving the public wildlife viewing? The wildlife are out there; let them find them and look at them." One's reaction to this approach to managing wildlife-viewing recreation depends on his or her philosophical orientation, which in turn is influenced by the individual's disciplinary tradition. Recreation management is based on a service-oriented philosophy that differs from the traditional protectionist view of the wildlife profession. Integration of these philosophical orientations (service and protection) is the first step toward developing an approach for wildlife-viewing management that considers both resource capabilities and the experiences desired by recreationists.

A Protection Philosophy of Wildlife

The wildlife profession has emphasized a resource-protection philosophy in recreation management since the early 1900s. As stated by Alexander (1962), "The goal of those of us in the wildlife field is, fundamentally, to save and perpetuate many things for many people." This philosophy can be traced to the beginning of the twentieth century when the human exploitation of wildlife threatened the very existence of many species in the United States. One of the most significant threats to wildlife arose through uncontrolled hunting and fishing. As a consequence, a significant effort of wildlife management was directed toward regulation of these pursuits. Over the twentieth century, the public's motivation for hunting and angling shifted from subsistence to recreation, yet the protection emphasis toward these activities remained unchanged. The wildlife tradition often regards recreation as a remnant of the very threat that formed the impetus for the profession.

A few of the guiding principles of recreation management that have emerged from the wildlife profession can be summarized in the following way.

• *It is important to ensure healthy wildlife populations are present for recreational pursuits such as viewing, hunting, and fishing.* Relative to viewing, this is the essence of the previously mentioned official's philosophy: "If wildlife are present, the rest will take care of itself." In the case of hunting, the goal is to ensure an adequate annual surplus is available for harvest.

• *Recreational hunting is a "tool" to be used in controlling wildlife populations.* If hunting is a tool, decisions about how to provide hunting recreation (e.g., season timing and length, number of hunting licenses, gender harvested) are driven primarily by concerns of administrative efficiency. More specifically, decisions are sought that will allow for the harvest of surplus animals with minimal administrative and enforcement costs, produce the least resistance from stakeholders, and have a positive effect on license revenues.

• *In ironic contrast, wildlife-viewing recreation is driven by an impact-intolerance rule.* That is, wildlife viewing is deemed acceptable only when it does not interfere appreciably with the "natural" conditions of wildlife. The irony is curious because the rule has certainly not been applied to hunting. Wildlife management in North America has focused on creating high, perhaps even *abnormally* high, populations of game species for the purposes of recreational hunting. These populations, in turn, have had a dramatic effect on the ecosystems of which they are a part. It is unlikely that the resultant species abundance, composition, and distribution could be considered "natural."

• *Regulatory mechanisms are used primarily to restrain people from adversely impacting wildlife with little concern for enhancing their experience.* Recreation management is seen principally as controlling the threat from recreational impact.

• *To "educate" people that there are "right" and "wrong" ways to recreate.* Mass recreation is viewed as inferior and destructive compared to ideal forms of recreation, which are high-solitude and low-impact recreation. This view is forged in the romantic lore of the wildlife-management profession. As Leopold (1938) declared in an essay critical of the "masses" participating in outdoor recreation, "Recreational development is a job, not of building roads into lovely country, but of building receptivity into the still unlovely human mind."

A Service Philosophy of Recreation Management

In contrast to the wildlife profession's protection philosophy, the recreation profession is built largely on the concept of providing service to people. Management direction is guided by serving public interests. The principles of this service philosophy can be summarized as follows:

• *To provide a range of quality recreation opportunities that meet the diverse demands of the public.* Two important but distinct concepts are introduced here: (1) the *types* of recreation opportunities desired by a person, and (2) the recreation *quality* that a person experiences. *Type* of recreation refers to the characteristics of the engagement that people seek and, clearly, not all people are seeking the same type of engagement. For example, some might seek a wildlife-viewing experience that allows them to see animals from an automobile while others wish to hike far from the road. Recreation type can be characterized by the motivations of the visitor (i.e., the desired experience), the resource, social, and managerial setting in which the experience occurs, and the activity that facilitates the experience.

Quality refers to how people evaluate the opportunities they experience. For example, did the types of recreation activities encountered detract from or enhance their visit to a particular site? Did the visitors feel crowded by the number of other recreationists? Did the behavior of other people limit their chances of seeing or photographing wildlife? Overall, how satisfied were they with the experience?

Wildlife managers can directly influence both recreation opportunities available and their quality through their policies and actions. For example, allowing snowmobiles into an area may provide winter access for recreationists who physically are not able to reach areas where wildlife can be found. Such a policy, however, may limit opportunities for a quiet and peaceful wildlife-viewing experience. Similarly, actions that encourage viewing at wayside attractions (e.g., paving the pullout) may increase the number of people who use the site, and decrease opportunities for solitude.

It is important that professionals not judge different *types* of desired visitor opportunities as having inherently higher or lower *quality* than other experiences. For example, experienced birders may regard the opportunity for a solitude experience, in a remote location with no development, as highly desirable and refer to it as "high quality." Conversely, these same individuals may regard opportunities designed to accommodate large numbers of recreationists at visitor centers as undesirable and refer to that experience as "low quality." However, depending on the desired type of experience, *either of these might be considered high-quality recreation opportunities.* There are low-quality experiences at visitor centers and high-quality experiences at visitor centers, but the visitor-center experience is not inherently of lower quality than the remote-viewing experience. If both types of experience are in demand by the public, and are appropriate, safe, and within resource-impact tolerances, management should attempt to provide for both.

• *Provision of wildlife is not an end in itself, but a means to an end.* The ends or products of wildlife-associated recreation management are the benefits provided to people. The ultimate test of a natural-resource policy depends on how it benefits people (Wagar 1966). This is *not* to suggest that the only benefits to consider are recreational, since "non-use" management benefits (e.g., protecting an endangered species) may outweigh recreational benefits. Nor does this imply that mass recreation should be favored over low-density recreation. The value (or benefits) of a single low-density experience may be many times greater than the high-density experience. But management decisions should be accompanied by an analysis of the benefits that are provided to people.

Biology is important for wildlife-associated recreation planning because the number and types of wildlife both facilitate and constrain available opportunities. Wildlife numbers and species facilitate specific recreation pursuits. On fall weekends, Rocky Mountain National Park, Colorado, for example, provides a high-visitor-density viewing opportunity due to the large number of elk that can easily be seen from the park's roads.Constraint relates to the sustainability of a desired biological condition, within acceptable levels of impacts, under different flows and types of recreational use.

While the biological situation constrains and facilitates, *it does not dictate the type of recreation that should occur at an area.* As noted by a state fish and wildlife manager at a meeting regarding hunting: "Biology sets the sideboards on what types of hunting we can provide. But within those constraints, there are a wide variety of types of hunts possible. The hunts might vary by levels of solitude, types of weapons, season lengths, and season timing. Biology won't tell us the right combination of hunts to offer; that should be based on hunter demand."

• *While demand drives recreation management, there are limits on what are acceptable and appropriate types of recreation for an area.* Behaviors such as poaching and vandalism, for example, are clearly unacceptable and are declared so through law. Certain types of recreation may, at times, also be incompatible. Hunting and wildlife viewing in the same location at the same time, for example, are typically not compatible (Vaske et al. 1995).

Approach of This Book

The philosophy of service creates the undertone of this text. It encourages wildlife professionals to expand their view of recreation management. It does not ask them to abandon a protection philosophy; yet it will challenge them to bring a protection and a service philosophy into a more thoughtful balance.

This text emphasizes an approach to managing wildlife-viewing recreation that has been identified in the literature as Experience-based Recreation Management (EBM). It is critical to note that the EBM approach is not attainable without some adoption of a service philosophy.

EBM was born in the late 1970s from a need in recreation management to become more responsive and accountable to the public. It is not the *only* framework available for recreation management, but elements of EBM (and a service philosophy) are found in all of the contemporary approaches to outdoor recreation management.

The book has been developed around four goals. The initial goal is to describe the overall context of wildlife viewing. To accomplish this goal, Dan Witter gives a description of the magnitude of growing interest in wildlife-viewing recreation and the need for management frameworks (Chapter 2). Manfredo, Pierce, and Teel then examine the societal factors that encouraged the growth of wildlife viewing and all forms of non-consumptive recreation in North America in the twentieth century Chapter 3).

The next goal is to describe the essence of the EBM approach to recreation management. This is accomplished in two chapters. In one, Manfredo and Driver advocate recognition of the benefits that arise from managing for recreation experiences (Chapter 4). In the other, Manfredo, Pierce, Vaske, and Whittaker describe the notion of EBM (Chapter 5).

The third goal is to show how EBM gives direction to the many elements of the management planning process. First, Fulton, Whittaker, and Manfredo give an overview of the planning process (Chapter 6), while Lauber, Decker, and Chase discuss the use of stakeholders in implementing a management planning effort (Chapter 7). Vaske, Manfredo, Whittaker, and Shelby then describe how to develop standards that will guide the EBM approach (Chapter 8). Whittaker, Vaske, and Manfredo discuss management strategies that link on-the-ground actions to experience-management objectives (Chapter 9). In Chapter 10, Donnelly, Whittaker, and Jonker discuss wildlife viewing as a critical element of planning for international ecotourism. Other chapters address additional functions essential to the planning process. Fix, Loomis, and Manfredo give an overview of the types of economic analysis that will guide management and planning activities (Chapter 12). Bright and Pierce offer basic principles of communicating to viewers (Chapter 13), while Duda and Bissell discuss the basics of marketing (Chapter 14). Lastly, Dean illustrates the utility of geomatics in the management planning process (Chapter 15).

As a final goal, the book offers a perspective on how biological management tools might be integrated with an EBM approach to managing wildlife-viewing recreation. First, Gill provides a philosophical perspective and a description of the strategies for managing wildlife in a way that facilitates viewing recreation

(Chapter 11). Then, in the final chapter (Chapter 16), Matt and Aumiller describe a unique management effort at McNeil River Alaska that illustrates the enormous potential of well-managed wildlife-viewing recreation. This chapter is important because it illustrates the compatibility of the protection and service philosophies.

To conclude, I must confess that there are certain times when the book writer or editor pauses to reflect upon the effort they have initiated. In private moments, they may say to themselves, "What effect might this effort have?" For me, the answer to this question was stimulated by Mihaly Csikszentmihalyi (1982) who wrote:

> *It is useful to remember occasionally that life unfolds as a chain of subjective experiences. Whatever else life might be, the only evidence we have of it, the only direct data to which we have access, is the succession of events in consciousness. The quality of these experiences determines whether and to what extent life was worth living (p. 13).*

If this book can have some small part to play in facilitating quality in life's experiences, we will have met the ultimate aim of our endeavor.

Literature Cited

Alexander, H. E. (1962). "Changing concepts and needs in wildlife management." *Annual Conference of the Southeast Association of Game and Fish Commissions, Proceedings* 16, 161-67.

Csikszentmihalyi, M. (1982). "Toward a psychology of optimal experience." *Review of Personality and Social Psychology*, 3, 13-36.

Jepson, T. (1994). *Wild Italy: The Sierra Club Natural Traveler*. San Francisco, CA: Sierra Club Books.

Lee, R., and I. Devore, Eds. (1968). *Man the Hunter*. Chicago, IL: Aldine Publishing.

Leopold, A. (1938). "Conservation esthetic." *Bird-Lore* 40(2),101-9.

Vaske, J. J., K. Wittmann, S. Laidlaw, and M. P. Donnelly. (1995). *Human-Wildlife Interactions on Mt. Evans*. Project Report for the Colorado Division of Wildlife. Human Dimensions in Natural Resources Unit Report Number 18. Fort Collins, CO: Colorado State University.

Wagar, J. A. (1966). "Quality in outdoor recreation." *Trends* 3(3), 9-12.

Emergence and Importance of Wildlife Viewing in the United States

Daniel J. Witter

Cultural Importance of Wildlife Viewing

People rely on fish and wildlife to maintain and improve their emotional and physical condition—a reliance that is primordial and profound. Fish and wildlife have been the origin of food, shelter, commerce, art, and spiritual identity across epochs and cultures. Wild animals have inspired people's folkways, prompted human migration and political arrangements, and even precipitated conflict (Ambrose 1996). Some conclude that our appreciation of nature and wildlife is so deep rooted that humans possess "biophilia," an innate affinity for nature (Kellert 1997).

The technological revolution of the past two centuries has altered human perceptions of fish and wildlife as purely consumables, at least in the world's post-industrial, service- and convenience-oriented economies. Three of four U.S. citizens say they have fulfilled most or all of their material needs, and are searching for fulfilling recreational diversions (Russell 1990). In societies focussed on survival and basic creature comforts, wild animals are raw materials for food and shelter. In higher-technology societies focused on gratification, agricultural industry and technology permit a range of options from managing wildlife for regulated recreational and commercial consumption to total preservation.

In the U.S., diverse viewpoints on fish and wildlife have given rise to a range of private groups and public agencies promoting different conservation perspectives. These include state fish and wildlife agencies as outgrowths of anglers' and hunters' conservation activism that began in the early twentieth century. Today, agencies continue to advocate hunting, fishing, and trapping as effective management techniques and valued folkways (Trefethen 1975). In 2001, about 16 percent of U.S. citizens sixteen years old and older went fishing and 6 percent went hunting (U. S. Department of the Interior, Fish and Wildlife Service 2002). Participation in these activities varies by region of the country, and is highest in the West North Central Census Division and lowest in the Middle Atlantic Division (Table 2-1). These activities have generated billions of dollars in state permit revenues and Federal Sport Fish and Wildlife Restoration monies supporting most states' conservation infrastructures (U. S. Department of the Interior, Fish and Wildlife Service 1999).

Non-consumptive, appreciative, or aesthetic-oriented wildlife users have long been a force in wildlife conservation. A single term for people who enjoy wildlife in ways other than fishing and hunting is elusive, with wildlife viewers or wildlife watchers perhaps the most expressive and inclusive. Bird watchers were among the first to voice concern about over-exploitation of wildlife resources at the start of the twentieth century. Naturalists reported their field observations on the life histories of non-game species in early scientific publications, and formed nongovernment organizations encouraging habitat conservation and game regulations (Phillips 1931).

Popularity of Wildlife Viewing

Emergence of wildlife viewing as a pervasive cultural phenomenon in the U.S. can be traced to social and economic prosperity following World War II, which fostered consumerism and heightened Americans' pursuit of personal gratification. Greater personal income, more leisure time, and greater mobility in the 1950s resulted in an unprecedented rush or return to the outdoors by Americans (Outdoor Recreation Resources Review Commission [ORRRC] 1962). In 1987, in an updating of the 1962 ORRRC report, the President's Commission on Americans Outdoors concluded that wildlife watching had become one of the nation's most popular forms of outdoor recreation (Lohmann 1987).

Across epochs and cultures, fish and wildlife have provided food, shelter, commerce, art, and emotional satisfaction to humans. (Art panel by Charles W. Schwartz, Missouri Department of Conservation)

Late nineteenth-century naturalists, including bird watchers, were among the first to voice concern about over-exploitation of United States' wildlife resources. (Art panel by Charles W. Schwartz, Missouri Department of Conservation)

In 2001, 66 million people, or 31 percent of the United States' population sixteen years old and older observed, fed, or photographed wildlife. Among this group, 21.8 million (10 percent of the population) took trips for the primary purpose of enjoying wildlife, while 62.9 million (30 percent of the population) stayed within a mile of their homes to participate in wildlife-viewing activities. Participation in wildlife viewing varies by region of the country; it is highest in the West North Central Census Division, and lowest in the West South Central (Table 2-1).

By one estimate, the number of people observing, photographing, or feeding wildlife in the United States had peaked in 1980, with 59 percent of the population participating. In fact, the number of participants in 1996 was substantially less than even 1991. There were 17 percent fewer people observing, photographing, or feeding wildlife in 1996 than in 1991, with 18 percent fewer residential participants and 21 percent fewer nonresidential participants. However, the number of days of nonresidential wildlife watching in 1996 (314 million) did not statistically change from 1991. Expenditures for wildlife watching in 1996 increased about $5 billion over 1991, to a total of $30 billion (U. S. Department of the Interior, Fish and Wildlife Service 1997), rising to $40 billion in 2001 (U. S. Department of the Interior, Fish and Wildlife Service 2002).

Table 2-1. U.S. National and Regional Participation in Fishing, Hunting, and Wildlife Viewing, 1996 and 2001, U.S. population 16 years old and older. Numbers in thousands (USDI Fish and Wildlife Service 1997, 2002).

	1996		2001	
	Number	Percent	Number	Percent
United States				
Total population	201,472	100	212,298	100
Sportsmen	39,694	20	37,805	18
Anglers	35,246	17	34,067	16
Hunters	13,975	7	13,034	6
Wildlife viewers	62,868	31	66,105	31
Non-residential	23,652	12	21,823	10
Residential	60,751	30	62,928	30
New England				
Total population	10,306	100	10,743	100
Sportsmen	1,673	16	1,504	14
Anglers	1,520	15	1,402	13
Hunters	465	5	386	4
Wildlife viewers	3,710	36	3,875	37
Non-residential	1,443	14	2,070	19
Residential	3,586	35	3,765	35
Middle Atlantic				
Total population	29,371	100	29,308	100
Sportsmen	4,192	14	3,810	13
Anglers	3,627	12	3,250	11
Hunters	1,453	5	1,633	5
Wildlife viewers	8,185	28	8,740	29
Non-residential	2,960	10	3,287	11
Residential	8,023	27	8,453	29
East North Central				
Total population	33,121	100	33,684	100
Sportsmen	6,912	21	6,400	19
Anglers	6,006	18	5,655	17
Hunters	2,712	8	2,421	7
Wildlife viewers	11,731	35	11,631	34
Non-residential	4,501	14	3,894	12
Residential	11,297	34	11,196	33
West North Central				
Total population	13,875	100	14,617	100
Sportsmen	3,977	29	4,239	29
Anglers	3,416	25	3,836	27
Hunters	1,917	14	1,710	12
Wildlife viewers	5,089	37	6,206	43
Non-residential	1,927	14	2,439	17
Residential	4,900	35	5,939	41

	1996		2001	
	Number	*Percent*	*Number*	*Percent*
South Atlantic				
Total population	36,776	100	38,650	100
Sportsmen	7,282	20	6,957	18
Anglers	6,636	18	6,451	16
Hunters	2,050	6	1,875	5
Wildlife viewers	11,252	31	11,395	29
Non-residential	3,992	11	4,453	12
Residential	10,964	30	10,826	28
East South Central				
Total population	12,459	100	13,022	100
Sportsmen	2,907	23	2,865	22
Anglers	2,514	20	2,543	20
Hunters	1,301	10	1,164	9
Wildlife viewers	3,904	31	4,514	35
Non-residential	1,118	9	1,475	11
Residential	3,795	30	4,390	34
West South Central				
Total population	21,811	100	23,448	100
Sportsmen	5,093	23	4,924	21
Anglers	4,616	21	4,375	19
Hunters	1,812	8	1,988	9
Wildlife viewers	5,933	27	5,747	25
Non-residential	2,096	10	1,930	8
Residential	5,773	26	5,491	23
Mountain				
Total population	11,966	100	13,129	100
Sportsmen	2,761	23	2,757	21
Anglers	2,411	20	2,443	18
Hunters	1,061	9	1,020	8
Wildlife viewers	4,099	34	4,619	35
Non-residential	1,967	16	4,080	31
Residential	3,855	32	4,282	33
Pacific				
Total population	31,787	100	33,454	100
Sportsmen	4,897	15	4,349	13
Anglers	4,501	14	4,111	12
Hunters	1,203	4	837	2
Wildlife viewers	8,966	28	9,377	28
Non-residential	3,648	11	4,678	14
Residential	8,558	27	8,503	25

See map on following page for geographic divisions.

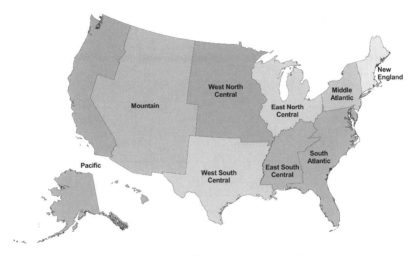

Regional geographic divisions. (Bureau of census)

Gallup polls in Missouri tracked citizen participation in aesthetic-oriented or nature activities over seven consecutive years in the 1990s. Consistently high participation was found for "reading about nature" (about 90 percent) and "feeding birds and wildlife at home" (about 80 percent). Especially revealing was the extent to which electronic media brought nature into the home. Practically all households in Missouri reported "nature-oriented TV viewing" (about 90 percent) (Missouri Dept of Conservation 2000), a finding confirmed in similar studies in the U.S. (Duda et al. 1998).

Conservation Potential of Wildlife Viewing

Most wildlife watchers are not fundamentally opposed to hunting, fishing, and trapping (Witter and Shaw 1979). In fact, 41 percent of wildlife watchers went hunting and fishing in 1996 (U. S. Department of the Interior, Fish and Wildlife Service 1997). But the 59 percent of wildlife watchers who did not remain politically and financially uninvolved with the conservation efforts of those state fish and wildlife agencies that rely on revenues from fishing and hunting permit sales.

Still other private citizens oppose fish and wildlife management and associated harvest, characterizing the activities as anachronistic and cruel. About 10 to 30 percent of the U.S. citizenry oppose hunting for any reason, and 30 to 60 percent oppose trapping for any reason, with disapproval varying by region of the country (Duda et al. 1998, Missouri Department of Conservation 1997). Sentiment against recreational angling is practically nonexistent in the United States and is a cultural curiosity more than a real phenomenon.

In 1975, the National Survey of Fishing and Hunting was renamed the National Survey of Hunting, Fishing, and Wildlife-associated Recreation. Among important survey findings: about as many women as men participate in wildlife watching, and nature study and wildlife viewing hold great appeal to youngsters (USDI Fish and Wildlife Service 1997). (Missouri Department of Conservation)

Despite a few divisive issues among some American fish and wildlife interest groups, two issues unite them. First, fish and wildlife remain profoundly fascinating as objects of aesthetic wonder for wildlife viewers, hunters, fishers, trappers, animal-rights advocates, hikers, campers, boaters, walkers, joggers, and "armchair outdoor-appreciators." Simply put, fish and wildlife appreciation contributes immensely to the quality and promise of twenty-first-century life.

Second, the loss of fish and wildlife habitat continues as a result of from human population pressures including urban and suburban sprawl; residential, industrial, and agricultural wastes and residues; issues of air and water quality and quantity; and even climate change (U. S. Department of the Interior, Geological Survey 1998). Given the importance of wildlife to so many, and continuing losses of wildlife habitat, wildlife-viewing programs represent common ground where diverse wildlife interests can agree on conservation action.

The benefits of linking wildlife watchers with traditional fish and wildlife interests have been recognized since the earliest days of fish and wildlife management. In characteristically visionary fashion, Aldo Leopold and colleagues (1930, pp. 17-18) saw the benefits of consolidated conservation efforts among fishers, hunters, and wildlife viewers. Note the authors' use of the separate words "wild life" implying all wild plant and animal resources—a subtle but fuller and richer connotation lost in contemporary usage of "wildlife."

In the long run lop-sided programs dealing with game only, songbirds only, forests only, or fish only, will fail because they cost too much, use up too much energy in friction, and lack sufficient volume of support. No game program can command the good will or funds necessary to success, without harmonious co-operation between sportsmen and other conservationists. To this end sportsmen must recognize conservation as one integral whole, of which game restoration is only a part. In predator-control and other activities where game management conflicts in part with other wild life, sportsmen must join with nature lovers in seeking and accepting the findings of impartial research.

But half a century later, the joint effort Leopold called for was not yet realized: ". . . most game management agencies must reevaluate their goals . . . to optimize support from a broad range of people interested in wildlife. . . . for all its past utility, the production of game is no longer a sufficient basis for providing the sole financial support for wildlife management" (Schick et al. 1976, p. 64).

During the late 1970s and early 1980s, social researchers across the U.S. confirmed strong potential for cooperative efforts among harvest-oriented and aesthetic-oriented wildlife users (e.g., Shaw et al. 1978; More 1979; Witter and Shaw 1979; Kellert 1980; Witter et al. 1980). Conservation staff at the federal and state levels also recognized the powerful implications of garnering the support of wildlife watchers. Questions were added to the 1975 National Survey of Hunting, Fishing, and Wildlife-Associated Recreation to quantify Americans' interests in wildlife watching and photography—the first such line of inquiry in this national survey series that had begun in 1955 (U. S. Department of the Interior, Fish and Wildlife Service 1977). The 1980 National Survey placed even greater emphasis on wildlife viewing, and collected information on wildlife watching at home and on trips, as well as wildlife enjoyment incidental to other outdoor activities (Shaw et al. 1985). These studies confirmed that many citizens enjoyed wildlife viewing and generously purchased memberships and products of private conservation organizations. The political and financial resources of wildlife viewers, however, remained largely untapped by conservation agencies. Again, the challenge was stated: "One of the most pressing needs facing the wildlife management profession is to broaden its financial and political base beyond its traditional constituency of sportsmen" (Shaw et al. 1978, p. 255).

This challenge was finally met in uniquely important and conclusive fashion. In 1976, the Missouri Department of Conservation presented the state's citizenry with an initiative proposal for a comprehensive plan for conservation of that state's fish, wildlife, and forests. This "Design for Conservation" called for state acquisition of roughly 300,000 acres (124,500 ha) for conservation, and expansion of programs, including non-game and education. The financial base for this plan was a 0.125 (one-eighth) percent sales tax, or 1¢ tax on every $8 spent, earmarked for conservation. In describing the collection of signatures, Assistant Director Allen Brohn (1977, p. 66) explained, "Hunting and fishing clubs did well getting signatures, but the best petition carriers often were college students, birders, and hikers." The initiative passed and went into effect on July 1, 1977. Missouri's conservation sales tax has since generated about $1.2 billion in support of conservation services to the state's 5.6 million citizens.

Almost three-quarters of a century have passed since Aldo Leopold called for a conservation coalition among wildlife watchers and sportsmen and -women. What other progress has been made?

Wildlife Viewing Programs Today

Encouragingly, every state fish and wildlife agency now has some form of wildlife-viewing program (Pierce and Manfredo 1997). In contemporary agency parlance, wildlife viewing is only one part of what has come to be called "wildlife diversity" programming. Wildlife diversity initiatives encompass wildlife viewing, non-game wildlife, urban wildlife, and threatened and endangered wildlife, as well as responsibilities in the broad areas of conservation, education, and recreation, and may also include the state's natural areas or heritage programs. Wildlife diversity programs may be fully integrated with game or sport-fish management, or they may be distinct from, but presumably complementary to, traditional game and fish emphases (Wildlife Diversity Committee 1999). Practically speaking, however, the act of viewing wildlife remains at the core of most wildlife diversity programs.

Expenditures for wildlife diversity initiatives comprise a relatively small portion of state fish and wildlife programs' total annual budgets. Combining all states, wildlife diversity funding accounted for about 9 percent of the total amount spent by states on fish and wildlife management in 1998 (Richie and Holmes 1999).

Private conservation groups such as National Audubon Society, National Wildlife Federation, Defenders of Wildlife, and Sierra Club have been preeminently successful in responding to the expectations of their memberships for diverse wildlife programming and services. Efforts to expand wildlife diversity programs in state and federal agencies do not threaten the long-standing financial relationship between nongovernment organizations and their memberships (Witter and Shaw 1979). Private organizations successfully garner members' financial support through sales of magazines, videos, clothing, jewelry, home furnishings, and outdoor equipment. Moreover, they offer services not likely within the scope of most state government agencies—political lobbying and fund-raising at national and international levels, extraordinary eco-tours or guided travel such as out-of-country viewing safaris and cruises, and purchase of conservation easement or title to critical or significant wildlife habitat.

In the public sector, individual states' expenditures on wildlife diversity in 1998 ranged from a low of $50,000 in Rhode Island to a high of $24.3 million in Illinois (Illinois accounting for 18 percent of all wildlife diversity spending in 1998). Fourteen states spent less than half a million dollars on wildlife diversity programs, twenty-eight states spent less than one million dollars, and just nine states spent two million dollars or more. The total amount of wildlife diversity revenue for state fish and wildlife agencies during fiscal year 1998 was $134.3 million compared to $66.3 million in 1992. This doubling of funding in six years resulted from 1998 increases in state general funds ($13.3 million more), vehicle plates or watchable wildlife "vanity" plates ($12.8 million more), trust funds ($9.4 million more), Pittman-Robertson Wildlife Restoration funds ($8.2 million more), Wallop-Breaux Sport Fish Restoration funds ($7.6 million more), and state sales tax ($3.9 million more) (Richie and Holmes 1999).

State wildlife diversity programs are funded from a variety of state and federal sources, but mainly from one or a combination (Table 2-2). In 1992, 37 percent of all wildlife diversity funding came from two agency funds. First was the sale of state hunting and fishing permits, and second, federal payments from the Sport Fish and Wildlife Restoration Programs. In 1998, these sources comprised 29 percent of all wildlife diversity funding (Richie and Holmes 1999), a decrease from 1992 due to reduced use of proceeds from hunting and fishing licenses.

Revenue is diminishing from voluntary state income-tax check-offs, at one time considered among the most promising mechanisms for funding wildlife diversity programs. In 1992, thirty-five states received money from tax check-offs. These funds comprised the second largest source of diversity funding nationwide, accounting for about 14 percent ($9.7 million). In 1998, an equal number of states received funding from this source, but it provided only 6 percent of diversity funding ($8.1 million). Connecticut is the only state to have established a check-off since 1992 (Richie and Holmes 1999). Diminishment of tax check-offs likely stems from

Table 2-2. State Wildlife Diversity Funding by Source, 1998 (Richie and Holmes 1999).

Funding Sources	Number of States	Total Dollars	1998%	1992%
General Funds	24	20,763,856	15.4	11.0
Vehicle License Plates	15	16,633,751	12.3	5.6
Hunting & Fishing Licenses	28	16,629,711	12.3	27.9
Pittman-Robertson	24	12,680,035	9.4	6.7
Trust Funds	6	10,767,700	8.0	2.1
Other State Funds	19	9,867,240	7.3	NA
Wallop-Breaux	7	9,258,611	6.9	2.3
Tax Checkoff	35	8,060,512	6.0	14.3
Sales Tax	3	7,692,426	5.7	5.6
Other Federal Funds	38	6,489,376	4.8	NA
State Lottery	3	6,237,500	4.6	6.2
Endangered Species Act (Sect. 6)	43	4,686,675	3.5	7.3
Interest Income	20	3,677,274	2.7	0.9
Private Donations	35	1,153,433	0.9	0.7
Merchandise	16	264,267	0.2	0.8
Royalties	9	35,899	<0.1	0.1
Total Revenue for All States	**49**	**134,898,266**	**100.0**	**100.0**

proliferation of competing tax check-offs, and the mechanics of completing tax forms, rather than a reflection of disinterest in wildlife (Duda et al. 1998).

Great promise for funding wildlife diversity programs is offered by federal bills that would generate monies with an excise tax on outdoor equipment analogous to the Wildlife and Sport Fish Restoration funds, or place a surcharge on selected commodities (Luntz Research Companies 1999). The Conservation and Reinvestment Act, the most promising of the bills, would underwrite wildlife diversity programs with a levy on U.S. off-shore oil revenues. Whatever the funding source, strong support for the Conservation and Reinvestment Act among both harvest- and aesthetic-oriented wildlife groups shows a desire to base wildlife diversity programs on financial grounds other than traditional fish and wildlife license sales (Teaming with Wildlife 1999).

Framework for a Wildlife Diversity Program

The Wildlife Diversity Committee of the International Association of Fish and Wildlife Agencies (1999) suggested a framework or model for a wildlife diversity program. The model rests on two prerequisites. First is the premise that a state fish and wildlife agency must be fully committed to its wildlife diversity program for the program to succeed, with agency leaders affirming with employees and the public that wildlife diversity is equal in stature to the agency's game and sport-fish programs. Second is the expectation that agency staff consciously reflect on key program elements to determine if and how these functions impact individualization of a state's wildlife diversity program. These elements are:

(1) agency mission,
(2) agency authorities or legal mandates,
(3) strategic and operational planning,
(4) public and private partnerships,
(5) public outreach and education,
(6) wildlife population research, inventory, and monitoring,
(7) natural resource management, protection, and law enforcement,
(8) training and human resources,
(9) marketing and public relations, and
(10) recreational experiences sought by wildlife viewers
(Wildlife Diversity Committee 1999).

Wildlife Planning and Management Based on Recreational Experiences Sought

Notable progress is evident in fish and wildlife agencies' efforts to serve wildlife watchers. Credit for these advances must in large measure be given to bold administrators who endorsed diverse wildlife programming. They translated their commitment to wildlife diversity into staffing and programs such as urban-wildlife biologists, wildlife-viewing coordinators, trail-development specialists, nature centers, wildlife-viewing guides, special events for wildlife watchers (e.g., Eagle Day, Prairie Day, Evening with Wildlife)—even editorial policies in state fish and wildlife magazines that recognized a more inclusive definition of fish and wildlife management.

Provision of diverse wildlife programming represents a maturation of fish and wildlife agency administration—both a cultural and deeply personal way of behaving and thinking (Muth et al. 1998). First, for the agency, there is a public-service mentality that embraces diverse wildlife programming, generating trust between agency and citizenry. Second, for the citizenry, there is a deep appreciation for fish and wildlife expressed through citizens' political and financial trust and support of agency efforts. In total, there is a coalescing of conservation commitment between agency and public.

This uniting of agency and public commitment to fish and wildlife conservation is the very essence of a term now widely used

Nature study and wildlife viewing break down common barriers to recreational participation such as gender and age. (Art panel by Charles W. Schwartz, Missouri Department of Conservation)

in fish and wildlife management—"human dimensions" (Witter and Jahn 1998). Agencies' studies of human dimensions of fish and wildlife management will be most productive if directed at discovering and understanding the *experiences* people seek through wildlife viewing; determining what motivates people to view wildlife, or what emotional, educational, and recreational gratification participants anticipate experiencing through wildlife viewing (Thorne et al. 1992, Wallace and Witter 1991). This understanding allows agencies to respond with services and facilities providing opportunities for these experiences, which in turn deepens the citizenry's commitment to conservation.

In the following chapters, Experience-based Management (EBM) is proposed for wildlife viewing. EBM is based on a theory of human motivation that defines recreation demand in terms of the desired psychological outcomes, activities, and settings associated with participation. The importance of wildlife viewing to Americans is clear, and the promise of funding and agency support for wildlife-viewing programs is stronger than ever. Resource managers' commitment to providing creative programs of wildlife viewing based on EMB could yield some of the most noteworthy professional innovations and conservation success stories yet produced by the fish, wildlife, and outdoor recreation disciplines.

Summary Points

• Fish and wildlife are deeply fascinating as objects of aesthetic wonder to many Americans, as illustrated by the fact that about one-third of the United States' population sixteen years old and older observed, fed, or photographed wildlife in 2001. Practically the entire U.S. population occasionally watches nature-oriented TV programming.

• This attraction to wildlife represents great potential for state fish and wildlife agencies to broaden their political and financial support bases to include wildlife watchers.

• Private wildlife organizations have long recognized and served citizens' interests in wildlife diversity, including wildlife viewing, non-game and urban wildlife, and threatened and endangered species.

• Wildlife viewing engages diverse aesthetic- and harvest-oriented wildlife users, and consolidates wildlife watchers' social and political activism with habitat and management initiatives of fish and wildlife agencies.

* Experience-based Planning and Benefits-based Management will help resource managers provide wildlife diversity programs.

Literature Cited

Ambrose, S. E. (1996). *Undaunted Courage*. New York: Touchstone.

Brohn, A. J. (1977). "Missouri's design for conservation." *Proceedings of the International Association of Fish and Wildlife Agencies*, 67:64-67.

Duda, M. D, S. J. Bissell, and K. C. Young (1998). *Wildlife and the American Mind*. Harrisonburg, VA: Responsive Management. 804 pp.

Kellert, S. R. (1980). "Americans' attitudes and knowledge of animals." *Transactions of the North American Wildlife and Natural Resources Conference* 45:111-24.

Kellert, S. R. (1997). *Kinship to Mastery: Biophilia in Human Evolution and Development*.Washington DC:Island Press. Washington DC

Leopold, A., W. P. Taylor, R. Bennitt, and H. H. Chapman (1930). "An American game policy." *Proceedings of the 17th American Game Conference. Washington DC: American Game Association*. Reprinted as "The American game policy in a nutshell" *in The American Game Policy and its Development, 1928-30* (1971)Washington DC: Wildlife Management Institute.

Lohmann, D. (1987). *Americans Outdoors*.Washington DC: Island Press.

Luntz Research Companies (1999). *Conservation of Land, Water, and Open Spaces is Congress's Chance to Shine on Environmental Issues*. Arlington, VA.

Missouri Department of Conservation (2000). *Missouri's Annual Conservation Monitor, 1994-2000, Four Years of Gallup Polls of Missourians' Conservation Interests and Opinions*. Public Profile 2-97. Jefferson City, MO.

More, T. A. (1979). *The Demand for Nonconsumptive Wildlife Uses: A Review of the Literature*.General Technical Report NE-52. U.S.

Department of Agriculture Forest Service, Northeastern Research Station, Newtown Square, PA.

Muth, R. M., D. A. Hamilton, J. F. Organ, D. J. Witter, M. E. Mather, and J. J. Daigle (1998). "The future of wildlife and fisheries policy and management: assessing the attitudes and values of wildlife and fisheries professionals."*Transactions of the North American Wildlife and Natural Resources Conference* 63: 604-27

Outdoor Recreation Resources Review Commission (1962). *Projections to the Years 1976 and 2000: Economic Growth, Population, Labor Force and Leisure, and Transportation.*Washington DC: U.S. Government Printing Office.Washington DC 434 pp.

Phillips, J. C. (1931). "Naturalists, nature lovers, and sportsmen." *Auk* 48:40-46.

Pierce, C. L., and M. J. Manfredo (1997). "A profile of North American wildlife agencies' viewing programs." *Human Dimensions of Wildlife* 2:3:27-41.

Richie, D., and J. Holmes (1999). *State Wildlife Diversity Program Funding: 1998 Survey.*Washington DC: International Association of Fish and Wildlife Agencies. Russell, C. (1990). "Everyone's gone to the moon." *American Demographer* 12:2:2.

Schick, B. A., T. A. More, R. M. Degraaf, and D. E. Samuel (1976). "Marketing wildlife management." *Wildlife Society Bulletin* 4:2:64-68.

Shaw, W. W., W. Mangun, and J. R. Lyons (1985). "Residential enjoyment of wildlife recreation by Americans." *Leisure Science* 7:361-75.Shaw, W. W., D. J. Witter, D. A. King, and M. T. Richards (1978). "Non-hunting wildlife enthusiasts and wildlife management." *Proceedings of the Western Association of Fish and Wildlife Agencies*, pp. 255-63.

Teaming with Wildlife (1999). Internet web site on Conservation and Reinvestment Act, www.teaming.com.Washington DC: International Association of Fish and Wildlife Agencies.

Thorne, D. H., E. K. Brown, and D. J. Witter (1992). "Market information: matching management with constituent demands."*Transactions of the North American Wildlife and Natural Resources Conference* 57:164-73.

Trefethen, J. B. (1975). *An American Crusade for Wildlife.* New York: Winchester Press. 409 pp.

U.S. Department of the Interior, Fish and Wildlife Service (1977). *1975 National Survey of Fishing, Hunting, and Wildlife-associated Recreation.* Washington DC: U.S. Government Printing Office.

U. S. Department of the Interior, Fish and Wildlife Service (1997). *1996 National Survey of Fishing, Hunting, and Wildlife-associated Recreation.*Washington DC: U.S. Government Printing Office.

U. S. Department of the Interior, Fish and Wildlife Service (2002). *2001 National Survey of Fishing, Hunting, and Wildlife-associated Recreation, State Overview (Preliminary Findings).*Washington DC: U.S. Government Printing Office.

U. S. Department of the Interior, Fish and Wildlife Service (1999). *Sport Fish and Wildlife Restoration: Program Update March 1999.*Washington DC: U. S. Department of the Interior, Division of Federal Aid.Washington DC.

U. S. Department of the Interior, Geological Survey (1998). *Status and Trends of the Nation's Biological Resources,* volume one.Washington DC : U.S. Government Printing Office.

Wallace, V. K., and D. J. Witter (1991). "Urban nature centers: what do our constituents want and how can we give it to them." *Legacy* 2:2: 20-24

Wildlife Diversity Committee (1999). *A Functional Model for a Wildlife Diversity Program.*Washington DC: International Association of Fish and Wildlife Agencies.

Witter, D.J., and L.R. Jahn (1998). "Emergence of human dimensions in wildlife management." *Transactions of the North American Wildlife and Natural Resources Conference* 63: 200-214.

Witter, D. J., and W. W. Shaw (1979). "Beliefs of birders, hunters, and wildlife professionals about wildlife management."*Transactions of the North American Wildlife and Natural Resources Conference* 44:298-305.

Witter, D. J., J. D. Wilson, and G. T. Maupin (1980). "Eagle days in Missouri, characteristics and enjoyment ratings of participants." *Wildlife Society Bulletin* 8:1:64-65.

Participation in Wildlife Viewing in North America

Michael J. Manfredo, Cynthia L. Pierce, and Tara L. Teel

Purpose

PROVIDING WILDLIFE-VIEWING OPPORTUNITIES requires that managers consider both recreationists' demand for opportunities and the supply of resources available. This chapter looks at the growing trend of participation and interest in wildlife viewing, and the reasons why this trend is occurring. It also provides an overview of the natural resources, funding, and management efforts devoted to the provision of wildlife-viewing recreation.

Demand for Wildlife-viewing Opportunities

A significant reason for producing this book is, of course, that there are current high levels of public participation in and demand for wildlife-viewing recreation in North America, and a likelihood of increased participation over time. This conclusion is supported by long-term monitoring of trends in participation. Studies such as the 2001 National Survey of Fishing, Hunting, and Wildlife-Associated Recreation indicate the magnitude of current participation (U. S. Department of the Interior, Fish and Wildlife Service and U. S. Department of Commerce, Bureau of the Census, 1997). As noted elsewhere in this volume, in 2001 a third of the U.S. population (33 percent or 66 million people) sixteen years old and older participated in wildlife-watching recreation. Similar findings are reported in the 1994-1995 National Survey on Recreation and the Environment (NSRE), which reported that 62.6 million Americans participated in wildlife viewing (Cordell and Teasley 1998). Also, the National Survey on the Importance of Wildlife to Canadians reported that nearly twelve million Canadians watched wildlife in 1987 (Federal-Provincial Task Force for the 1987 National Survey on the Importance of Wildlife to Canadians 1992).

In addition, studies have generally revealed a trend toward growth in interest and participation in viewing wildlife. Estimates from the 2001 National Survey of Fishing, Hunting, and Wildlife-associated Recreation show a 5 percent increase in participation over 1996 estimates. Also, the 1994-1995 NSRE estimate of 62.6 million people represents an almost 17 percent increase from 1985-87 figures (Cordell et al. 1990; Cordell and Teasley 1998). Earlier versions of this survey (generally referred to as the National Recreation Survey series) revealed that participation in bird watching more than doubled from 1965 to 1983 (Kelly 1987). The 1987 estimate of twelve million Canadian wildlife-viewing participants represents an 18 percent increase since 1981 (Federal-Provincial Task Force for the 1987 National Survey on the Importance of Wildlife to Canadians 1992).

Projections suggest that viewing participation will continue to increase by up to 142 percent over the next half century (Walsh et al. 1989). Wildlife viewing is predicted to be one of the most rapidly growing outdoor recreation activities in the United States (Cordell et al. 1990). Cordell et al. (1990) conclude that, by the year 2040, participation in wildlife observation and photography will increase by 74 percent, or 121 million trips. In Canada, the total number of participants in primary nonconsumptive activities (defined as those activities involving trips for the specific purpose of watching, photographing, feeding, or studying wildlife) is predicted to increase 21 percent over 1987 estimates by the year 2006 (Federal-Provincial Task Force for the 1987 National Survey on the Importance of Wildlife to Canadians 1992).

The Basis for Growth in Wildlife-viewing Recreation

Growing public interest in wildlife viewing is not a sudden fad. It is quite understandable in the context of North America's changing society. It is apparent that the growth of interest in wildlife viewing is just part of a larger trend toward increased participation in outdoor

Wildlife viewing is currently very popular in North America, and participation is rapidly increasing. (Photo by Doug Whittaker)

Box 1. Defining Wildlife Viewing

Wildlife viewing encompasses a wide variety of behaviors. It might include the home owner who sits at the dining-room window watching birds at a feeder, the person who watches one of the many television programs about wildlife, the hunter who takes to the field to pursue deer but enjoys the many other sights and sounds of wildlife, and the avid viewer who spends thousands of dollars to travel to McNeil River in Alaska to watch bears in their natural habitat. Studies suggest that how wildlife viewing is defined greatly affects estimates of participation. This is revealed in U.S. Fish and Wildlife study reports that define four classes of viewing activity based on the breakdown of two variables: primary versus secondary viewing, and residential versus non-residential viewing. Primary viewing describes activities where the primary purpose is wildlife viewing. Secondary includes situations in which wildlife viewing occurs but was not the primary motive for the trip(s). Residential viewing involves activities that are within 1 mile of home, and secondary deals with situations involving trips of at least 1 mile away from home.

This book focuses on a subset of the many ways in which people enjoy wildlife viewing. More specifically, it is focused on:
• *Experiences which have viewing wildlife as the primary objective.* This would include trips where wildlife viewing is the primary trip purpose or trips where a portion of time is devoted to engaging in recreation at a destination where the primary purpose is viewing wildlife.
• *Management of destination sites* that are publicly or privately managed locations where viewing experiences are provided to people.
• *Recreation behavior,* which has been defined as human behavior that is intrinsically rewarding, freely chosen, and occurs in non-obligated time.

recreation which began in the mid-twentieth century (see Table 3-1; Cordell et al. 1999). With the exception of hunting, participation in most nature-based recreational activities (especially those that occur away from developed areas) has risen dramatically in recent years (Cordell and Teasley 1998, Flather and Cordell 1995). For example, the percentage of people participating in hiking has increased from 16 percent in 1985 to 24 percent in 1994 (Cordell et al., 1990; Cordell and Teasley, 1998).

Table 3-1. Summary of Trends in U.S. Participation in Selected Outdoor Recreational Activities: 1965-1994.

Type of activity	Percent [a] participating in 1965	Percent participating in 1982-3[b]	Percent participating in 1994[c]
Bicycling	16	32	29
Bird watching	5	12	27
Golf	9	13	15
Hiking	7	NA[d]	24
Mountain climbing	1	NA	5
Off-road driving	NA	11	14
Outdoor sports events	30	40	48
Sightseeing	49	46	57
Walking / jogging	48	53	67

[a]*Source:* U.S. Bureau of Outdoor Recreation (1972). [b]*Source:* Van Horne et al. (1985). [c]*Source:* Cordell et al. (1996). [d]NA: Data not available.

The results of the 1994-95 National Survey on Recreation and the Environment showed that the growth of participation in the majority of outdoor recreational activities exceeded the growth of the population (Cordell et al. 1999). Nearly 95 percent of the U.S. population (or 189 million people) participated in some form of outdoor recreation, according to 1995 estimates (Cordell and Teasley 1998). This represents a 6 percent increase over the last thirteen years, from 89 percent in 1982 (Cordell et al. 2000). The reasons for this growth are found in changes in the basic fabric of society, some of which we explore in the following sections.

Economic Change. One of the most important influences on the growth of outdoor recreation participation in North America has been economic change. Three factors regarding this change have been particularly important. The first factor has been the enormous growth in economic prosperity and productivity in North America. The Gross Domestic Product (GDP), expressed in dollars per person-hour, adjusted to 1985 dollars, increased from about $2.50 in the early 1900s to almost $24 in the 1990s (Vietor 1996). As productivity increased, so did income level. In 1920, adjusted hourly income was approximately $2.75 and by 1980 it was about $9.25 (U. S. Bureau of Labor 1999). This dramatic increase in wealth and its distribution throughout all levels of society provided the means for the citizenry to exist beyond subsistence and therefore pursue recreational enjoyment.

The second factor was the diminished reliance upon public domain natural resources for things such as food and fiber, which opened the way for recreational use of public lands. As we explore this point, it is important to remember that the roots of the natural-resource profession were established at the turn of the century, at a time when settlement was expanding into all corners of the continent. The political leaders of the day, such as Theodore Roosevelt and Gifford Pinchot, were faced with the task of building an economically sound nation, and the country's abundant natural resources were seen as means to that end. Sustained yield was the production principle of the conservation philosophy, and it was this philosophy that guided early management of most of the publicly owned land in the U.S.

Although use of natural resources in the U.S. did increase over the twentieth century—for example, between 1938 and 1970, the annual rate of usage of water, timber, metals, non-metallic minerals and fuels in the U.S. more than doubled (Vietor 1996)—reliance on the public domain resources diminished. There were three main reasons for this. First, over the duration of the twentieth century, the U.S. increasingly became a service economy that did not depend upon extraction of raw materials. Second, the nation increased its reliance on imports of natural resources. Lastly, there was increased use of resources from private land. For example, timber harvest on U.S. Forest Service lands reached a high of twelve billion board feet per year in the 1980s (Sedjo 1991), but, by 1997, annual harvest had diminished to four billion board feet (U. S. Forest Service 1998).

The third economic factor contributing to the increase in participation in natural-resource recreation has been the expanding number of available opportunities. Demands for natural-resource activities have resulted in the growing viability of recreation and tourism providers (English and Marcoullier 1999; Stenger 1999). Today, outdoor recreation is a significant component of the United States' economy and is becoming the dominant use of public lands. The National Recreation Lakes Study reported that recreation constitutes 10 percent of all consumer spending and contributes over $350 billion annually to the GDP (National Recreation Lakes Study 1998). It is estimated that $180 billion of that recreation GDP is generated by visitation to federally managed public lands (National Recreation Lakes Study 1998).

The growth in recreation has been a global phenomenon. According to the World Tourism Organization (1999), tourism receipts in 1996 totaled $423 billion and have been growing, on average, 9 percent annually over the past sixteen years. Expenditures are projected to be $1,550 billion by 2010 (World Tourism Organization 1999). Furthermore, receipts from international tourism outstripped exports of petroleum products, motor vehicles,

*Participation in most nature-based recreation activities such as
mountain biking has risen dramatically in recent years.
(Photo by Tara Teel)*

telecommunications equipment, textiles, or any other product or
service in 1996 (World Tourism Organization 1999). While outdoor
recreation does not account for all these expenditures, it is certainly
a significant contributor to the totals.

Wildlife viewing has been a lucrative object of marketing by
recreation and tourism businesses. Opportunities to watch wildlife
are often one of the prime draws to an area for visitors and an
essential part of many outdoor experiences. These wildlife-viewing
opportunities may provide a significant economic contribution to
areas and businesses catering to these interests.

Demographic Change. Demographic changes have also
contributed to changes in participation in outdoor recreation in
North America. Over the twentieth century, the U.S. population
increased three-fold, expanding from 76 million people in 1900 to
263 million people in 1995 (U. S. Bureau of the Census 1997). Even
if the proportion of individuals participating in outdoor recreation
had remained the same, the potential for increased demand in
outdoor recreation activities in sheer numbers alone is dramatic.

In addition to this population growth, people appear to be
moving away from rural areas to urban centers. At the turn of the
twentieth century, only 40 percent of the U.S. population resided
in urban environments, while 77 percent resided in urban areas in
the 1990s (U. S. Bureau of the Census 1997). The trend toward
urbanization has removed people from day-to-day experiences with
the land base and natural resources. For many people, outdoor
recreation has became the primary means by which they can explore
natural environments. As noted by James Lyons (Under Secretary
for Natural Resources and the Environment during the Clinton
administration) in a public speech during Outdoor Recreation Week,
"recreation is the window through which most Americans see their
national forests" (Lyons 1998).

Available Leisure Time and Mobility. Another factor contributing to the growth in recreation has been the increased demarcation between work time and leisure time. In rural, pre-industrialized societies, leisure and work were intermingled, with no clear definition of when work started and leisure began. With urbanization and industrialization, the workweek attained a clear boundary from leisure (Kaplan 1975). As North America moved through the twentieth century, the workweek gradually decreased to a point where the forty-hour workweek became a cultural standard, leaving significant blocks of time for leisure (Jensen 1985).

In addition, the United States has become an increasingly mobile society. Between 1920 and 1941, the number of private automobiles increased from eight to twenty-two per one hundred people (Clawson and Harrington 1991), while in 1995 it was seventy-eight per one hundred people (U. S. Department of Transportation 1995). The length and quality of the road system increased, as did the durability and amenities of the automobile, facilitating accessibility to natural areas. As noted by Clawson and Harrington (1991), the automobile made the outdoors closer for long vacations as well as weekend recreational engagements.

Another important trend has been the increasing availability of airline travel. Personal/pleasure airline travel increased by nearly 300 percent between 1977 and 1995 (National Business Travel Association Industry 2000). Once considered a privilege of the elite, airline travel is now accessible to a broad spectrum of economic classes; airfares declined by over 30 percent for domestic travel and 43 percent for international travel from 1978 to 1998 (Federal Aviation Administration 1998).

The growth in television and other media has dramatically increased people's enjoyment and awareness of outdoor-recreation opportunities and, particularly, wildlife. As an example, 15.7 million Canadians, or 78.4 percent of the population, reported watching films or television programs on wildlife in 1987, a significant increase from estimates for 1981 (Federal-Provincial Task Force for the 1987 National Survey on the Importance of Wildlife to Canadians 1992). In 1988, the Discovery Channel, a United States-based cable-television network dedicated to providing viewers with educational nature programming, was named the fastest-growing cable-television network in history with 32 million subscribers (Discovery.com, Inc. 2000). By 1996, the network reached 101 million households around the world, and in 2000 had 178 million subscribing households.

This trend toward increasing use of technology to heighten awareness of outdoor-recreation opportunities will continue to be

Urbanization is associated with growth in outdoor recreation and shifting natural-resource values. (Photo by Matt Weithaus)

enhanced in the twenty-first century as computer technology becomes more accessible and sophisticated. A report by the U.S. Internet Council in 2000 indicates that 136.9 million North Americans use the internet and that growth of internet use outside that region is accelerating rapidly (U.S. Internet Council 2000). Individuals can quickly access information online about wildlife, viewing opportunities, organizational activities, and may even make virtual visits to natural areas. This information can be used as a source for planning future trips or may be sought merely for education or entertainment value.

Twentieth-century Change in Public Values. The changing economic, demographic, and technological context of society has led the way for changes in the basic thought processes of the public. This is revealed in the work of Ingelhart (1990), who proposed that, following World War II, there was a large-scale shift from "materialistic" to "post-materialistic" values. Materialistic values are oriented toward improving and protecting economic well-being and physical security, and are often associated with the accumulation of goods. Post-materialistic values are oriented toward self-esteem, self-expression, belonging, and quality of life. Ingelhart (1990) suggested that populations first must satisfy their basic materialistic needs for food, goods, and security, and can then move on to higher-order post-materialistic needs that include expectations for quality of life (this is similar to Maslow's [1970] need hierarchy). Ingelhart (1990) provides cross-cultural data that show a strong

Improvements in readily available transportation have made the outdoors available to most Americans. (Photo by Nina Roberts)

relationship between increased economic prosperity and the shift from materialistic to post-materialistic values. Materialistic values would view wildlife and other natural resources as a source for meeting physical needs. Once these needs are satisfied adequately, more appreciative natural-resource values may emerge.

As materialistic needs have been replaced by post-materialistic needs in North America, the role of leisure has changed. The Puritan work ethic, associated with materialistic times in the early European settlement of North America, views leisure as nonproductive and trivial (see Chapter 4). Now that post-materialistic values are prevalent, leisure-time pursuits have become increasingly important because they serve the high-order needs that motivate the lives of people. Csikszentmihalyi and Kleiber (1991) suggest that leisure-time pursuits offer a critical means by which individuals can fulfill their highest-order needs such as self-actualization or realizing their true potential. It is not surprising, therefore, that in recent national surveys, about two-thirds of respondents report that they rate their leisure-time activities as "important"" or "more important" than their work (Roper Poll 1990; Godbey et al. 1992).

A related trend that has affected recreation participation is the shift from the utilitarian values of the early twentieth century toward the more protectionist views toward natural resources later in the century. This shift is evident in national polls that show a dramatic increase in concern for the environment among North Americans since the 1960s (Dunlap and Mertig 1991). This trend was also evidenced by Manfredo and Zinn (1996), who tested for Coloradans' generational differences regarding value orientations toward wildlife utilization, wildlife welfare, and wildlife rights. Their

Web-based nature experiences are rapidly on the rise. (Photo by Tara Teel)

results showed a clear pattern: younger generations placed greater emphasis on protectionist values (wildlife welfare and rights) and older generations placed greater emphasis on utilitarian values (Manfredo and Zinn 1996).

The latter study is particularly relevant given research by Fulton et al. (1996), which shows that participation in hunting and angling is based on strong utilitarian value orientations. It would be reasonable to hypothesize that the decline in hunting participation in North America is related to the diminishing prevalence of these utilitarian value orientations. By contrast, participation in wildlife viewing was positively associated with what these researchers labeled an appreciative value orientation, an orientation widely held among the Colorado public they surveyed (Fulton et al. 1996). In other words, the growth in wildlife-viewing participation is consistent with the broader shift of values in North America toward more protectionist positions.

The Supply of Wildlife Viewing Opportunities

The supply of outdoor-recreation opportunities is determined primarily by two factors: availability of natural resources and management efforts (e.g., facilities, programs, enforcement, educational materials, and policy). In the case of wildlife viewing, natural resources are abundant, but management efforts to provide opportunities to allow the enjoyment of those resources are lacking.

Availability of Natural Resources. It is undeniable that the availability of natural resources for wildlife viewing throughout North America is substantial. First, an enormous land base is accessible to outdoor recreationists. Federal agencies in the United States manage 652 million acres of land and water, most of which are available for public recreation (Betz et al. 1999). Add to that a substantial land base managed by state and local entities.

Importantly, the management of these lands throughout the twentieth century, during which sustainability has been emphasized, has left these resources in generally good condition.

Second, the efforts of fish and wildlife professionals have led to widespread restoration and protection of fish and wildlife species in North America (Harrington 1991). At the turn of the century, market hunting, predator control, and habitat loss were drastically affecting the viability of many wildlife species, leading to extirpation and even total extinction. The turnabout and conversion toward restoration occurred with:

- the establishment of the Fish and Wildlife Service in 1885 and of state fish and wildlife agencies, which controlled the taking of wildlife;
- key federal legislation such as the Lacey Act (1900, regulating cross-state transport of wildlife), the Migratory Bird Treaty Act (1918, granting federal prerogative in management of migratory birds), the Pittman-Robertson Act (1937, providing funding for wildlife management), the Dingell-Johnson Act (1950, providing funding for fish management), and the Endangered Species Act (1973, providing for the protection and recovery of threatened and endangered species) (Anderson 1999); and
- a decline and even reversal in the amount of land used for agricultural production.

As a result, wildlife populations in North America are being sustained at high levels, relative to conditions at the turn of the century. This is particularly true for game species, which have been the primary target of management enhancement efforts. High-profile examples include the growth in populations of white-tailed deer and wild turkeys in the north and south and the dramatic increase in elk numbers in the Rocky Mountain states (Harrington 1991). These charismatic species are also popular for wildlife viewing. This apparent species abundance indicates the potential for creation of wildlife-viewing opportunities, but it is not enough to simply provide animal populations.

Management Effort. Much of growth in outdoor-recreation management at the federal level originated with a burst of legislation beginning in the 1960s, providing increased funding and personnel resources for natural-resource management. This included legislation such as the Land and Water Conservation Fund Act of 1964, which provided a significant source of acquisition and development funding, and a host of special land and water classification acts such as the National Trails System Act (1968), the Wild and Scenic Rivers Act of 1968, and the Wilderness Act of 1964. As a result of this emphasis, there was phenomenal growth in funding and the number of available facilities, management professionals, and university-based programs that train professionals (Cordell et al. 1990). Recent studies indicate that these trends in growth in facilities and funds are continuing as we begin the twenty-first century (Cordell et al. 1990).

Certainly the growth of resources available to state fish and wildlife agencies has been substantial. For example, in 1938, only $300,000 was available to the Idaho Department of Fish and Game, while revenue sources totaled $52 million in 1998 (Idaho Department of Fish and Game 2000). This growth mirrors the increases in funding for fish and wildlife management at the federal level. For example, the distribution of appropriations to the U.S. Fish and Wildlife Service from the Land and Water Conservation Fund totaled only $148,000 in 1967, but rose to $94.7 million in 1998 (U.S. Department of the Interior, Office of Budget 2000). Additionally, more than $1.5 billion was raised and spent on wildlife restoration in the last fifty years of the twentieth century as a result of the Pittman-Robertson Act (Duda et al. 1998). The states matched this figure by over $500 million during that period.

In addition to federal aid (e.g., that provided by the Pittman-Robertson Act), numerous other sources of revenue currently exist to enable state wildlife agencies to fund their growing programs. These sources, the majority of which have gained great public acceptance, include voluntary income-tax check-offs, entrance fees to wildlife-management areas, fees paid by developers and development-related interests, lottery-related funding mechanisms, voluntary license plates with fish or wildlife symbols, RV sales taxes, and hunting and fishing licenses (Duda et al. 1998; Anderson 1999).

The Watchable Wildlife Initiative, represented by the logo shown in this picture, is a cooperative initiative among NGOs and agencies to provide wildlife-viewing opportunities. (Photo by Pete Fix)

Despite this diversity of funding sources, wildlife agencies continue to rely heavily upon the sale of hunting and fishing license fees (Mangun and Mangun 1991; Anderson 1999). However, with increasing interactions between agencies and new and existing constituencies, and with greater attempts to develop broader-based funding programs that utilize a marketing approach, the growth of monetary resources should continue (Duda et al. 1998).

Although there has been a tremendous growth in resources for recreation management as a whole since the 1960s, relatively few resources are devoted to management for wildlife-viewing recreation among state fish and wildlife agencies. In a 1995 survey of state and provincial fish and wildlife agencies in North America, Pierce and Manfredo (1997) found that, on average, the budget for wildlife-viewing programs is less than 4 percent of that for either hunting or fishing programs. Only five of the agencies had full-time coordinators of watchable-wildlife programs while fifty-three indicated that administration of their watchable-wildlife programs was a part-time responsibility (Pierce and Manfredo 1997). When asked about the constraints to program development, 97 percent indicated lack of resources, while 55 percent indicated lack of agency acceptance (Pierce and Manfredo 1997). Efforts to seek sources of funding for wildlife-viewing programs, such as Missouri's sales tax, Teaming with Wildlife, and the Conservation and Reinvestment Act (CARA), give some indication that resources for viewing management may increase in the twenty-first century (Pierce and Manfredo 1997, Wildlife Management Institute 1998).

While funding has been a constraint, professionals charged with managing for provision of wildlife-viewing opportunities are becoming organized. This growth in professionalism may improve the credibility of and justification for increased recognition of viewing programs. A significant effort to promote wildlife-viewing recreation has been the establishment of the Watchable Wildlife Initiative in 1989 (Vickerman 1989). This initiative is a cooperative partnership among government and nongovernmental agencies who are dedicated to working together to provide watchable wildlife associated recreation. The partnership is dedicated to developing a network of nature viewing sites, a wildlife viewing site signing system, a series of guide books, and increasing public awareness of and support for natural resource management (Anderson, 1999).

Why Wildlife-viewing Recreation Programs?

Management efforts to provide wildlife viewing recreation do not appear to have kept up with the demand for opportunities. This is due to a number of factors. One reason is the lack of agency resources being devoted to viewing programs. Traditionally, fish and wildlife agencies have been funded largely by hunter and angler license fees. The use of these funds for management of wildlife viewing has been met with opposition from traditional stakeholder groups, such as hunters and anglers, who have been skeptical of viewing programs because they are seen as competition for wildlife resources. To some, wildlife viewing is associated with an anti-hunting movement that threatens the viability of consumptive forms of recreation. Another reason relates to the protection philosophy of the wildlife profession. As noted in Chapter 1, the traditional philosophy of wildlife management, which views recreation as a threat to the management of sustainable wildlife populations, has placed a low emphasis on providing ways to maximize the recreational benefits and services to stakeholders.

Challenging these assumptions, supporters of wildlife viewing recreation programs have pointed to the multiple positive outcomes they will provide. First, active programs of management show accountability to the public. Public participation in and demand for wildlife viewing recreation is substantial and is not being met with adequate response. Increased programmatic efforts would show responsiveness to the public, which would in turn increase credibility. Second, as participation in this form of recreation increases, the agencies' stakeholder constituency will expand. This new constituency can provide a significant source of agency funding and a strong political voice for wildlife. Third, increased contact with wildlife may lead to a public that is better-educated about wildlife issues. As the trend toward participatory decision-making increases, it will be critical to have an informed and interested public involved in the process. Fourth, and perhaps most important, wildlife viewing provides positive benefits to individuals, communities, and ecosystems, which is ultimately the goal of management (see Chapter 4).

Summary Points

The purpose of this chapter was to explore the demand for and supply of wildlife viewing recreation in North America. Our findings can be summarized in the following.

• There is a considerable amount of participation in wildlife viewing recreation in North America. In addition there is considerable latent demand for this type of recreation.

• The growth in wildlife viewing participation mirrors the growth in recreation participation that occurred in North America in the twentieth century. The growth is rooted in changes in economics, demographics, available leisure time and accessibility, and social value trends.

- There is an abundant natural resource supply from which wildlife viewing opportunities can be developed, including significant land and wildlife resources.
- While few management or monetary resources are currently devoted to management of wildlife viewing recreation, it is likely that this will change in the twenty-first century.
- Wildlife viewing opportunities provide benefits to both the recreationists and the wildlife management agencies and organizations.

Literature Cited

Anderson, S. H. (1999). *Managing our Wildlife Resources,* 3rd edition. Upper Saddle River, NJ: Prentice Hall.

Betz, C. J., D. B. English, and H. K. Cordell (1999). "Outdoor recreation resources." In S. McKinney (Ed.), *Outdoor Recreation in American Life: A National Assessment of Demand and Supply Trends.* Champaign, IL: Sagamore Publishing.

Clawson, M., and W. Harrington (1991). "The growing role of outdoor recreation." In K. D. Frederick and R. A. Sedjo (Eds.), *America's Renewable Resources: Historical Trends and Current Challenges.* Washington DC: Resources for the Future.

Clawson, M., R. B. Held, and C. H. Stoddard (1960). *Land for the Future.* Baltimore, MD: Johns Hopkins University Press for Resources for the Future.

Cordell, H. K., J. C. Bergstrom, L. A. Hartmann, and D. B. English (1990). *An Analysis of the Outdoor Recreation and Wilderness Situation in the United States: 1989-2040.* General Technical Report RM-189: A technical document supporting the 1989 USDA Forest Service RPA Assessment. Fort Collins, CO: U.S. Department of Agriculture Forest Service, Rocky Mountain Forest and Range Experiment Station.

Cordell, H. K., B. McDonald, B. Lewis, M. Miles, J. Martin, and J. Bason (1996). "United States of America." In G. Cushman, A. J. Veal, and J. Zuzanek (Eds.), *World Leisure Participation: FreeTime in the Global Village.* Cambridge, England: CAB International.

Cordell, H. K., and J. Teasley (1998). "Recreational trips to wilderness: Results from the USA National Survey on Recreation and the Environment." *Science and Research,* 4, 23-27.

Cordell, H. K., N. G. Herbert, N. G., and F. Pandolfi (1999). "The growing popularity of birding in the United States." *Birding,* 31, 168-76.

Cordell, H. K., B. L. McDonald, J. A. Briggs,R. J. Teasley, R. Biesterfeldt, J. Bergstrom, and S. H. Mou (2000). Emerging markets for outdoor recreation in the United States: Based on the National

Survey on Recreation and the Environment [On-line]. Available: http://www.outdoorlink.com/infosource/nsre/index.htm.

Csikszentmihalyi, M., and D. A. Kleiber (1991). "Leisure and self-actualization." In B. Driver, P. Brown, and G. Peterson (Eds.), *Benefits of Leisure*. State College, PA: Venture Publishing.

Discovery.com, Inc. (2000). Press event 2000 [On-line]. Available: http://www.discovery.com/pressevent2000/timeline.html.

Duda, M. D., S. J. Bissell, and K. C. Young (1998). *Wildlife and the American Mind*. Harrisonburg, VA: Responsive Management.

Dunlap, R. E., and A. G. Mertig (1991). "The evolution of the United States environmental movement from 1970-1990: An overview." *Society and Natural Resources*, 4(3), 209-18.

English, D. B., and D. Marcoullier (1999). "Local jobs and income from outdoor recreation." In S. McKinney (Ed.), *Outdoor Recreation in American Life: A National Assessment of Demand and Supply Trends*. Champaign, IL: Sagamore Publishing.

Federal Aviation Administration. (1998). Twenty years of deregulation: 1978 to 1998 [On-line]. Available: http://www.api.faa.gov/dereg\Dereg.pdf.

Federal-Provincial Task Force for the 1987 National Survey on the Importance of Wildlife to Canadians. (1992). *The Importance of Wildlife to Canadians in 1987: Trends in Participation in Wildlife-related Activities, 1981 to 2006*. Ottawa, Ontario, Canada: Canadian Wildlife Service.

Flather, C. H., and H. K. Cordell (1995). *Outdoor Recreation: Historical and Anticipated Trends*. Washington, DC: Island Press.

Frederick, K. D., and R. A. Sedjo (Eds.) (1991). *America's Renewable Resources: Historical Trends and Current Challenges*. Washington DC: Resources for the Future.

Fulton, D. C., M. J. Manfredo, and J. Lipscomb (1996). "Wildlife value orientations: A conceptual and measurement approach." *Human Dimensions of Wildlife*, 1(2), 24-47.

Godbey, G., A. Graefe, and S. James (1992). *The Benefits of Local Recreation and Park Services: A Nationwide Study of the Perceptions of the American Public*. State College, PA: Pennsylvania State University.

Harrington, W. (1991). "Wildlife: Severe decline and partial recovery." In K. D. Frederick and R. A. Sedjo (Eds.), *America's Renewable Resources: Historical Trends and Current Challenges*. Washington DC: Resources for the Future.

Idaho Department of Fish and Game. (2000). 1938-1998 Revenue sources. Unpublished data.

Inglehart, R. (1990). *Culture Shift in Advanced Industrial Society*. Princeton, NJ: Princeton University Press.

Jensen, C. R. (1985). *Outdoor Recreation in America* (4th edition). Minneapolis, MN: Burgess Publishing Company.

Kantor, A., and M. Neubarth (1996). "Off the charts: The internet 1996." *Internet World, 7*, 44-51.

Kaplan, M. (1975). *Leisure: Theory and Policy*. New York: John Wiley & Sons, Inc.

Kelly, J. R. (1987). *Recreation Trends: Toward the Year 2000*. Champaign, IL: Sagamore Publishing.

Lyons, J. R. (1998). Outdoor recreation week: Outdoor recreation on the national forests [On-line]. Available: http://www.fs.fed.us/ intro/speech/lyonsrec.html.

Manfredo, M. J., H. C. Zinn (1996). "Population change and its implications for wildlife management in the New West: A case study of Colorado." *Human Dimensions of Wildlife*, 1(3), 62-74.

Mangun, W. R., and J. C. Mangun (1991). "An intergovernmental dilemma in policy implementation." In W. R. Mangun (Ed.), *Public Policy Issues in Wildlife Management*. Westport, CT: Greenwood Press.

Maslow, A. (1970). *Motivation and Personality*. New York: Harper and Row.

National Business Travel Association Industry. (2000). Passenger enplanements [On-line]. Available: http://www.nbta.org/industry/ stats.htm#passengers.

National Recreation Lakes Study. (1998). *Fact Sheet*. 1951 Constitution Ave., Washington D.C.

Pierce, C. L., and M. J. Manfredo (1997). "A profile of North American wildlife agencies' viewing programs." *Human Dimensions of Wildlife*, 2(3), 27-41.

Roper Poll. (1990). *The American Enterprise*, 1, 118-20.

Sedjo, R. A. (1991). "Forest resources: Resilient and servicable." In K. D. Fredrick and R. A. Sedjo (Eds.), *America's Renewable Resources: Historical Trends and Current Challenges*. Washington DC: Resources for the Future.

Stenger, R. (1999). "Recent trends in consumer spending and industry sales." In S. McKinney (Ed.), *Outdoor Recreation in American Life: A National Assessment of Demand and Supply Trends*. Champaign, IL: Sagamore Publishing.

U. S. Bureau of the Census (1997). *Statistical Abstract of the United States: 1997* (117th edition). Washington, DC: U.S. Government Printing Office.

U. S. Bureau of Labor (1999). Bureau of Labor statistics data: National employment, hours, and earnings [On-line]. Available: http:// 146.142.4.24/cgi-bin/surveymos.

U. S. Bureau of Outdoor Recreation (1972). *The 1965 Survey of Outdoor Recreation Activities.* Washington, DC: U.S. Government Printing Office.

U.S. Department of Agriculture Forest Service (1998). National forest timber harvest [On-line]. Available: http://www.fs.fed.us/land/fm/salefact/salefact.htm.

U. S. Department of the Interior, Fish and Wildlife Service and U. S. Department of Commerce, Bureau of the Census (1997). *1996 National Survey of Fishing, Hunting, and Wildlife-associated Recreation.* Washington, DC: U.S. Government Printing Office.

U. S. Department of the Interior, Fish and Wildlife Service and U. S. Department of Commerce, Bureau of the Census (2002). *2001 National Survey of Fishing, Hunting, and Wildlife-associated Recreation (Preliminary Findings).* Washington, DC: U.S. Government Printing Office.

U.S. Department of the Interior, Office of Budget (2000). Distribution of appropriations from the Land and Water Conservation Fund (in thousands of dollars) [On-line]. Available: http://www.ios.doi.gov/budget/LWCFApprop.html.

U.S. Department of Transportation (1995). Number of vehicles [On-line]. Available: http://www.bts.gov/btsprod/nts/chp1/tb11x25.html.

U.S. Internet Council (2000). *State of the Internet 2000.* 1301 K Street NW, Suite 350, Washington, DC.

Van Horne, M., L. Szwak, and S. Randall (1985). "Outdoor recreation activity trends: Insights from the 1982-83 nationwide recreation survey." In *Proceedings of the 1985 National Outdoor Recreation Trends Symposium, Volume II.* Atlanta, GA: U. S. National Park Service.

Vickerman, S. (1989). *Watchable Wildlife: A New Initiative.* Portland, OR: Defenders of Wildlife.

Vietor, R. (1996). "Economic performance." In S. Kutler, R. Dallek, D. A. Hollinger, T. K. McCraw, and J. Kirkwood (Eds.), *Encyclopedia of the United States in the Twentieth Century* (Vol. 3) New York: Charles Scribner's Sons.

Walsh, R. G., K. H. John, J. R. McKean, and J. G. Hof (1989). "Comparing long-run forecasts of demand for fish and wildlife recreation." *Leisure Sciences,* 11, 337-51.

Wildlife Management Institute (1998). " 'Teaming With Wildlife' funding closer." *Outdoor News Bulletin,* 52, 2.

World Tourism Organization (1999). Tourism highlights 1999 [On-line]. Available: http://www.world-tourism.org/esta/monograf/highligh/hl.pdf.

Benefits: The Basis for Action

Michael J. Manfredo and B. L. Driver

Introduction: Why Consider Benefits?

The purpose of this chapter is to describe the benefits of wildlife viewing. To some people, a journey into this arena may seem removed from the "realities" of on-the-ground management. Yet, recognition of benefits may be the most important task of planning and management. This is true because the targeted benefits of a program or plan should be the starting point for all other management activities. Moreover, a clear statement of benefits offers the only defensible justification for why we embark on a wildlife-viewing program instead of directing our efforts elsewhere. The purpose of this chapter is to describe some of the possible benefits from wildlife-viewing recreation and to describe how information on benefits can be used in the planning and management process.

As a starting point for the discussion, wildlife viewing is considered in the broader context of leisure in North America. This is important because of biases that often distort our perceptions of the worth of recreation to individuals and societies. Several writers have noted that our views of leisure and recreation have been colored by a powerful and pervasive work ethic that has been dominant in American and other cultures. To illustrate, at the height of the industrial revolution, Veblen (1899) defined leisure as involving a "nonproductive" consumption of time. Goodale and Cooper (1991, p. 30) suggest that "the Calvinist association of leisure with time not working, and so idleness, and consequently vice and sin . . . can still be found today in the implication of triviality that is so often suggested by the equation of leisure with free time." Continuing, these authors state, "The most widely shared conception . . . equates leisure with free time   . . . from work, obligation, or any other necessary activity. Leisure in this sense is generally thought of as earned by effort, work, discipline, denial, sacrifice and the like."

These work-based, cultural-bound, leisure-is-trivial views stand in sharp contrast to one which posits that recreation contributes to individual and societal well-being in many important and unique ways. This alternative view of the social merits of leisure can be traced back to Aristotle, who considered leisure and its associated freedom of choice as essential to the development of virtue and character formation (Goodale and Cooper 1991). Recognition of the social worth of recreation was also the basis for the parks and recreation movement in the United States, Great Britain, and Canada in the late 1800s and early 1900s. That social worth has been recently emphasized by Driver (1999), who proposed that leisure services (considered broadly) add as much aggregate value to human welfare as do any other social service, including health and educational services. He further suggested that the total recreation and tourism industry is probably the biggest economic sector of the U.S. economy.

Whether Driver's propositions are true or not, it seems critical that, in an era of increasing governmental accountability and cost-effectiveness, we consider the benefits of recreation quite carefully. It is particularly important that we attain a better understanding of the ways in which wildlife viewing contributes to meeting basic personal and societal needs and contributes to the quality of human life. Moreover, a fundamental challenge is for us to consider the ways in which we can overtly manage recreation resources so they explicitly lead to benefits to individuals and to society, while sustaining and improving the basic biophysical resources from which those benefits flow.

The Emergence of a Benefits Approach

The growing emphasis on the benefits of recreation emerged from leisure scientists and practitioners who asked two very basic questions: why do people recreate and how can management best accommodate their needs? One approach to answering this question has been to apply motivation theory (Csikszentmihalyi 1975; Tinsley et al. 1977; Neulinger 1981; Manfredo et al. 1996). The theoretical tradition associated with Experience-based Management (EBM, see Chapter 5) proposed that people would choose to participate in recreation based on the likelihood that it would fulfill the forces that motivated it. Borrowing from expectancy-valence motivation theory, it was proposed that motivational forces included both short-term outcomes (e.g., having a chance to be with family members) and long-term outcomes (e.g., family solidarity). Work by Driver and Brown in 1975 articulated the managerial relevance of this behavioral model by proposing that it reveals four levels of recreation demand: for activities, for settings, for satisfying recreational experiences, and for other types of benefits. A significant thrust of research following this model in the 1970s and 1980s focused on identifying experiences, settings, and activities desired by recreationists. From this effort was born the model of EBM.

Box 1. Benefits-based Management

Benefits-based Management was introduced as an approach to recreation management in the 1990s and while now widely applied is still being refined. Like EBM, BBM advocates a paradigm shift in management by urging policymakers and managers to base decisions on the outputs (consequences to people) of recreation management instead of primarily on the inputs (facilities, regulations, enforcement). Both EBM and BBM are derived from the same behavioral model of recreation and BBM "builds out" from the concepts in EBM, i.e., EBM is a core component of the BBM system. A few key features of the BBM approach include the following.

1. Decisions about recreation resource management are based on the positive (beneficial) and negative impacts or outcomes that management actions are overtly directed to produce or avoid. Thus, targeted benefits are explicitly identified, and management actions are selected which are expected to assure realization of those benefits and prevention of unwanted outcomes.

2. BBM extends the focus of EBM on beneficial on-site experiences to additionally consider: (1) the positive and negative impacts on the on-site users beyond the satisfying experiences they realize, such as physiological benefits, and any desirable later changes in behavior prompted by those on-site experiences; (2) the experiential, social, and economic benefits and any negative impacts to what are called "off-site" users (e.g., nearby private landowners, members of local communities, and more remote taxpaying owners of the public lands who never visit the areas managed); and (3) "benefits" that accrue to the biophysical environment from management.

3. Like EBM, BBM employs extensive use of stakeholder groups to assist in the determination of which benefits should be the targets of management and which likely negative impacts should be minimized. The stakeholder groups are broadly inclusive, involving not only recreationists but members of local communities, business leaders, partner agencies, etc.

4. Both EBM and BBM require use of demand studies; social, economic, and environmental assessments; special science-based studies; and good professional judgment to supplement information obtained from stakeholders.

box continues

> EBM is an integral, and the most important, part of BBM; it must be understood before BBM can be. It is a rather small step then from implementing EBM to implementing BBM. BBM offers managers a broad decision context for applying EBM, so the reader is encouraged to explore application of EBM in the BBM framework. Readers interested in learning more about Benefits-based Management are referred to Driver and Bruns (1999) and O'Sullivan (1999).
>
> B. L. Driver

By the early 1990s, growing attention was being given to the benefits of recreation, i.e., the "fourth" level of demand in the Driver and Brown (1975) model (Driver et al. 1991a). Increasingly, professionals were urged to recognize benefits of leisure in the policy-making and management process. The most recognized management model that explicitly considers leisure benefits is Benefits-based Management (BBM) which builds from the EBM approach (Driver and Bruns 1999). See the box, which briefly describes BBM.

Defining the Term "Benefit"

Driver and Bruns (1999) have defined three types of benefits of recreation management and use. The first is the realization of a satisfying psychological experience, which is the centroid of EBM, as explained in Chapter 5 . The second type of benefit is the realization of an improved condition; a gain or a change in condition or state that is more desirable or is an improvement over an existing state. Examples include increased learning about natural ecological processes from studying and observing wildlife, an improved physiological condition such as reduced tension or better physical fitness, greater family cohesion as a result of wildlife-related family outings, enhanced commitment to environmental stewardship, economic growth of a local community from expenditures by wildlife-viewing tourists, and any improvement made by managers in the biophysical environment. The third type of benefit is defined as the maintenance of a desired condition and thereby the prevention of a worse condition, which means, in contrast to the second type, that no change in condition occurs. Examples include maintenance of good family relationships, maintenance of mental or physical health, or maintenance of any aspect of the biophysical environment, such as biodiversity.

Benefit Identification

The concept of benefit is closely related to a judgment or evaluation of what is "better" or "worse" and "good" or "bad." To conclude that a consequence is beneficial, requires a human judgment. Evaluation can certainly be the tricky part of identifying a benefit. This is obvious from years of research in the behavioral sciences that indicate:

People vary in their evaluative judgments of consequences. It is probably human nature to believe that one's perception of a benefit is universally true. However, it is not at all uncommon for people to disagree over the evaluation of an outcome. To illustrate, in a recent study, subjects were presented a list of consequences of wolf reintroduction in Colorado and asked to rate these as good or bad (Pate et al. 1996). Opponents and supporters of reintroduction differed in the "goodness" ratings of a wide number of outcomes including "returning the natural environment to the way it once was," "preserving the wolf as a species," and "ranchers losing money."

Evaluations vary in degree or strength. That is, some positive consequences are more positive than others and some negative consequences are more negative than others. For example, while a person may agree that it is desirable to maintain a healthy functioning ecosystem, they may think it is much more desirable to ensure economic stability. This is important because, while a person may identify a particular outcome as beneficial, they may actually oppose pursuing it if it conflicts with attainment of an even higher-ranked outcome.

An individual's judgment that an outcome is good or bad is based on beliefs associated with the outcome. One's evaluation of a particular consequence is part of a web of other evaluations, which explains why a particular outcome receives its evaluation. For example, a belief about ecosystem health might be supported because it is associated with beliefs about the type of environment needed for human survival. Similarly, beliefs that economic stability is important may be related to a person's desire for a good life for her child. It is important for managers to consider and articulate the benefits that are the targets of management programs. Yet, as indicated above, managers cannot assume stakeholders' orders of preferences for benefits or whether or not certain outcomes are even considered to be beneficial. As a consequence, a key component of the planning process is to determine stakeholders' preferences for the benefits that are to be pursued. This can be achieved through techniques such as stakeholder meetings, survey assessments, and focus group exercises.

The other challenge to considering benefits in the recreation planning process is in linking certain management strategies to beneficial consequences. This challenge requires us to address the question, "If we take this action, what benefits will arise?" Past research can certainly provide guidance in answering this question. However, this task will ultimately require stakeholder input, the best judgment of management (and/or "expert groups"), and programs of evaluation.

The Benefits of Wildlife Viewing

The benefits of wildlife viewing, and leisure more generally, are certainly diverse. When considering benefits in the planning process, managers are encouraged to begin their thinking at this broad level and then to focus on benefits most applicable to their specific situation. To assist this process, Table 4-1 lists a wide array of individual and social benefits of recreation that have been identified in the literature. For a more extensive list the reader is directed to "The Benefits Catalogue" (Canadian Park/Recreation Association 1997).

Recognizing the broad array of potential benefits, we sought to highlight a small sample of benefits likely to arise from participation in wildlife viewing. We selected these benefits based on available research about wildlife viewers (e.g., Manfredo and Larson 1993; McFarlane 1996; Martin 1997) and our professional judgment. This sample of benefits is certainly not exhaustive and focuses only on a few of the high-profile benefits possible from participation in viewing. Our descriptions of these benefits were guided, in part, from contributions to the text "Benefits of Leisure" (Driver et al. 1991a).

Family Bonding Benefits

Research shows that family bonding ranks as one of the top priorities of American adults, and leisure is seen as a critical means by which this bonding can occur (Orthner and Mancini 1991). In a study that examined the primary leisure objectives of Americans, spending time with family and companionship were the two most common objectives (United Media 1982). It should not be surprising, then, that participants in wildlife viewing have reported that "being with family" is an important outcome they seek from participation in this recreational activity (Manfredo and Larson 1993; Duda et al. 1998). Furthermore, abundant research confirms the importance of family bonding in other wildlife-associated recreation, most specifically hunting (Sofranko and Nolan 1972; Purdy and Decker 1986). We found no research that looked specifically at the

Table 4-1. Specific Types and General Categories of Benefits That Have Been Attributed to Leisure

I. Personal Benefits
A. Psychological
  1. Better Mental Health and Health Maintenance
    Holistic sense of wellness
    Stress management (prevention, mediation, and restoration)
    Catharsis
    Prevention of and reduced depression/anxiety/anger
    Positive changes in mood and emotion
  2. Personal Development and Growth
    Self-confidence
    Self-reliance
    Self-competence
    Self-assurance
    Value clarification
    Improved academic/cognitive performance
    Independence/autonomy
    Sense of control over one's life
    Humility
    Leadership
    Aesthetic enhancement
    Creativity enhancement
    Spiritual growth
    Adaptability
    Cognitive efficiency
    Problem solving
    Nature learning
    Cultural/historic awareness/learning/appreciation
    Environmental awareness/understanding
    Tolerance
    Balanced competitiveness
    Balanced living
    Prevention of problems to at-risk youth
    Acceptance of one's responsibility
  3. Personal Appreciation/Satisfaction
    Sense of freedom
    Self-actualization
    Flow/absorption
    Exhilaration
    Stimulation
    Sense of adventure
    Challenge
    Nostalgia
    Quality of life/life satisfaction

table continues

3. Personal Appreciation/Satisfaction (continued)
      Creative expression
      Aesthetic appreciation
      Nature appreciation
      Spirituality
      Positive change in mood/emotion
B. Psycho-physiological
      Cardiovascular benefits, including prevention of strokes
      Reduced or prevented hypertension
      Reduced serum cholesterol and triglycerides
      Improved control and prevention of diabetes
      Prevention of colon cancer
      Reduced spinal problems
      Decreased body fat/obesity/weight control
      Improved neuropsychological functioning
      Increased bone mass and strength in children
      Increased muscle strength and better connective tissue
      Respiratory benefits (increased lung capacity, benefits to people with
         asthma)
      Reduced incidence of disease
      Improved bladder control of the elderly
      Increased life expectancy
      Management of menstrual cycles
      Management of arthritis
      Improved functioning of the immune system
      Reduced consumption of alcohol and use of tobacco

II. Social/Cultural Benefits
      Community satisfaction
      Pride in community/nation (pride in place/patriotism)
      Cultural/historical awareness and appreciation
      Reduced social alienation
      Community/political involvement
      Ethnic identity
      Social bonding/cohesion/cooperation
      Conflict resolution/harmony
      Greater community involvement in environmental decision making
      Social support
      Support democratic ideal of freedom
      Family Bonding
      Reciprocity/sharing
      Social mobility
      Community integration
      Nurturance of others
      Understanding and tolerance of others
      Environmental awareness, sensitivity
      Enhanced world view
      Socialization/acculturation

Cultural identity
Cultural continuity
Prevention of social problems by at-risk youth
Developmental benefits of children

III. Economic Benefits
Reduced health costs
Increased productivity
Less work absenteeism
Reduced on-the-job accidents
Decreased job turn-over
International balance of payments (from tourism)
Local and regional economic growth
Employment opportunities
Contributions to net national economic development

IV. Environmental Benefits
Maintenance of physical facilities
Stewardship/preservation of options
Husbandry/improved relationships with natural world
Understanding of human dependency on the natural world
Environmental ethic
Public involvement in environmental issues
Environmental protection
Ecosystem sustainability
Species diversity
Maintenance of natural scientific laboratories
Preservation of particular natural sites and areas
Preservation of cultural/heritage/ historic sites and areas

Source: Driver and Bruns (1999). Pp. 352 & 353.

relationship between wildlife viewing and family bonding. However, past research suggests that "families who play together [do] stay together" because this increases marital satisfaction, improves husband-wife communication, and contributes to marital stability (Orthner and Mancini 1991). Family leisure experience also seems to have long-lasting impacts on children in later life, including their choice of and commitment to certain preferred recreational activities (McGuire Dottavio and O'Leary 1987). This implication is partially confirmed in McFarlane (1996), who found an association between specialization and age of introduction as well as family involvement in bird watching. More broadly, Csikszentmihalyi and Kleiber (1991) stressed the importance of family leisure experiences in early life, suggesting that this may dictate whether or not, in later life, a person can readily achieve self-actualization in their leisure experiences.

Family bonding can be one of the most important benefits of wildlife-viewing participation. (Photo by Kristie Maczko)

Managers can enhance the family experience in a number of ways. For example, family-bonding benefits might be used as a promotional message in marketing wildlife-viewing programs. People will choose to participate in wildlife viewing if they realize it will assist them in achieving the family-bonding outcomes that they seek. Managers can also enhance the family experience, for example, by focusing some facilities on family-group opportunities, and planning for activity packages (which include viewing) that would engage family groups (horse riding/viewing, biking/viewing). Another means for encouraging family participation in wildlife viewing would be through programming that targets family bonding. This might include, for example, parent/child tours and special viewing-related tasks for children and parents with rewards such as a badge, certificate, or wildlife photographs.

Community Satisfaction Benefits

Community satisfaction is an integrative concept and comprises a host of things such as availability of public services, educational opportunities, environmental quality, economics, medical facilities, social opportunities, recreational opportunities, and other amenities. Several studies have shown that community satisfaction is an important predictor of quality of life (Allen 1991; Marans and Mohia 1991).

The importance of opportunities to view wildlife to community satisfaction and quality of life was partially revealed in a study by Fulton et al. (1993). These researchers asked a random sample of residents how the quality of life in Colorado ranked compared to other places they might live and then asked them to rate a list of quality-of-life attributes. Factors which distinguished quality of life in Colorado for the greatest number of people were scenic beauty, opportunities to view wildlife, and opportunities to participate in outdoor recreation (other than hunting, fishing, and viewing).

Other research shows similar findings. Montana residents indicated that viewing and interest in wildlife were the primary reasons they enjoyed living in the state (Dolsen et al. 1996). Residents of Anchorage, Alaska, reported that the presence of wildlife and the risks that they pose are a valued symbol of the Alaskan lifestyle (Whittaker and Manfredo 1997). New York apple growers indicated the importance of seeing deer even if those deer caused economic loss through crop damage (Purdy 1987). In addition, most South Dakota residents reported that healthy wildlife populations are important to the economy and well-being of their lives (Gigliotti 1999). While these types of results might vary by region and by rural versus urban residence, it is clear that, for many people, opportunities to view wildlife are quite important in defining satisfaction with their community and the quality of their life.

Economic Benefits

The economic benefits of wildlife viewing have been revealed in a number of different studies, such as the 1996 National Survey(s) of Fishing, Hunting, and Wildlife-Associated Recreation, which found very large personal expenditures associated with seeing wildlife. The 1996 survey attributed $29 billion to spending for wildlife watching, $9 billion of which was due to trip-related expenses (USDI Fish and Wildlife Service and U.S. Department of Commerce 1997). In a study that focused more specifically on recreational trips on National Forest lands, Maharaj and Carpenter (1999) reported that a total of 53 million days of use for wildlife viewing in 1996. This compares to 26.1 million days for hunting. The estimated retail sales associated with these uses were $2.16 billion for wildlife viewing and $1.77 billion for hunting.

McCool (1996) suggested that tourism expenditures from wildlife viewing can play an important role in sustaining the viability of rural communities in the western United States. Many of these communities have suffered economic loss as the policies of public-land management have reduced their emphasis on consumptive uses (e.g., timber harvest, mining, and grazing) and increased emphasis on recreation, wildlife, and preservation.

Manfredo et al. (1988) explored the economic potential of wildlife viewing for coastal communities in the Pacific Northwest where, in the late 1980s, declining salmon stocks created hardships for commercial fishers and recreational charter operators. That study indicated a high potential for ecotourism alternatives, including whale-watching and general sightseeing, for charter operators who had depended heavily on salmon fishing.

The economic benefits of wildlife viewing hold particular promise for developing countries, which struggle to maintain a balance of conservation and development activities (Bissonette and Krausman 1995). Navrud and Mungatana (1994), for example, indicated that the recreational value of wildlife viewing in Kenya was $7.5-15 million U.S.. They indicated that wildlife preservation had considerable potential to contribute to the overall social welfare of the nation. Other studies have focused specifically on the potential economic contributions from wildlife viewing that can accrue to local communities in North America.

The ability of active management to influence tourism expenditures is clear from examination of the gateway communities that surround many of the national parks in the western United States. Because of the preservation-oriented management philosophy of the parks, there are abundant opportunities to view wildlife and, in fact, the opportunity to see wildlife is one of the primary reasons people visit the parks (e.g., USDI National Park Service 1995). The economic contributions of the national park experience to surrounding communities is critical to their viability, because the diversity of businesses in the gateway communities are affected by wildlife-viewing participation. Expenditures arise from the need for services that support the viewing experiences in the parks (e.g., motels, restaurants, sporting goods). Furthermore, there has been a steady growth of private business enterprises that provide or package wildlife-viewing experiences. These include outfitted

Outdoor recreation provides economic benefits to rural communities. This IMAX, located in West Yellowstone, Montana, offers Yellowstone National Park visitors a complementary experience to touring the park. (Photo by Mike Manfredo)

tours to see wildlife, businesses that offer substitutes to seeing wildlife in the park (e.g., zoos), and businesses that offer complements to seeing wildlife in a park (e.g., movies on wildlife, visitor centers). Recent trends reveal a rapidly expanding interest in wildlife viewing and birding in particular (Baicich et al. 1999). Projections of recreation demand in the United States suggest that wildlife viewing will continue to grow over the next half century (Cordell 1999), and it seems likely that the economic benefits derived from wildlife viewing will also continue to mount. As agencies develop plans for managing wildlife viewing, economic contributions will be a critical concern.

Learning Benefits
It seems obvious that people visit sites, attend interpretive programs about wildlife, watch wildlife on television, and read wildlife books in order to learn. Indeed, the limited research available confirms the supposition that learning is an important motivation for wildlife viewing (Manfredo and Larson 1993; Daigle et al. 2002). Furthermore, research indicates that the public believes that providing education about wildlife is an important goal for wildlife-management agencies (Fulton et. al, 1993; Whittaker and Manfredo 1997; Duda et al. 1998).

Prior research suggests that, when learning is the primary motivation for wildlife-viewing participation and there is a structured learning environment, factual learning and even concept learning can occur (Hair and Pomerantz 1987; Roggenbuck et al. 1991; Bogner 1998). For example, research suggests that visits to interpretive centers, reading magazines about wildlife, or involvement in environmental-education programs or museums directly affect factual learning. These educationally focused activities have also been linked to increased awareness of wildlife issues and positive attitudes toward wildlife (Hair and Pomerantz 1987; Preston and Fuggle 1987).

We know of no research that examines the learning benefits of less structured types of wildlife viewing. However, Roggenbuck et al. (1991) suggest there is a hierarchy of learning outcomes associated with recreation participation, with lower-level learning including learning of skills, behaviors, facts, and visual memories. Higher-order learning would include such things as concepts and schemata (larger frameworks for understanding). Researchers who looked at recreation specialization (Bryan 1977) and involvement (McIntyre 1989) have proposed that increased participation and attachment in leisure have a profound effect on these types of learning. Consequently, we contend that higher levels of learning

Learning is a key motivation for and benefit derived from wildlife-viewing recreation. (Photo by Don Rodriguez)

would be dependent on a number of factors including whether or not learning is a strong motivation for participation, the repetition of participation, and the level of involvement a person has in the activity.

Personal Identity Benefits

People present themselves to others, in part, through their recreational identity. Their shirts, hats, belt buckles, pictures on their walls, photographs that they show friends, their club membership, the magazines to which they subscribe, the recreation equipment they possess, and the natural-resource values they profess each proudly proclaim a recreational pursuit. Upon meeting others, conversations often move to a leisure pursuit as a means of conveying a certain image. The phrases "I'm a birder," "Im a fly angler," "I'm a mountain biker," or "I'm a snowboarder" not only project an image but a lifestyle and a type of personality. Research has explored the importance of leisure to one's identity through topics such as enduring involvement (McIntyre 1989), commitment (Buchanan 1985; Barro and Manfredo 1996), and specialization (Bryan 1977). Each of these concepts suggests that as people become attached to specific forms of recreation, this has a profound affect on their lives. It affects how they perceive themselves, the social group with which they interact, and their ability to achieve self-actualizing experiences.

The topic of specialization has been directly applied in describing wildlife viewers by McFarlane (1996), Martin (1997), and Hvenegaard (2002). Specialization describes a process by which participants become more attached to and gain better skills in a particular form of leisure activity. As participants become more specialized, they become more specific in their setting preferences and more particular in their equipment. In addition, increased specialization leads to shifting motivations for participation, refinement in skills and knowledge, more certain and well-formed attitudes toward management, and an association with a social group of people with similar interests. Managers can be more responsive to the public by explicitly recognizing highly involved or highly specialized wildlife viewers, and thereby enhance those viewers' opportunities to benefit through self-identity, self-development and self-actualization.

Stress Reduction Benefits

Many studies have shown that recreation provides very important mental-health benefits, helping people release tensions and otherwise cope with everyday life stresses (Driver et al. 1991b; Ulrich et al. 1991). Stress is a process whereby a person responds physiologically, psychologically, and behaviorally to a situation that threatens well-being (Baum et al. 1982). Although there is evidence that a short-term stressful situation can improve human performance, either acute or chronic stress can lead to long-term decrements in performance and well-being. Many studies have shown that leisure encounters, particularly those in natural environments, can be quite stress mediating (Kaplan and Kaplan 1989; Hartig et al. 1991; Ulrich et al. 1991). Measured responses include reduction in heart rate, blood pressure, and muscle tension; reduced stress-related hormones in the urine and blood; and other physiological and psychological indications of greater overall relaxation. In addition, a large number of self-report studies have shown that tension release and escaping from stressors at home are very important motivational forces, prompting people to seek recreational pursuits as a coping mechanism. Much of that research used the Recreation Experience Preference (REP) scales which are described in Table 4-2 (Driver et al. 1991b; Manfredo et al. 1996). Using the REP scales, Manfredo and Larson (1993) found that wildlife viewers rated stress-related motivations as highly important reasons for their viewing activities. Although there is very little additional research that focuses on wildlife viewing, stress reduction has been found to be an important motivation in a wide variety of other outdoor recreation activities, including hunting and fishing (Spaulding 1970; Knopf 1973; Wellman 1979; Manfredo 1984; Driver et al. 1991b).

Table 4-2. Motivational Dimensions (Desired Psychological Outcomes) for Recreation As Measured by The Recreation Experience Preference Scales[1].

Enjoy Nature[2]
    Scenery
    General nature experience

Physical Fitness

Reduce Tension
    Tension release
    Slow down mentally
    Escape role overloads
    Escape daily routines

Escape Physical Stressors
    Tranquility/solitude
    Privacy
    Escape crowds
    Escape noise

Outdoor Learning
    General learning
    Exploration
    Learn geography of area

Independence
    Independence
    Autonomy
    Being in control

Risk Taking

Risk Reduction
    Risk moderation
    Risk prevention

Family Relations
    Family kinship
    Escape family

Introspection
    Spiritual
    Personal values

Being With Considerate People

Achievement/Stimulation
    Reinforcing self-confidence
    Social recognition
    Skill development
    Competence testing
    Seeking excitement
    Endurance
    Telling others

Physical Rest

Teach/Lead Others
    Teaching/sharing
    Leading others

Meet New People
    Meet new people
    Observe other people

Creativity

Nostalgia

Agreeable Temperatures

[1]Taken from Driver, Tinsley and Manfredo, 1991
[2]Major motivational domains are shown in bold. Sub-dimensions of these domains are shown indented and listed below domains.

Physical Health Benefits

Improvements in one's physical health from exercising during recreation are the benefits of leisure that have the best scientific documentation (Paffenbarger et al. 1991). However, generalizations about these benefits are certainly contingent on factors related to the type of participation (endurance, weight bearing, strength inducing, competitive, therapeutic, etc.) and on the extent, duration, and frequency of physical activity. Furthermore, these variables are known to interact with other personal characteristics such as eating behavior, use of tobacco, substance abuse, etc. Findings suggest, however, that to the extent that one's participation style is highly active and participation is frequent, a number of benefits do occur, especially those related to hypertensive, atherosclerotic diseases such as coronary heart disease, sudden cardiac death, and stroke. Regular physical activity not only reduces these effects, but can also promote improved body-weight management and lead to increased bone mass and bone and muscle strength, and better muscle tone. It can also lead to improved structure and function of connective tissues; eg., ligaments, tendons, and cartilage (Paffenbarger et al. 1991). Furthermore, evidence suggests that regular physical activity will reduce symptoms of moderate depression and anxiety by improving self-image, social skills, mental health, and total well-being. Overall, regular physical activity has found to be related to longer life (Paffenbarger et al. 1991). Wildlife viewing and many other recreation activities, if infrequent, are unlikely to produce all these effects. But Manfredo and Larson (1993) found that physical exercise was a valued outcome for the highly involved wildlife viewer. Our belief is that any exercise experienced during recreation of any form synergistically contributes to the above-described physical health benefits. Furthermore, recent research shows that even modest levels of physical exercise promote mental and physical health benefits (U.S. Department of Health and Human Services 1996).

Self-Actualization and Flow Benefits

The concept of self-actualization is most readily identified with the work of Maslow (1970), who proposed that it represents the highest level of human need. He define self-actualization as the "full use and exploration of talents, capacities and potentialities" (1970:150). This concept and that of "peak experience" led Csikszentmihalyi (1975) to develop the notion of flow experience, which he defined as the optimal human experience. Flow occurs when there are a matching of challenges and skills; clear goals and immediate feedback; a depth of concentration that prevents worry and the intrusion of unwanted thoughts into consciousness; and transcendence of the self and a loss of time perception. Csikszentmihalyi and Kleiber (1991) indicate that flow experiences are frequently attained in many types of recreational engagements. Although we know of no research to confirm this conclusion, we would propose that flow is certainly an outcome that is possible in a variety of wildlife-viewing experiences. It might occur for the artist who is carefully concentrating on her surroundings in an attempt to render the best interpretation via her chosen medium. Or it might occur for the photographer who is carefully stalking a herd of elk in the day's fading light in an attempt to attain mastery in his photographic pursuits. In general, it would seem that flow experiences and self-actualization are more readily available through the more involved and specialized wildlife-viewing pursuits.

Nature-based Spiritual Benefits

For an increasing number of North Americans, there is a reverence and spirituality associated with the natural world. Outdoor recreation, including wildlife viewing, offers participants an important means for attaining these types of spiritual experience (Driver et al. 1996).

The importance of outdoor spiritual experiences shifted during the twentieth century. At the beginning of the century, spiritual thought in North America was guided by the Judeo-Christian religions, which generally proposed a view of human domination over nature. Over the duration of the century, there was a shift away from this view of nature. The inception of this change can be traced to the romantics and transcendentalists of the 1800s, who argued against the scientific and technological forces that they felt dehumanized people and exploited and degraded nature. From this period emerged the writings of transcendentalists Ralph Waldo Emerson and Henry David Thoreau who emphasized the spirituality of nature. Emerson wrote, "in the wilderness, I find something more dear and connate than in streets or villages in the woods we return to faith and reason" (quoted in Nash 1967). This spiritual view of natural resources was coupled with a preservation emphasis that grew stronger throughout the twentieth century due, in large part, to its association with the adoption of protectionist attitudes toward the environment (Dunlap and Mertig 1991). Today, nature-based spirituality is recognized as a key concern of managers of public natural resources, including wildlife (Driver et al. 1996). Jack Ward Thomas (1996), Chief of the Forest Service from 1993 to 1996, suggested that an emphasis on spirituality is consistent with the notion of an "ecological conscience," which was advocated by Aldo

Leopold. Thomas suggested that there is a need for policymakers to attain a greater understanding of public land experiences that "serve to renew and fulfill the human spirit."

Wildlife viewing might emphasize two aspects of spirituality. One is to allow people opportunities to explore the personal, spiritual meanings they associate with wildlife. The other is to examine the spiritual relevance of wildlife to human cultures over time. Throughout human existence, wildlife played a central role in religion. Wildlife have, for example, served as deities to be worshiped; sacrifices to appease gods; and oracles of wisdom, meaning, and guidance. As we learn more about this aspect of the relationship between humans andwildlife, we attain a better cross-cultural understanding.

McDonald and Schreyer (1991) suggested that spirituality can be managed in a variety of ways. It might include recognizing and managing for sacred places and wildlife, providing information about the spiritual meaning of wildlife to various cultures over time, or programming that emphasizes the spiritual meaning of places or wildlife to people.

Environmental Protection and Sustainability Benefits
Wildlife viewing is frequently recognized as a mechanism that can provide assistance in achieving ecosystem sustainability (Jordan 1995). One way this can occur is by offering an economically viable alternative to more obtrusive or destructive uses of wildlife resources. Environmentally sensitive forms of recreation have become known as ecotourism (Ceballos-Lascurain and Johnsingh 1995). The World

Wildlife viewing can promote ecosystem sustainability benefits.
(Photo by Julie Whittaker)

Conservation Union (IUCN) has defined ecotourism as "environmentally responsible travel and visitation to relatively undisturbed natural areas, to enjoy, study and appreciate nature (and accompanying cultural features—both past and present) that promotes conservation, has low visitor impact, and provides for beneficially active socio-economic involvement of local populations" (Ceballos-Lascurain 1995). Ecotourism has been proposed as a particularly useful tool for developing countries that face serious threat from resource exploitation (Western and Pearl 1989; Jordan 1995). A number of studies have been conducted that illustrate the economic, social, and biophysical benefits of wildlife-viewing-based ecotourism (see Chapter 10).

Another way by which viewing contributes to attainment of goals for ecosystem sustainability is by building public support for wildlife. Participation in wildlife viewing leads to increased knowledge about wildlife and its needs, about wildlife management and the needs of wildlife-management agencies, and about the practice of sustainable ecosystem management. Furthermore, as fee-generation programs are established, viewing participation will provide important sources of revenue that can facilitate an ecosystem-management paradigm.

Benefits of More Responsive, Effective, and
Accountable Management

EBM and, later, BBM were developed to help managers meet public expectations. Therefore, probably one of the biggest benefits of any agency adopting the EBM or BBM approach is that the agency will be more responsive, effective, and accountable.

Conventional approaches to the management of natural resources, including wildlife resources, have tended to focus primarily on the inputs to the managerial system and on the biophysical structure of the area(s) being managed. Thereby, management has focused on investment and maintenance capital; personnel and skills needed; the biophysical resources including summer and winter habitat, numbers, ages, and sex ratios of specific species of wildlife; facilities; programs; and marketing. Too often, if not generally, this orientation to inputs and biophysical structure is erroneously viewed as the end of management.

In sharp contrast, the approach proposed here views management of inputs and of system structure only as necessary means to attain the end of capturing desired outcomes or impacts. Specifically, it views the goal of management to be one of optimizing net benefits that accrue to individuals and groups of individuals, such as family units and local communities, and to the biophysical resources being managed. The process by which this can be attained is by interweaving careful management planning and plan implementation (see Chapter 6). A summary of the key elements of this approach includes the following points:

- The intended benefits of management are declared through broad statements of purpose, such as goals. Most arguments over natural-resource management strategies are rooted in fundamental disagreements over the basic ends sought by stakeholders. While it is not a guarantee against further arguments, the first step of planning should be to open a dialogue about stakeholders' perceptions of the benefits that management should pursue. Should family bonding be a central focus? Are learning and self-development a critical foundation of the management program? Should the targets be high community satisfaction and high economic growth? What is the potential conflict and priority among these pursuits?

- Alternative recreation-allocation strategies should be compared and evaluated as to the benefits they provide. For example, this analysis might ask how well the proposed mix of opportunities meets the demands of the public, results in net economic benefits to participants (see Chapter 12), leads to desired economic impacts to local communities, results in undesirable impacts to local landowners, and is perceived to affect perceptions of community satisfaction.

- More targeted and specific statements of intent, such as objectives and standards, should be clearly linked to the broader statements of intent and should focus on the desired outcomes of participation. For example, under EBM, the immediate benefits of participation are articulated in the description of the "bundle" of desired psychological outcomes (experiences) sought. By setting an objective to provide a particular type of opportunity, managers are making a commitment to provide opportunities for the benefits described in this type of experience. For example, when applying the typology used by Manfredo and Larson (1993), if managers commit to providing opportunities for Creative experiences, they are explicitly especially targeting benefits related to self-expression, self-actualization, flow, and creativity.

- As managers explore the types of information, facilities, programs, and on-site management they will provide, they should be guided by the types of experiences and benefits they are attempting to produce and the negative impacts they wish to avoid or minimize. Recognition of the benefits to be produced and the negative impacts to be avoided should not just be window dressing for decisions already made. This recognition of benefit targets should stimulate the creativity and invention of managers as they lay strategies for management.

• Attainment of benefits should not be assumed to occur as a result of management; they should be measured, which will require an active and clear system of standards, monitoring and evaluation (see Chapter 8). Adjustments in management objectives or goals should be guided by evaluative efforts.

Summary Points

The purpose of this chapter was to introduce readers to the potential benefits of wildlife viewing and to argue that these benefits should be considered the targets or the outputs of wildlife-viewing programs. Key points we considered include the following.

• Current views urge us to understand the contributions of leisure in a new perspective that views leisure as being integral to personal growth, social well-being, and quality of life.

• There is rapidly spreading interest in, and application of EBM and its expansion to BBM. These models explicitly urge managers to recognize and overtly plan for the benefits of recreation. Recreation leads to a broad array of direct benefits to individuals and to society and to indirect, nonrecreational benefits.

• Because the importance of specific benefits varies among people, managers must take steps to assess stakeholders' orderings of preferred benefits and negative outcomes by active engagement with stakeholders to identify which benefits should be targeted for managerial actions.

• A wide variety of benefits can be produced from the management and use of wildlife-viewing programs. We described several such benefits that have been documented by research, including family bonding, community satisfaction, economic stability, learning, definitions of self, stress release, self-actualization, health, spirituality, and ecosystem health.

• Managers are encouraged to consider the benefits of wildlife viewing directly in the planning and management process.

Literature Cited

Allen, L. (1991). "Benefits of leisure services to community satisfaction." In B. Driver, P. Brown, and G. Peterson (Eds.), *Benefits of Leisure*. State College, PA: Venture Publishing.

Baicich, P., G. Butcher, and P. Green (1999). "Trends and issues in birding." In H. Cordell (Ed.), *Outdoor Recreation in American life: A National Assessment of Demand and Supply Trends*. Champaign, IL: Sagamore Publishing.

Barro, S., and M. Manfredo (1996). "Constraints, psychological investment, and hunting participation: Development and testing of a model." *Human Dimensions of Wildlife*, 1: 42-61.

Baum, A., J. Singer, and C. Baum (1982). "Stress and the environment." In G. Evans (Ed.), *Environmental Stress.* New York: Cambridge University Press.

Bissonette, J., and P. Krausman, Eds. (1995). *Integrating People and Wildlife for a Sustainable Future.* Bethesda, MD: The Wildlife Society.

Bogner, F. (1998). "The influence of short-term outdoor ecology education on long-term variables of environmental perspectives." *Journal of Environmental Education,* 29: 17-19.

Bryan, H. (1977). "Leisure value systems and recreational specialization: The case of trout fishermen." *Journal of Leisure Research,* 9: 174-87.

Buchanan, T. (1985). "Commitment and leisure behavior: A theoretical perspective." *Leisure Sciences,* 7: 401-20.

Canadian Parks/Recreation Association (1997). *The Benefits Catalogue.* Gloucester, Ontario: Canadian Parks/Recreation Association. (Accessible by e-mail at cpra@rtm.activeliving.ca).

Ceballos-Lascurain, H. (1995). "Overview on ecotourism around the world: IUCN's ecotourism program." In Bissonette, J., and P. Krausman, Eds, *Integrating People and Wildlife for a Sustainable Future.* Bethesda, MD: The Wildlife Society.

Ceballos-Lascurain, H., and A. Johnsingh (1995). "Ecotourism: An introduction." In Bissonette, J., and P. Krausman, Eds, *Integrating People and Wildlife for a Sustainable Future.* Bethesda, MD: The Wildlife Society.

Cordell, H., Ed. (1999). *Outdoor Recreation in American Life: A National Assessment of Demand and Supply Trends.* Champaign, IL: Sagamore Publishing.

Csikszentmihalyi, M. (1975). *Beyond Boredom and Anxiety: The Experience of Play in Work and Games.* San Francisco, CA: Jossey-Bass.

Csikszentmihalyi, M., and D. Kleiber (1991). "Leisure and self-actualization." In B. Driver, P. Brown, and G. Peterson (Eds.), *Benefits of Leisure.* State College, PA: Venture Publishing.Daigle, J. J., O. Hrubes, and I. Ajzen (2002). "A comparative study of beliefs, attitudes, and values among hunters, wildlife viewers, and other outdoor recreationists." *Human Dimensions of Widlife* 7:1-19.

Dolsen, D., S. McCollough, G. Dusek, and J. Weigand (1996). "Beliefs about wildlife-related recreation in Montana." *Human Dimensions of Wildlife,* 1: 83-84.

Driver, B., and P. Brown (1975). "A sociopsychological definition of recreation demand, with implications for recreation resource planning." In *Assessing Demand for Outdoor Recreation.* Washington DC: National Academy of Sciences.

Driver, B., P. Brown, and G. Peterson, Eds. (1991a). *Benefits of Leisure.* State College, PA: Venture Publishing.

Driver, B., H. Tinsley, H., and M. Manfredo (1991b). "The paragraphs about leisure and recreation experience preference scales: Results from two inventories designed to assess the breadth of the perceived psychological benefits of leisure."In B. Driver, P. Brown, and G. Peterson (Eds.), *Benefits of Leisure.* State College, PA: Venture Publishing.

Driver, B., D. Dustin, T. Baltic, G. Elsner, and G. Peterson, Eds. (1996). *Nature and the Human Spirit: Toward an Expanded Land Management Ethic.* State College, PA: Venture Publishing.

Driver, B. L. (1999). "Management of public outdoor recreation and related amenity resources for the benefits they provide." In H. K. Cordell (Ed.), *Outdoor Recreation in American Life: A National Assessment of Demand and Supply Trends.* Champaign, IL: Sagamore Publishing.

Driver, B. L., and D. Bruns (1999). "Concepts and uses of the benefits approach to leisure." In E.L.Jackson and T. L. Burton (Eds.), *Leisure Studies: Prospects for the Twenty-first Century.* State College, PA: Venture Publishing, Inc.

Duda, M., S. Bissell, and K. Young (1998). *Wildlife and the American Mind: Public Opinion on and Attitudes toward Fish and Wildlife Management.* Harrisonburg, VA: Responsive Management.

Dunlap, R., and A. Mertig (1991). "The evolution of the United States environmental movement from 1970-1990: An overview." *Society and Natural Resources,* 4: 209.

Fulton, D., M. Manfredo, and L. Sikorowski, L. (1993). *Coloradans' Recreational Uses of and Attitudes toward Wildlife.* (Project Report No. 6). Project Report for the Colorado Division of Wildlife. Fort Collins, CO: Colorado State University, Human Dimensions in Natural Resources Unit.

Gigliotti, L. (1999). "Environmental and wildlife attitudes of South Dakota residents." *South Dakota Conservation Digest,* 66: 16-19.

Goodale, T., and W. Cooper (1991). "Philosophical perspectives on leisure in English-speaking countries." In B. Driver, P. Brown, and G. Peterson (Eds.), *Benefits of Leisure.* State College, PA: Venture Publishing.

Hair, J., and G. Pomerantz (1987). "The educational value of wildlfife." In D. Decker and G. Goff (Eds.), *Valuing Wildlife: Economic and Social Perspectives.* Boulder, CO: Westview Press.

Hartig, T., M. Mang, and G. W. Evans (1991). "Restorative effects of natural environment experiences." *Environment and Behavior* 23(1): 3-26.

Hvenegaard, G. (2002). "Birder specialization differences in conservation involvement, demographics, and motivations." *Human Dimensions of Wildlife* 7:21-36.

Jordan, C. (1995). *Conservation: Replacing Quantity with Quality as a Goal for Global Management.* New York: John Wiley and Sons.

Kaplan, R., and S. Kaplan (1989). *The Experience of Nature.* New York: Praeger.

Kellert, S., and E. Wilson, E. (1993). *The Biophilia Hypothesis.* Washington, DC: Island Press.

Knopf, R. C., B. L. Driver, and J. R. Bassett (1973). "Motivations for fishing." In *Transactions of the Thirty-eighth North American Wildlife and Natural Resources Conferences.* Washington, DC: Wildlife Management Institute.

Lee, R., and I. Devote, Eds. (1968). *Man the Hunter.* Chicago, IL: Aline Publishing.

Maharaj, V., and J. Carpenter (1999). *The Economic Impacts of Hunting, Fishing, and Wildlife Viewing on National Forest Lands.* Report prepared for the Wildlife, Fish, and Rare Plants staff of the USDA Forest Service by the American Sportfishing Association. Washington, DC.

Manfredo M. (1984). "The comparability of onsite and offsite measures of recreation needs." *Journal of Leisure Research,* 16: 245-49.

Manfredo M., M. Lee, and K. Ford (1988). "Alternative markets for charter boat operators affected by declining salmon allocations in Oregon." *Coastal Management,* 16: 215-27.

Manfredo M., and R. Larson (1993). "Managing for wildlife viewing recreation experiences: An application in Colorado." *Wildlife Society Bulletin,* 21: 226-36.

Manfredo M., B. Driver, and M. Tarrant (1996). "Measuring leisure motivation: A meta-analysis of the recreation experience preference scales." *Journal of Leisure Research,* 28: 188-213.

Marans, R. W., and P. Mohai (1991). "Leisure resources, recreation activity, and the quality of life." In B. Driver, P. Brown, and G. Peterson (Eds.), *Benefits of Leisure.* State College, PA: Venture Publishing, Inc.

Martin, S. (1997). "Specialization and differences in setting preferences among wildlife viewers." *Human Dimensions of Wildlife,* 2: 1-18.

Maslow, A. (1970). *Motivation and Personality.* New York: Harper & Row.

McCool, S. (1996). "Wildlife viewing, natural area protection, and community sustain ability and resiliency." *Natural Areas Journal,* 16: 147-51.

McDonald, B., and R. Shreyer (1991). "Spiritual benefits of leisure participation and leisure settings." In In B. Driver, P. Brown, and G. Peterson (Eds.), *Benefits of Leisure.* State College, PA: Venture Publishing.

McFarlane, B. (1996). "Socialization influences of specialization among birdwatchers." *Human Dimensions of Wildlife,* 1: 35-50.

McGuire, F., F. Dottavio, and J. O'Leary (1987). "The relationship of early life experiences to later life leisure involvement." *Leisure Sciences,* 9: 251-57.

McIntyre, N. (1989). "The personal meaning of participation: Enduring involvement." *Journal of Leisure Research,* 21: 167-79.

Nash, R. (1967). *Wilderness and the American Mind.* Third Edition. New Haven, CT: Yale University Press.

Navrud, S., and E. Mungatana (1994). "Environmental valuation in developing countries: The recreational value of wildlife viewing." *Ecological Economics,* 11: 135-51.

Neulinger, J. (1981). *The Psychology of Leisure.* Springfield, IL: Charles C. Thomas.

Orthner, D., and J. Mancini (1991). "Benefits of leisure for family bonding."In B. Driver, P. Brown, and G. Peterson (Eds.), *Benefits of Leisure.* State College, PA: Venture Publishing.

O'Sullivan, E. (1999). *Setting a Course for Change: The Benefits Movement.* Ashborn, VA : National Recreation and Parks Association.

Paffenbarger, R. Jr., R. Hyde, and A. Dow (1991). "Health benefits of physical activity." In B. Driver, P. Brown, and G. Peterson (Eds.), *Benefits of Leisure.* State College, PA: Venture Publishing.

Pate, J., M. Manfredo A. Bright, and G. Tischbein (1996). "Coloradans' attitudes toward reintroducing the gray wolf into Colorado." *Wildlife Society Bulletin,* 24: 421-28.

Preston, G., and R. Fuggle (1987). "Awareness of conservation among visitors to three South African nature reserves." *Journal of Environmental Education,* 18: 25-30.

Purdy, K. (1987). "Landowners'willingness to tolerate white-tailed deer damage in New York: An overview of research and management response." In D. Decker and G. Goff (Eds.), *Valuing Wildlife: Economic and Social Perspectives.* Boulder, CO: Westview Press.

Purdy, K., and D. Decker (1986). *A Longitudinal Investigation of Social-psychological Influences on Hunting Participation in New York.* (Study I: 1983-1985) (Series No. 86-87). Ithaca, NY: Cornell University, Human Dimensions Research Unit.

Roggenbuck, J., R. Loomis, and J. Dagostino (1991). "The learning benefits of leisure."In B. Driver, P. Brown, and G. Peterson (Eds.), *Benefits of Leisure.* State College, PA: Venture Publishing.

Sofranko, A., and M. Nolan (1972). "Early life experiences and adult sports participation." *Journal of Leisure Research,* 4: 6-18.

Spaulding, I. (1970). *Variation of Emotional States and Environmental Involvement during Occupational Activity and Sport Fishing.* (VIR-1 Experiment Station Bulletin No. 402.) Kingston, RI: VIR-1 Experiment Station.

Thomas, J. (1996). "Foreword."In B. Driver, D. Dustin, T. Baltic, G. Elsner, and G. Peterson (Eds.), *Nature and the Human Spirit: Toward an Expanded Land Management Ethic.* State College, PA: Venture Publishing.

Tinsley, H., T. Barrett, and R. Kass (1977). "Leisure activities and need satisfaction." *Journal of Leisure Research,* 9: 110-120.

Ulrich, R., U. Dimberg, and B. Driver (1991). "Psychological indicators of leisure benefits." In B. Driver, P. Brown, and G. Peterson (Eds.), *Benefits of Leisure.* State College, PA: Venture Publishing.

United Media Enterprises Report on Leisure in America. (1982). *Where Does the Time Go?* New York: United Media Enterprises.

U.S. Department of Health and Human Services (1996). *Physical Activity and Health: Report of the Surgeon General.* Atlanta, GA: Centers for Disease Control and Prevention, National Center for Chronic Disease Prevention and Health Promotion.

U.S. Department of the Interior, Fish and Wildlife Service, and U.S. Department of Commerce, Bureau of the Census (1997). *1996 National Survey of Fishing, Hunting, and Wildlife-associated Recreation.* Washington, DC: Government Printing Office.

U.S. Department of the Interior, National Park Service (1995). *Rocky Mountain National Park Visitor Use Survey.* Statistics compiled by N.P.S. Socio-economic Studies Division, Denver, CO.

Veblen, Thorstein. (1899). *The Theory of the Leisure Class.* Reprinted 1953, New York: The New American Library.

Wellman, J. (1979). "Recreational response to privacy stress: A validation study." *Journal of Leisure Research,* 11: 61-73.

Western, D., and M. Pearl (Eds,) (1989). *Conservation for the Twenty-first Century.* Oxford, England: Oxford University Press.

Whittaker, D., and M. Manfredo (1997). *Living with Wildlife in Anchorage: A Survey of Public Attitudes.* (Reference Report No. 35). Project Reference Report for the Alaska Department of Fish and Game. Fort Collins, CO: Colorado State University, Human Dimensions in Natural Resources Unit.

An Experience-based Approach to Planning and Management for Wildlife-viewing Recreation

Michael J. Manfredo, Cynthia Pierce, Jerry J. Vaske, and Doug Whittaker

Introduction

ONE OF THE PRIMARY PURPOSES of this book is to propose Experience-based Management (EBM) as a concept for guiding the management of wildlife-viewing recreation. This chapter introduces the need for EBM and describes how it guides management. EBM proposes a definition of recreation opportunity that includes an experiential component, a setting component, and an activity component. The first step of management is to define the range of recreation opportunities demanded by the public. This typology of opportunities guides the process of resource allocation and on-the-ground management activities. An example wildlife-viewing typology based on EBM is introduced from research in Colorado.

The Need for Experience-based Management

At the core of the recreation-planning and -management process are three questions: what types of opportunities should be provided, how should they be provided, and are these intended opportunities being delivered? As simple as these may seem, their answers rest upon an even more basic question: how should opportunities be defined, characterized, and classified? Experience-based Management provides the response to this latter question (see Box 1).

EBM originated with a recognition of the inadequacies of the traditional activity approach to recreation, which are illustrated by the enormous expansion of recreation facilities in the 1960s. This expansion was guided by Statewide Comprehensive Outdoor Recreation Planning, implemented by state and local park agencies as a prerequisite to obtaining federal Land and Water Conservation development funds (Knudson 1984). These early plans identified statewide recreation *need* based on balancing recreation *supply* and recreation *demand*. Recreation supply was expressed in terms of facilities per population and demand in terms of preferences for activities. One problem with this approach is the narrow conceptualization of supply. It suggested that recreation requires

---

Box 1. The Recreation Opportunity Spectrum

The EBM approach to recreation management served as the basis for the Recreation Opportunity Spectrum (ROS), which is a broad-based typology of recreation opportunities. The ROS was developed in the late 1970s for use in integrated land-management planning. The typology identifies six different opportunities to be considered in planning, which are labeled Primitive, Semi-Primitive Non-Motorized, Semi-Primitive Motorized, Roaded Natural, Rural, and Urban. This ROS typology and associated standards (Brown et al. 1978; Driver and Brown 1978) is used in the Recreation Opportunity Planning System, the primary system used by the U.S. Forest Service and U.S. Bureau of Land Management and a host of other management agencies. It has been used within the U.S. and at an international level (Driver et al. 1987).

---

facilities, yet clearly some activities depend on the *absence* of facility development. Another problem was the lack of precision in representing the demands of recreationists. For example, although hiking near a wildlife visitor center and hiking to a remote location to learn about bear behavior could be classified as the same activity (hiking), it is unlikely that participants would consider these two opportunities interchangeable. The two activities also require different management strategies. Providing opportunities for hiking and wildlife viewing at the developed visitor center necessitates very different management actions than for the backcountry bear viewing.

Another weakness of the activity approach is that it tells little about the "products" of recreation engagement. More specifically, it does little to indicate why people participate or what they derive from that participation. Without an indication of the "outputs" of participation, it is difficult to provide a clear justification for investing in recreation instead of alternative resource uses, or selecting one type of recreation over another.

Experience-based Management

Experience-based Management proposes a more holistic approach to defining recreation. The approach is derived from a model of recreation choice (see Box 2). This model proposes that people choose to participate in a particular recreation *activity* and a specific type of *setting* in order to attain certain *desired psychological outcome*

Box 2. Conceptual Background of Experience-based Management

In the past two decades, a number of authors have proposed an experience-based approach (and the closely related benefits-based approach) to recreation management (Driver and Brown 1975; Manfredo et al. 1983; Driver 1985). The experience-based approach rests on an expectancy theory model of human behavior (Driver and Knopf 1977; Haas et al. 1981; Manfredo et al. 1996). Expectancy theory has been the most frequently researched motivation theory (Lee et al. 1989). The theory originated in industrial psychology and was used to explain a person's work performance. Expectancy theory proposes that work behavior is a function of ability and motivation (Lawler 1973). Although formulations of the theoretical model vary, most propose that motivation results from two expectancies: the likelihood that one's efforts lead to performances and that these performances influence valued outcomes.

Proponents of experience-based management adopted expectancy theory to explain people's recreation choices (Haas et al. 1981; Manfredo et al. 1983; Driver et al. 1991; Manfredo et al. 1996). Motivation for recreation was viewed as the expectation that effort leads to onsite performances, and the expectation that performances affect valued psychological outcomes. For example, effort would include things such as pre-planning activities, expenditures of money, travel to a site, and hiking a long distance to a particular destination. Performances represented onsite events such as seeing wildlife, camping, or being with companions. The desired psychological outcomes refer to the goal states preferred by an individual. Obtaining these goal states results in satisfied motivation states. For example, a risk-taking motivation might influence participation in mountain climbing. The goal is to satisfy the risk-taking motivation and mountain climbing offers such an opportunity. The type of recreation demanded defines a unique combination of motivations that propel participation. Quality of the experience is explained by the extent to which the participation satisfies the motivations that caused it. Stated otherwise, the quality of a recreation experience depends on the expectancies, motivations, and outcomes (e.g., seeing wildlife) that are important to the individual.

This conceptual approach spawned a large number of empirical investigations in the past twenty-five years (Manfredo et al., 1996). These investigations have studied, for example, hunters (Hautaluoma and Brown 1979), wilderness visitors (Brown and Haas 1980; Manfredo et al. 1983) , and skiers (Ballman et al. 1981). Manfredo and Larson (1993) illustrate the application of this approach in a study of recreation demand for wildlife viewing in the Denver, Colorado, metro area.

or satisfactions (i.e., *experiences*). The EBM approach considers all three elements—activity, setting preference, and satisfactions/experiences—when managing and planning for recreation (Table 5-1). More specifically, it suggests that a recreation opportunity is composed of an experience opportunity, a setting opportunity, and an activity opportunity.

Experience Opportunity. The experience opportunity is defined by the multiple satisfactions that a person derives from a recreation engagement. It has been described as the "bundle" of satisfactions or psychological outcomes sought from participation. For example, psychological outcomes that have been shown to be important to wildlife viewing include developing and experiencing relations with nature, stress release, family bonding, and exploration (Manfredo and Larson 1993).

The experience component is important because it gives a fundamental understanding of *why* an individual is participating in a recreation opportunity. With this understanding, managers are better positioned to make inferences about the types of management that will provide a given opportunity or that will negatively affect it. For example, allowing high densities of recreationists at a site can interfere with attainment of solitude-oriented experiences; providing in-depth information about wildlife can facilitate learning outcomes; and modifying the design of wildlife-viewing platforms can influence creativity outcomes (e.g., painting, photography).

Types of experiences can be distinguished by the emphases placed on desired psychological outcomes. Manfredo and Larson (1993) examined types of wildlife-viewing experience desired by residents of the Denver, Colorado, metro area. Cluster analysis of motivation questions identified four experience types that the researchers labeled "high involvement," "creativity," "generalist," and "occasionalist." These experiences and their associated setting and activity opportunity preferences are summarized in Table 5-2. The

Table 5-1. EBM Approach to Defining a Wildlife Viewing Recreation Opportunity

Components of an Opportunity	Component Description	Example
Experience Opportunity	The bundle of satisfactions or psychological outcomes sought from participation in a recreation engagement	Achievement Creativity Fitness Risk taking Social Escape
Setting Opportunity	Describes the environment in which will facilitate the type of recreation experience sought. It comprises resource, social, managerial attributes	
Resource Attributes	Elements of the biophysical environment that will facilitate an experience.	Wildlife themes, species Remoteness from roads and development Degree of naturalness Types of Eco-zones
Social Attributes	Elements of the social environment that facilitate or constrain a recreation experience	Number and density of recreationists present at one time Types of recreationists together at one time
Managerial Attributes	Presence of management to facilitate or constrain experience attainment.	Degree of development Amount of site hardening. Offsite versus onsite control
Activity Opportunity	The activity or mix of recreation activities which together facilitate the total recreation experience	Hiking Biking Camping Picnicking

typology was used later in a planning assessment conducted by the Colorado Division of Wildlife as illustrated in Box 3.

Setting Opportunity. The setting opportunity is defined as the entire environment in which a recreation opportunity occurs and comprises resource, social, and managerial attributes. Attributes of the setting yield answers to questions about what types of wildlife, scenery, development, levels of crowding, types of information, and management restrictions are appropriate and desirable for a given type of recreation opportunity.

Resource attributes include those elements of the biophysical environment that facilitate an experience. Wildlife attributes, such as the number and diversity of different wildlife species, and the frequency of wildlife sightings, are the central components of a wildlife-viewing opportunity. For example, species that had prominence in the media and were highly symbolic of Colorado (such as eagles, elk, bighorn sheep, and mountain goats) were also of greatest interest to viewers (Manfredo and Larson 1993). People also expressed an interest in having the opportunity to see rare or endangered species.

Although wildlife are an important attribute of a setting, other attributes are also important. Key questions include: Is it necessary to have a natural setting or are developed facilities acceptable? Should opportunities be remote or next to high-volume highways? Are eco-zone considerations important in providing the recreation opportunity? In Colorado, for example, we might consider three eco-zones (alpine, forest, plains), three levels of remoteness (viewing by roads, viewing within a quarter mile of roads, viewing more than a quarter mile from roads) and three levels of naturalness (low, medium, and high intrusions on the natural environment).

Social attributes deal with the social environment that will facilitate a particular recreation engagement. The social environment is often the greatest source of conflict and can be the most difficult to manage. For example, frequently cited problems among recreationists are "there are too many people" (crowding), "some people are acting inappropriately" (uninformed or illegal behavior), "there is a group of people that ruins our experience" (conflict such as motorized versus non-motorized recreation). Research shows that the acceptable number and behavior of others varies by the type of experience being pursued (Graefe et al. 1990).

Several authors have emphasized the importance of social attributes through the concept of social carrying capacity (Shelby and Heberlein 1986; Vaske et al. 1986). For a given type of recreation

text continues on page 79

Table 5-2. Summary of Empirically Derived Wildlife Viewing Typology for Denver Metro Residents. Description of Experience Opportunity Preference and Highlights of Attributes Associated with the Experience. (From Manfredo and Larson, 1993.)

	Recreation Opportunity Label			
	High Involvement	Creativity	Generalist	Occasionalist
Experience Opportunity	A wide range of desired outcomes is highly valued. Compared to other experiences, emphasis placed on developing spiritual values, teaching outdoor skills to others, nostalgia, privacy/solitude, friendship, stimulation, being near others who are considerate, developing skills and abilities.	High on experiencing nature, escaping life's demands, tranquility, nostalgia, exploration, family togetherness. Very high on creativity. Low on solitude/privacy.	High on experiencing nature, tranquillity, escaping life's demands, family togetherness, exploration.	Overall, low level of importance to outcomes associated with viewing. Highest on nature experience, nostalgia, tranquility, family togetherness.

Setting Opportunity	Wide interests, including rare and endangered species, eagles, and large mammals. Strong interest in information including information about threatened and endangered species, how to be successful at viewing, habits of wildlife, natural history, and management activities.	Unique due to emphasis placed on seeing animals in the wild and interest in seeing many different animals in a single outing.	Interested in rare and endangered species, symbolic species (e.g., eagles), and large mammals. Responsive to designated viewing areas, visitor centers, trails with signs, brochures at visitor centers.	Low specific interest. Items of greatest interest include rare and endangered species, symbolic species (e.g., eagles), and large mammals. Responsive to designated viewing areas, visitor centers, trails with signs, brochures at visitor centers.
Activity Opportunity	Viewing is combined with a wide array of activities, especially camping, hiking, picnicking, fishing.	Camping, hiking, picnicking. Unique due to emphasis placed on photography.	Emphasis on camping, hiking, picnicking.	Camping, hiking, picnicking.

Box 3. Wildlife Viewing Markets in Colorado

It is important to distinguish *viewer markets* and *opportunity preferences*. The term "viewer markets" describes the characteristics of segments of users based on their viewing interests and participation behavior. Opportunities (including setting, experience, and activity) describe the characteristics of a single engagement. A person's classification in one type of market does not suggest they have interest in only one type of opportunity. For example, a person classified in the "high involvement" user segment might engage in highly specialized viewing opportunities with friends who share that interest. But, when choosing a family outing, the same person might choose to participate in a low-specialization, general-interest type of experience.

The confusion between markets and opportunity preferences is exacerbated by the fact that the research method frequently used to determine a typology of experiences establishes groups of people. With that approach, study participants are asked to rate attributes of their favorite (or most frequent) wildlife-viewing experience. Responses are cluster analyzed to group respondents with like preferences. The result is a description of the range of preferred (or most frequent) experiences of the group studied.

Table 5-2 summarizes results of the Manfredo and Larson (1993) study, which was directed toward describing a range of preferred wildlife-viewing experiences. To illustrate the contrast between viewer markets and opportunity preferences, as part of Colorado Division of Wildlife's Long Range Planning Efforts, Manfredo et al. (1993), conducted a statewide survey to determine wildlife-viewing markets. A single summary question was developed to assess visitor types derived from Manfredo and Larson's (1993) study. (See Moscardo [2000] for an additional example.) The question asked subjects to indicate which of the following types described them best.

Type 1 (High Involvement) describes people who are highly interested in wildlife viewing. They take several trips throughout the year, and they enjoy opportunities to study wildlife and wildlife behavior and opportunities to teach and lead others.

Type 2 (Creative) are also very active and interested in wildlife. However, what they value most highly is the opportunity to photograph, paint, or sketch wildlife. These people often have a high-investment in equipment such as camera gear.

Type 3 (Generalist) is a person with general interest in seeing and learning more about wildlife. These people take trips to see wildlife sporadically throughout the year and do so to have a change of pace, to get out with friends or family just to see new scenery.

Type 4 (Occasionalist) is a person who has a slight level of interest in trips specifically to view wildlife. Occasionalists take wildlife-viewing trips only occasionally. The primary means by which they enjoy wildlife is when it is associated with other types of activities such as auto driving, camping, walking, or fishing.

Results indicated that the highest proportion of the public was either an Occasionalist (51 percent) or Generalist (35 percent), while 6 percent were in the Creative group and 8 percent in the High Involvement group.

opportunity, a social carrying capacity defines the maximum amount, type, and distribution of recreationists tolerable at a given location.

Managerial attributes describe the types of management interventions that are appropriate and/or desirable in providing a given type of opportunity and include the broad array of tools and techniques available for providing a wildlife-viewing experience (e.g., visitor centers, roadside rests, brochures, field guides, video tapes, guided tours, areas with limited entry permits).

As with the other attributes, certain types of management are appropriate only for certain types of experiences. When attempting to provide an experience that emphasizes high solitude and high naturalness, for example, it is inappropriate to plan highly developed or hardened sites (e.g., kiosks, visitor centers, asphalt on trails). In this case control can be attained without site manipulation, using, for example, informational techniques or permit systems (see Chapter 9).

Activity Opportunity. The activity opportunity refers to the recreation activity or activities that are associated with a given

text continues on page 82

Experience-based managerment urges managers to recognize the diversity of public demand for wildlife-viewing experiences. (Photos, clockwise from top left, Doug Whittaker, Rich Larson, Kristie Maske, Mike Manfredo, Julie Whittaker, Doug Whittaker)

opportunity. This component provides a link to the traditional "activity" classification used in recreation nomenclature. For example, the Manfredo and Larson (1993) study determined that the types of activities that Denver residents most frequently combine with wildlife viewing were camping, hiking, picnicking, and photography. Activity participation varied, of course, by the type of experience and setting desired by a person.

Benefits or Consequences Associated with Recreation. In the late 1980s, the concept of Experience-based Management was linked to the concept of management for benefits (see Chapter 4). The emphasis on benefits proposes that "better" decisions (i.e., more equitable, effective, and efficient decisions) can be made by recognizing the social and personal benefits associated with recreation opportunities. When evaluating alternatives, for example, strategies emphasizing a generalist wildlife-viewing experience may produce a high volume of recreationists and have a large economic impact on the community. Conversely, a strategy emphasizing highly specialized viewing experiences may have lower economic impact, but be less likely to affect the character of a community (e.g., need for infrastructure, social and economic characteristics of a community, and sense of community cohesion).

The benefits-based management concept has been particularly useful because it places EBM within the context of the planning process. Contemporary models of natural-resource planning emphasize the importance of creating alternative strategies, examining the consequences of these alternatives, and selecting the alternative that yields the benefits determined to be important by stakeholders (see Chapter 6). In considering benefits, managers are urged to look beyond an economic assessment of alternatives. Although an economic assessment is important, other benefits (e.g., cultural traditions, community stability, effects on family stability and solidarity, and personal health benefits), and costs associated with an alternative should be considered. For example, when planning for wildlife-viewing opportunities for international tourists, it would be important to determine costs associated with development, investment, infrastructure, education, and training.

Adopting EBM

The most useful tools of the EBM approach are the *concepts* about how to define recreation "products" and how to use that definition in the planning/management process. EBM application begins with development of a classification of the recreation opportunities. The classification serves the following functions that are described in greater detail in other chapters of this book:

- It accurately reflects the diversity of *types* of recreation desired by the public (described by an experiential, setting, and activity component) and yields information about how to maintain *quality* for a given type of recreation.
- It is useful to management in making on-the-ground decisions. Managers must be able to clearly identify the combination of resources, recreation density, and types of developments that create the opportunities for the quality experiences people are seeking.
- It leads the way to a better understanding of the short- and long-term consequences of participation, and allows managers to develop programs that emphasize the societal benefits of participation (e.g., economic impacts and impacts on families and local communities).
- It is the building block for managers who need to make determinations about:
  - the types of opportunities that are in high demand and those in lower demand;
  - opportunities that are valued highly, and those that have lower value;
  - opportunities that are, on the ground, readily available and those that are relatively scarce; and
  - a clear allocation and action strategy that involves balancing opportunity availability, opportunity preference, and engagement consequences (i.e., benefits of participation in opportunities).

To illustrate the possibilities, Table 5-3 (see page 87) shows a typology of viewing opportunities that was developed at a specific site located at a Front Range plains environment in Colorado (Manfredo et al. 1992). The typology used the findings from Manfredo and Larson (1993) as the core concept, but adapted it to fit existing participation and to address timely management issues. The typology also incorporated concepts of specialization (Bryan 1977), normative definitions of crowding (Vaske et al. 1986), and the importance of prior experience in shaping onsite preferences and behavior (Schreyer et al. 1984).

A ready-made classification of wildlife-viewing opportunities is not available or even recommended. Development must be accomplished through the efforts of local planners and managers. In developing a typology, they are encouraged to use available research, their own experience and knowledge of an area, and the participation of stakeholders. The classification should be refined through the trial and error of its use.

Conclusion

EBM stresses the importance of accurately and parsimoniously defining the public's recreation demands. This emphasis can be applied further in development of planning, management, marketing, and communication, described in subsequent chapters of this volume. The history of EBM originates from a desire to meet recreationists' needs and a recognition of the complexity of their experiences. This approach should prove useful in planning and management for wildlife-viewing experiences.

Summary Points

• Experience-based Management aims to identify and meet the diverse interests of publics, while operating within the capabilities of the resource.

• Early approaches to meeting recreational needs defined opportunities in terms of the types of activities that could be provided. However, this approach has proven inadequate because it focuses on producing things (e.g., facilities, trails) rather than experiences, and fails to recognize individual motivations for recreation participation, variation within similar activities, and the context in which activities occur.

• Experience-based Management proposes that recreation opportunities be described in terms of experience, setting, and activity. Experience refers to the desired outcomes and satisfactions an individual hopes to realize from the recreational opportunity. Setting includes the resource, social, and managerial attributes associated with the environment in which the recreation occurs. The activity is the type of recreation in which an individual is participating (e.g., hiking, biking).

• Adopting EBM begins with classifying opportunities. This classification reflects the types, quality, and consequences of recreational opportunities, and helps in making allocation decisions.

• Using an example from Colorado (Table 5-3), we can see how wildlife viewers can be defined in terms of their interests and how this will help determine the types of activities, settings, and experiences to provide. This is only one example, and it will be important for local managers and planners to work through the process for their own particular situation and needs.

Literature Cited

Ballman, G. E., T. B. Knopp, and L. C. Merriam, Sr. (1981). *Managing the Environment for Diverse Recreation: Cross Country Skiing in Minnesota*. (Agricultural Experiment Station Bulletin 544, Forestry Series 39). St. Paul, MN: University of Minnesota.

Brown, P.J., B. L. Driver, and C. McConnell (1978). "The opportunity spectrum concept and behavioral information in outdoor recreation supply inventories: Background and application." In *Integrated Inventories of Renewable Natural Resources: Proceedings of the Workshop.* (USDA Forest Service General Technical Report RM-55). Fort Collins, CO: U.S. Department of Agriculture, Forest Service, Rocky Mountain Forest and Range Experiment Station.

Brown, P. J., and G. E. Haas (1980). "Wilderness recreation experience: The Rawah case." *Journal of Leisure Research,* 12(3), 229-41.

Bryan, H. (1977). "Leisure value systems and recreation specialization: The case of trout fishermen." *Journal of Leisure Research,* 12, 229-41.

Driver, B. L. (1985). "Specifying what is produced by management of wildlife by public agencies." *Leisure Sciences,* 7(3), 281-95.

Driver, B. L., and P. J. Brown (1975). "A sociopsychological definition of recreation demand, with implications for recreation resource planning." In *Assessing Demand for Outdoor Recreation.* Washington DC: National Academy of Sciences.

Driver, B. L., and R. C. Knopf (1977). "Personality, outdoor recreation and expected consequences." *Environment and Behavior,* 9(2), 169-93.

Driver, B. L., and P. J. Brown (1978). "The opportunity spectrum concept and behavioral information in outdoor recreation resource supply inventories: A rationale." In *Integrated Inventories of Renewable Natural Resources: Proceedings of the Workshop.* (USDA Forest Service General Technical Report RM-55). Fort Collins, CO:U.S. Department of Agriculture, Forest Service, Rocky Mountain Forest Range Experiment Station.

Driver, B. L., P. J. Brown, T. G. Gregoire, and G. H. Stankey (1987). "The ROS planning system: Evolution and basic concepts." *Leisure Sciences,* 9, 203-14.

Driver, B. L., H. E. A. Tinsley, and M. J. Manfredo (1991). "The paragraphs about leisure and recreation experience preference scales: Results from two inventories designed to access the breadth of the perceived psychological benefits of leisure." In B. L. Driver, G. L. Peterson, and P J. Brown (Eds.), *Benefits of Leisure* (pp. 263-86). State College, PA: Venture Press.

Graefe, A. R., F. R. Kuss, and J. J. Vaske (1990). *Visitor Impact Management: The Planning Framework.* Washington, DC: National Parks and Conservation Association.

Haas, G. E., B. L. Driver, and P. J. Brown (1981). "Measuring wilderness recreational experiences." In *Proceedings of the Wilderness Psychology Group Annual Conference.* Durham, NH: University of New Hampshire, Department of Psychology.

Hautaluoma, J., and P. J. Brown (1979). "Attributes of the deer hunting experience: A cluster analytic study." *Journal of Leisure Research*, 10, 271-87.

Knudson, D. M. (1984). *Outdoor Recreation*. New York: Macmillan.

Lawler, E. E. (1973). *Motivations in Work Organizations*. Monterey, CA: Brooks/Cole.

Lee, T. W., E. A. Locke, and G. P. Latham (1989). "Goal setting theory and job performance." In L.A Pervin (Ed.), *Goal Concepts in Personality and Social Psychology* (pp. 291- 326). Hillsdale, NJ : Lawrence Erlbaum Associates.

Manfredo, M. J., B. L. Driver, and P. J. Brown (1983). "A test of concepts inherent in experience based setting management for outdoor recreation areas." *Journal of Leisure Research*, 15, 263-83.

Manfredo, M. J., M. Paulson, J. Wurtz, and A. Bright (1992). *Development of a Recreation and Tourism Assessment System for the Rocky Mountain Arsenal*. (Final Report for the Cooperative Agreement 14-16-0009-1552, Work Order No. 27). Fort Collins, CO: Colorado Fish and Wildlife Research Co-op Unit, Colorado State University.

Manfredo, M. J., D. Fulton, F. Ciruli, S. Cassin, J. Lipscomb, L. Sikorowski, and S. Norris (1993). *Summary of Project Report: Colorado's Recreational Uses of and Attitudes toward Wildlife*. (Human Dimensions in Natural Resources Unit Report Number 6). Fort Collins, CO: Human Dimensions in Natural Resources Unit, Colorado State University.

Manfredo, M. J., and R. A. Larson (1993). "Managing for wildlife viewing recreation experiences: An application in Colorado." *Wildlife Society Bulletin*, 21, 226-36.

Manfredo, M. J., B. L. Driver, and M. A. Tarrant (1996). "Measuring leisure motivation: A meta-analysis of recreation experience preference scales." *Journal of Leisure Research*, 28(3), 188-213.

Moscardo, G. (2000). "Understanding wildlife tourism market segments: An Australian marine study." *Human Dimensions of Wildlife*, 5:2: 36-53.

Schreyer, R., D. Lime, and D. Williams (1984). "Characterizing the influence of past experience on recreation behavior." *Journal of Leisure Research*, 16(1), 34-50.

Shelby, B., and T. A. Heberlein (1986).*Carrying Capacity in Recreation Settings*. Corvallis, OR: Oregon State University Press.

Vaske, J. J., B. Shelby, A. Graefe, and T. Heberlein (1986). "Backcountry encounter norms: Theory, method and empirical evidence." *Journal of Leisure Research*, 18, 137-53.

Table 5-3. Example of A Wildlife Viewing Opportunity Typology Developed for a Site Along the Front Range, Colorado.

<u>Activities</u>
Highly Specialized Hiking/Wildlife Viewing
Hiking and wildlife viewing. Low confinement with opportunities for stalking.

Highly Specialized Wildlife Photography
Photography, sketching, and painting with low confinement with opportunities for stalking.

Nature Touring
Visiting nature center, taking bus or train tour, visiting roadside information stops. Observes and enjoys nature, but does so at a distance. Does not enter or intrude upon the natural environment.

General Interest Wildlife Viewing/Hiking
Taking short to moderate distance hikes on nature/interpretive trails. This might include organized nature tours and/or group nature tours. Observes and enjoys nature with moderate entrance into the natural environment.

General Interest Wildlife Viewing/Biking
Bike touring combined with wildlife viewing. Observes and enjoys nature with low entrance into the environment. An emphasis of this experience is on mobility and exercise.

<u>Experience Type</u>
Highly Specialized Hiking/Wildlife Viewing
Learning and teaching others are the immediate and most easily recognizable motivations of this group. There is a strong and enduring interest in learning about nature. Socialization is also important due to the involvement with a particular referent group. Being part of a particular referent group allows expression and reinforcement of self,e.g., these individuals want to be recognized as naturalists or birdwatchers, many of their friends share this common interest and offer social reinforcement for the individuals involved in this form of recreation.

Highly Specialized Wildlife Photography
Participation in this form of recreation emphasizes satisfaction of motivations for creativity, which is of primary importance to the recreationist. Also of importance is the opportunity to develop skills, achieve, use equipment, explore and experience autonomy.

Nature Touring
Participation in this form of recreation allows realization of a broad range of satisfactions. These are satisfactions which might be found in a number of different forms of recreation, i.e. these goals are not necessarily dependent upon the specific qualities related to wildlife viewing; other experiences may

table continues

provide a somewhat reasonable substitute. The types of satisfactions realized include enhancement of family togetherness, in-group relations, general learning, physical rest, escape the stress of every day life, stimulation and change of pace.

General Interest Wildlife Viewing/Hiking
Participation in this form of recreation allows realization of a broad range of satisfactions. These are satisfactions which might be found in a number of different forms of recreation, i.e. these goals are not necessarily dependent upon the specific qualities related to wildlife viewing; other experiences may provide a somewhat reasonable substitute. The types of satisfactions realized by this group include enhancement of family togetherness, in-group relations, general learning, physical rest, escape the stress of every day life, stimulation and change of pace.

General Interest Wildlife Viewing/Biking
Participation in this form of recreation allows realization of satisfactions derived primarily from biking. These goals may not necessarily be dependent upon the specific qualities related to wildlife viewing, but this activity adds considerable quality to the experience. The types of satisfactions realized by this group include enhancement of family togetherness, in-group relations, general learning, exercise, stimulation and change of pace. Challenge, achievement skill and competence may also be important to this group.

Resource Requirements
Highly Specialized Hiking/Wildlife Viewing
Highly dependent on natural resource base. While preferences for viewing particular species (birds of prey, eagles, rare/endangered species) is similar to others who view wildlife, it is more specific and narrowly focused. For example, instead of just seeing an eagle they may want to see an eagle in a specific feeding or nesting situation. High naturalness of immediate surroundings would be desirable. Environment in which viewing occurs offers unconfined and spontaneous animal movement.

Highly Specialized Wildlife Photography
Highly dependent on natural resource base. While preferences for viewing particular species (birds of prey, eagles, rare/endangered species) is similar to others who view wildlife, it is more specific and narrowly focused. For example, instead of just seeing an eagle they may want to see an eagle in a specific feeding or nesting situation. High naturalness of immediate surroundings would be desirable. Environment in which viewing occurs offers unconfined and spontaneous animal movement.

Nature Touring
Moderate degree of naturalness of surroundings. Interest is not focused on a diversity of species or specific wildlife, although there is interest in a high profile or highly symbolic wildlife (e.g. eagles, bighorn sheep). Cultural or historical attractions, in addition to the naturalness, would add to the variety and to the overall quality of the experience.

General Interest Wildlife Viewing/Hiking
Moderate degree of naturalness of surroundings. Interest is not focused on a diversity of species or specific wildlife, although there is interest in a high profile or highly symbolic wildlife (e.g. eagles, bighorn sheep). Cultural or historical attractions, in addition to the naturalness, would add to the variety and to the overall quality of the experience.

General Interest Wildlife Viewing/Biking
Moderate degree of naturalness of surroundings. Interest is not focused on a diversity of species or specific wildlife, although there is interest in a high profile or highly symbolic wildlife (e.g. eagles, bighorn sheep). Cultural or historical attractions, in addition to the naturalness, would add to the variety and to the overall quality of the experience.

Likely Past Experience of Visitors
Highly Specialized Hiking/Wildlife Viewing
Moderate to high. Takes an average of 2 or more trips/month for the purpose of viewing wildlife.

Highly Specialized Wildlife Photography
Moderate to high. Takes, on the average, 2 or more trips per month for this form of recreation.

Nature Touring
Moderate to occasional participation in this form of recreation, however high repeat visitation to the same site is unlikely.

General Interest Wildlife Viewing/Hiking
Moderate to occasional participation in this form of recreation, however high repeat visitation to the same site is quite unlikely.

General Interest Wildlife Viewing/Biking
Occasional to high participation in biking. Moderate to low visitation to the same site.

Likely Level of Specialization of Visitors
Highly Specialized Hiking/Wildlife Viewing
Level of specialization likely for visitors is high. Many of these visitors are likely to be strongly committed and involved in this form of recreation. Their involvement has moderate to high probability of being related to central life interests.

Highly Specialized Wildlife Photography
Highly involved and committed to this form of recreation. This form of recreation is likely to be central to life interests of the participants.

Nature Touring
This is a generalist type of experience. Participants are typically not committed to, or highly involved in this form of recreation. There are virtually no technique requirements to participation.

table continues

General Interest Wildlife Viewing/Hiking
This is a generalist type of experience. Participants are typically not committed to or highly involved in this form of recreation, although they may be interested in becoming more specialized. Technique requirements for engaging in this experience are low and include only the physical capability to engage in short distance hikes or mechanized aids to mobility (e.g. wheel chairs).

General Interest Wildlife Viewing/Biking
This may range from generalist to specialist. However, participants are typically not committed to or highly involved in the wildlife viewing aspects of the experience, rather they are focused on biking. Technique requirements for engaging in this experience are moderate to high.

Social Norms
Highly Specialized Hiking/Wildlife Viewing
Highly focused and responsive to members of the same group. Strong preference for minimal outgroup contact necessitating standards and management practices which ensure opportunities for low density recreation.

Highly Specialized Wildlife Photography
Minimal outgroup contact is required. Participation occurs in small groups or alone.

Nature Touring
A high amount of outgroup contact is expected, and in moderate amounts is desirable. Social carrying capacity is typically limited by facility design (e.g. available parking, room in a visitor center, seats on tour bus or train).

General Interest Wildlife Viewing/Hiking
A moderate amount of outgroup contact is expected, although low amounts may be desirable. Outgroup contact with more mechanized forms of recreation like biking or bus tours are quite undesirable.

General Interest Wildlife Viewing/Biking
A moderate amount of outgroup contact is expected, although low amounts may be desirable.

Managerial Requirements
Highly Specialized Hiking/Wildlife Viewing
Strong interest in obtaining detailed information about species, where to view them, natural history of wildlife and management activities at the area. Likely to be responsive to reasoned cognitive appeals made by managers. Informational programs particularly for the likely visitor which is more experienced, would allow for continuous updating (a newsletter, a phone call-in for current conditions etc.). Controlled and/or limited access is acceptable and may be necessary to accommodate the interests of visitors pursuing this experience. Trails with minimal interpretive signing (since this is generally geared to first time users) and designated viewing areas (which permit off-trail hiking) assist in facilitating this opportunity. Viewers seeking

this experience are likely to have high involvement over a long term with managers. Often well formed opinions about management with a territorial, investment or ownership attitude being evident.

Highly Specialized Wildlife Photography
Strong interest in obtaining detailed information about species, where to view them, natural history of wildlife and management activities at the area. Visitors likely to be responsive to reasoned cognitive appeals made by managers. Informational programs would allow for continuous updating (a newsletter, a phone call-in for current conditions etc.). Controlled and/or limited access is acceptable and may be necessary to accommodate the interests of those seeking this experience. Trails with minimal interpretive and designated viewing areas which permit off-trail hiking assist in facilitating this opportunity. Viewers seeking this experience are likely to have moderate to high involvement over a long term with managers. Often posses well formed opinions about management with a territorial, investment or ownership attitude being evident.

Nature Touring
This form of recreation generally requires high site hardening and development by management. Developments such as visitor centers, bus tours, observation points typically involve sophisticated access development (roads, trails) with consideration of traffic flows (directional signs, appropriate designing), consideration of staging or queuing areas, considerable capital investment in facilities or machinery instrumental to the activity, high staffing requirements (interpretation, maintenance and enforcement) and perhaps involvement with concessionaires who might provide these services. There is a high demand for information/education/ interpretation for this experience. It is critical that information be quite general and be presented in a form that does not require extensive cognitive effort if it is to be effective. Information should be conveyed through symbols, pictures, dioramas, or in brief phrases and simple language.

General Interest Wildlife Viewing/Hiking
This form of recreation generally requires moderate site hardening and development by management. Reinforced trails and signing are the predominant facilities instrumental to the goals of the experience, however, parking and restroom facilities must be considered. Staffing is needed to ensure enforcement, moderate amounts of interpretation and maintenance. There is a high demand for information/ education/interpretation for this experience. It is critical that information be quite general and be presented in a form that does not require extensive cognitive effort if it is to be effective (effective in conveying anything from interpretive information to rules and regulations). Information should be conveyed through symbols, pictures, dioramas, or in brief phrases and simple language. Interpreters can also be instrumental.

table continues

General Interest Wildlife Viewing/Biking
This form of recreation generally requires moderate site hardening and development by management. Reinforced trails and signing are the predominant facilities instrumental to the goals of the experience, however, parking and restroom facilities must be considered. Staffing is needed to ensure enforcement, moderate amounts of interpretation, and maintenance. There is a moderate demand for information/education/interpretation for this experience. It is critical that information be quite general and be presented in a form that does not require extensive cognitive effort if it is to be effective (effective in conveying anything from interpretive information to rules and regulations). Information should be conveyed through symbols, pictures, dioramas, or in brief phrases and simple language. Interpreters can also be instrumental.

Temporal Requirements
Highly Specialized Hiking/Wildlife Viewing
Activities are typically timed to the movements of wildlife permitting the greatest opportunity for viewing. On a daily basis, this most often includes early morning and late evening hours.

Highly Specialized Wildlife Photography
Activities are typically timed to the movements of wildlife permitting the greatest opportunity for viewing. On a daily basis, this most often includes early morning and late evening hours.

Nature Touring
Non-specific

General Interest Wildlife Viewing/Hiking
Non-specific

General Interest Wildlife Viewing/Biking
Non-specific

A Planning Framework for Experience-based Wildlife-viewing Management

David C. Fulton, Doug Whittaker, and Michael J. Manfredo

The Challenge of Good Planning

Perhaps now more than ever before, the public is interested and involved in natural-resources planning and decision making. A diverse public is engaged in camping, hiking, hunting, fishing, wildlife viewing, and other outdoor activities and wants to express its competing desires and values to the agencies responsible for managing the resource. Due to federal and state laws such as the National Environmental Policy Act, agencies are also compelled to comply with the public's wish to be involved in decision making in meaningful ways. Most agency managers now recognize that not only does the public have a right to be involved, but collaborative involvement of the interested public will lead to better decisions.

Despite the emphasis on public involvement and planning in the natural-resources arena, natural-resource planning efforts often fail to meet four fundamental criteria of good environmental decision making identified by Driver (1999). These criteria are: (1) active involvement of stakeholders in policy development and decision making; (2) decisions that clearly express specific outcomes and adequately consider the positive and negative consequences of any decisions; (3) efficient, cost-effective decisions; and (4) decisions that protect or enhance the biophysical environment. Perhaps by examining the struggles of a typical planner in a fish and wildlife management agency, we can get a better feel for the challenge of doing "good" planning—planning that engages the public to develop specific outcomes that are cost effective and beneficial to the environment and people.

A Wildlife Planning Misadventure

Chris is in charge of planning efforts to develop a watchable-wildlife plan for a state fish and wildlife agency. As in most states, interest in viewing wildlife has grown dramatically in Chris's home state during recent years, with almost half of the state's residents participating in wildlife viewing each year. There is also growing recognition in the state that wildlife viewing means big business for local communities and the state's tourism industry, and a governor's task force has identified wildlife-related tourism as a key ingredient in improving the sagging economies of rural communities in the state. The array of stakeholders who are interested in the planning effort is immense, and it is widely anticipated by state decisionmakers that the plan Chris has responsibility for developing will provide key direction and innovations for effectively managing and enhancing the economic and social benefits of wildlife viewing in the state.

At the beginning of the planning process, Chris felt there was more than ample time to prepare a draft plan for review, but now the deadline is looming fast on the horizon without the level of preparation or detail that Chris had intended. During the past several months, Chris has been responsible for a variety of time-consuming activities that needed immediate attention, including the potential listing of a high-profile species, a cooperative habitat-protection program with ranchers, problem bears in campgrounds, a new budgeting system, and a host of personnel issues. Regardless, the first draft of the statewide wildlife-viewing plan is due in six weeks and stakeholders are calling, e-mailing, writing letters, and visiting the office. In just the last couple of weeks, Chris has communicated with state legislators, commissioners, ranchers, the agency director, several newspaper reporters, a representative of an anti-hunting group, several local hunting groups, local chambers of commerce, and a high-profile television personality. The topic even intrudes upon Chris's personal life with numerous calls at home from key stakeholders to discuss the impending draft plan.

Chris's original hope of conducting a long-range comprehensive analysis to drive the planning effort has vanished. Now, the focus is on a "minimalist" approach. This strategy asks what changes to prior plans and programs need to be made so that a new wildlife-viewing plan can be written without significant upheaval or dissent from stakeholders. Chris is now approaching this planning task in the time-honored way of doing business as usual by basing the planning decisions on what has been done by the agency in the past. A few key agency people will be assembled to develop a preferred planning alternative that can be taken to the Wildlife Commission, which has the ultimate decision authority for accepting or rejecting the plan. While good data may serve as a backdrop to discussions, the preferred alternative will be based primarily on the best guess of these managers and its acceptance will rest upon their credibility and a show of support from key stakeholders at the commission meeting. With any luck, the alternative will be accepted by the Wildlife Commission, and Chris's next assignment, development of a new predator population-control policy, can begin.

Lessons from Chris's Misadventure

Although Chris's story is purely fictional, it probably sounds all too familiar. This fictional case highlights the need for a clearly delineated planning process for involving stakeholders to make decisions about wildlife planning issues. Such a process must be based on the best available scientific information, but it must also be pragmatic in recognizing the time and budget constraints that are realities within any organization. As others have recognized, good public planning often involves a blending of planning approaches (Hudson 1979). In particular, public planning must chart a course between, on the one hand, decisions that are made without adequate consideration of desirable long-term goals or scientific information and, on the other, unrealistic decision-making processes that become bogged down in collecting and analyzing data and complex meetings without consideration of what information is actually necessary for decision making. What is needed is a planning approach that balances the advantages and disadvantages of *rational-comprehensive planning* and *incremental planning* (see Box 1).

Chris's planning efforts demonstrate the worst case of what can go wrong with both rational-comprehensive planning and incremental planning. Chris initiated the process with the intention of implementing a thorough rational-comprehensive planning effort that would consider all the technical and scientific information available, as well as involving all interested parties. Unfortunately, Chris failed to recognize a key criticism of rational-

To be effective, Experience-based Management planning efforts need to be integrated throughout other major planning processes.
(Photo by David Fulton)

Box 1. Comparison of Rational-Comprehensive and Incremental Planning (adapted from Lindblom, 1959)

Rational-Comprehensive	Incremental
• Identify and clarify values and objectives prior to generation and analysis of decision alternatives	• Identification of goals and analysis of options is not distinct but closely integrated
• Separation of "ends" and "means" in decision making	• "Means" and "ends" are not treated separately but together
• Test of an alternative is that it is the "best" means to achieve desired ends	• Test of a good alternative is agreement among analysts that it is good (no means/ends analysis)
• Comprehensive analysis with every factor taken into consideration	• Drastically limited analysis that may neglect important impacts and values considerations
• Emphasis on quantifiable measurement and scientific theory	• Successive comparison of alternatives decreases emphasis on quantification and theory
• Assumes ability to define a single public interest	• Explicitly recognizes multiple, competing public interests

comprehensive planning processes that others (Lindblom 1959; Braybrooke and Lindblom 1963; Bailey 1982) have identified: limits on knowledge, time, resources, and money often constrain the degree to which decision making can be comprehensive. All decisions involving complex problems are constrained to some degree. If we fail to recognize this, we might initiate a planning process that is unrealistic and impossible to complete. Instead, we must recognize the constraints and plan accordingly. Chris's failure to recognize the practical limitations of rational-comprehensive planning led to the circumstance rational-comprehensive planning strives to avoid—extreme incrementalism. While the critiques of rational-comprehensive planning must be acknowledged, "good" decisions cannot be made without at least some consideration of future goals and objectives and some analysis or assessment of the best means of achieving those goals and objectives. The severe time and resource constraints that Chris faced led to a decision-making process that failed to either adequately involve the public or ensure a technically competent decision.

The case highlights common pitfalls that produce less-than-optimal planning processes. First, while Chris had good intentions, the stakeholders were never truly engaged in developing policy or management decisions. Second, because stakeholders were not engaged in the planning process, the plan is not likely to define clear management outcomes that are supported by the stakeholders. Third, inadequate time and resources were devoted to analyzing the positive and negative consequences of the plan on the biophysical environment and people. The primary lesson of Chris's story is the need to balance the advantages of the rational-comprehensive and incremental approaches to ensure that planning achieves the four criteria of success identified by Driver (1999). In the following chapter, we introduce a planning process that embraces Etzioni's (1967) ideas of integrating incremental and rational-comprehensive approaches to planning and attends to the four criteria of good decision making delineated by Driver (1999).

The Philosophical Foundations of Outcomes-focused Planning

Several planning frameworks have been developed for use in managing natural-resource-based recreation (Stankey et al. 1985; Shelby and Heberlein 1986; Driver et al. 1987; Graefe et al. 1990; National Park Service 1997). While there are both important and subtle differences between the various frameworks developed by different researchers or different agencies, most of these frameworks favor: (1) collaborative processes that include substantial and meaningful involvement of stakeholders throughout the planning process; (2) approaches focused on defining clear outcomes for management via goals and objectives; and (3) science-based approaches for collecting and analyzing information important to decision making.

Experience-based Management (EBM) focuses on producing outcomes through management actions. These outcomes are manifested in crucial recreation experiences that provide the motivation for recreation participation and benefits that accrue to individuals, communities, societies, cultures, economies, or the environment due to management actions associated with producing recreational opportunities (Driver et al. 1991b; Driver 1999; Driver

Collaboration is an essential component of Experience-based Management. (Photo by David Fulton)

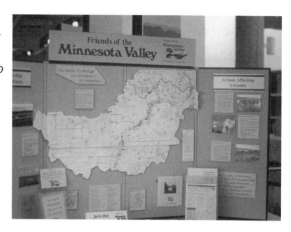

and Bruns 1999; Chapter 2 of this volume). The rational component of outcomes-focused planning emphasizes the use of scientific methods to evaluate the social and biophysical components of the systems in which planning is taking place. A collaborative approach emphasizes the fact that planning decisions, while informed by science, are ultimately social decisions that are directed by human values and preferences that are often in conflict.

Why Collaborative Planning?

Incorporation of public values to guide decision making is the foundation of good environmental decisions. This can only be accomplished through the direct and meaningful involvement of stakeholders in the decision-making process in a way the stakeholders deem to be fair (Wondolleck and Yaffee 2000, Lauber and Knuth 1999, Lawrence et al. 1997). Certainly, decisions concerning the management of wildlife resources must be based on solid biological and social scientific information (Manfredo et al. 1996), but plans for public resources such as wildlife are first and foremost a social agreement among the diverse stakeholders in society. Effective public involvement in planning requires the formation of a partnership with the stakeholders, or interests, who are affected by planning decisions (Bruns 1997). Planning that is supported by a broad, representative segment of the public does not occur by simply applying social or biophysical knowledge to solve problems; it requires direct communication among governmental agencies and the public to which these agencies are accountable.

This emphasis on direct public participation and consensus-seeking processes reflects a growing perspective within the natural-resource-planning community—a belief in the power of public and interest-group input and support in planning and decision making (Wondolleck and Yaffee 2000, Cortner and Moote 1999, Daniels and Walker 1996, Lee 1993, Crowfoot and Wondolleck 1990, Kartez 1989, Friedmann 1973, 1987). Moreover, it represents application of a growing body of evidence from political and social psychology concerning the need to address procedural justice concerns, or concerns that the process by which natural resource decisions are made is fair to all stakeholders (Lawrence et al. 1997; Tyler et al. 1997; Tyler and Degoey 1995; Lauber and Knuth 1999). One key aspect of procedural justice is having adequate access, or voice, in the process. Although reliable social information can be collected and utilized without direct public involvement, face-to-face, collaborative processes are essential to providing a voice to key stakeholders (Lawrence et al. 1997).

The collaborative approach emerged out of growing dissatisfaction with and distrust of government agencies in the 1960s and 70s, and its principal tenet is that stakeholders (interest groups, local communities, interested members of the public) ought to have greater input into decisions throughout the planning process (Friedman 1973; Culhane and Friesma 1979). Collaborative planning represents a shift to shared responsibility in defining goals and objectives, and selecting preferred alternatives among agency experts and the interest groups and stakeholders who use the resource (Decker and Chase 1997). A collaborative approach does *not*, however, entail dismissing reliable scientific information in favor of opinionated bargaining among competing interests. As Etzioni (1968) noted, a society that helps its members to achieve higher needs must balance consensus-seeking process with some form of social control that helps provide direction for the future. Commitment to using scientific and technical knowledge to evaluate consensus decisions serves as a mutually agreed-upon form of social control. The consensus decision is acceptable only so long as it also proves to be technically competent.

Integrating Collaboration and Science

EBM strives to integrate the tenets of goal-directed, science-based planning with the ideal of meaningful, collaborative involvement of stakeholders. This integration is facilitated by the systems outlook of EBM, which views the production of wildlife-viewing experiences, and subsequent benefits, as an interaction of inputs and outputs among physical, social, and managerial systems (Buckley 1968; Driver and Bruns 1999). Thus, the systems perspective of EBM encourages a more holistic approach to wildlife-recreation planning that considers and integrates social and biophysical parameters.

EBM is focused on managing for the experiences of the visitor as well as the condition of biophysical resources. For this reason, it emphasizes collaboration with users and affected communities along with developing scientifically valid information necessary for assessing opportunity demand and resource capability and supply, and evaluating potential alternatives (Driver and Bruns 1999). Such collaboration requires face-to-face social dialogue with a range of interests from affected communities including local governments, chambers of commerce, interest groups, and scientific experts from other agencies or organizations. Collaboration also includes community- and user-focused social science research directed at describing, in a reliable and valid fashion, the broader concerns and desires of the affected public. We want to emphasize again that the goal of collaboration in EBM planning processes is not to simply let stakeholders negotiate decisions based on opinions without the benefit of scientific or technical information. Instead, collaboration in EBM means ensuring that a full range of diverse values and interests are represented when deciding what future conditions are desired and when deciding which means are preferred for obtaining these conditions. Collaboration includes formal decision-making processes in which representatives from the broad range of stakeholder interests meet face-to-face to develop decisions that are acceptable to all. Such processes can be extremely beneficial if controversy or conflicts are anticipated (Wondolleck and Yaffee 2000, Susskind and Cruikshank 1988).

Although EBM recognizes the necessity of a collaborative approach to planning, the remainder of this chapter focuses on the underlying process of planning and not on the activities required for effective collaboration with stakeholders. The topic of stakeholder involvement is, however, thoroughly addressed in Chapter 7.

A Planning Model for Experience-based Management of Wildlife Recreation

Experience-based Management for wildlife recreation takes place within the context of a general comprehensive planning process (Figure 6-1) that thoroughly embraces the idea of a collaborative partnership among the planning agency and interested stakeholders (Figure 6-2, Boxes 2-5). As illustrated in Figure 6-2, the fundamental action in EBM planning is creating a partnership among management agencies, local communities (including, at least, local government, residents, interest groups, citizen organizations, and landowners), and the business community (including chambers of commerce, tourism associations, and businesses associated with wildlife-viewing recreation and tourism). This collaborative partnership provides an interactive forum that is necessary for assuring that the interests of all stakeholders are addressed in the planning process. A partnership is developed through various formal and informal activities such as the creation of a citizen/stakeholder planning committee that, together with agency representatives and technical experts, will provide oversight and direction for planning decisions. The development of a partnership among agencies and stakeholders creates an environment of trust and activism that guides planning activities based on community desires. Chapter 7 provides more detail on how to involve stakeholders throughout the planning process.

While comprehensive planning processes vary from organization to organization, almost all processes used in natural-resource settings share a common sequence of steps that begins with defining or clarifying an organization's mission or fundamental purpose. Among natural-resource agencies, the mission is typically a general statement emphasizing conservation of natural resources for long-term benefits. Pursuant to this fundamental mission, and depending on the scale of the planning effort, the planning process typically involves the development of (1) a vision of desired future conditions

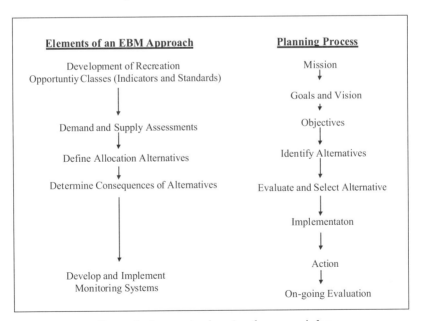

Figure 1. A generic planning framework for Experience-based Management

Box 2. Stages in Outcomes-focused Planning

Planning product	*How is it developed/decided?*
Mission	Formal political process when forming or directing the agency (legislative and/or executive branch of local, state, federal government).
Goals (Vision)	Long-range visioning involving collaboration of agency and stakeholders affected or interested in the agency's future management direction. Open dialogue to reach a consensus vision for the future. Social science research would include developing wildlife-viewing experience typology through focus groups and surveys of the users.
Objectives, standards, and indicators	Goals (consensus value statements) are reviewed in conjunction with scientific and technical knowledge to describe these goals in a measureable way consistent with the resources on the ground and demand for experiences. Integration of scientific research (social and biophysical) with consensus-seeking process. Social science research is directed at understanding what social and biophysical factors affect production of various viewing experiences and demand for these experiences.
Range of alternatives	Range of technically viable choices representing key preferences of major stakeholders and reflecting technical assessment of demand and supply of experiences and resources instrumental to the production of experiences. Integration of scientific research (social and biophysical) with consensus-seeking process is used to ensure an adequate range of alternatives.
Preferred alternative	Selection/creation of preferred alternative based on best available scientific and technical information and consideration of allocation preferences. Developing consensus on the preferred choice is instrumental to long-term political viability of the plan. Social science research and stakeholder processes are used to help determine social preferences.

Monitoring and evaluation	Long-term assessment of resource conditions and experience of users to determine if management actions are achieving desired consequences. Monitoring framework is based on system of indicators and standards using science-based methods for implementation.

explicated in a general set of goals, and (2) more specific objectives that give meaning to the goals. The specificity of such objectives is enhanced through indicators and standards that concretely define the quality of desired experiences and resource conditions. The model then requires (3) exploration of alternative actions for meeting the standards, including analyses of the consequences of these actions; and finally, (4) the evaluation and selection of the preferred actions. The preferred actions are then (5) implemented and (6) evaluated through an ongoing monitoring effort to determine if actions are helping meet the standards and objectives established in the plan. The "front end" of the plan (parts 1 and 2) defines what will be provided, while the "action" component (parts 3-6) explores the ways to make that happen.

The actual steps in a planning process are often matched to these elements in the plan, suggesting that planning is a serial process.

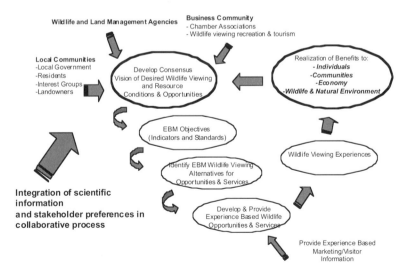

Figure 2. Collaborative process for Experience-based Management of wildlife-viewing opportunities. (Adapted from Bruns [1997])

Box 3. Description of Planning Missions, Goals, and Objectives

Mission Statements

The well-known mission of the National Park Service (NPS) is to "... promote and regulate the use of the ... national parks ... which purpose is to conserve the scenery and the natural and historic objects and the wildlife therein, and to provide for the enjoyment of the same in such manner and by such means as will leave them unimpaired for the enjoyment of all future generations" (National Park Service Organic Act). Similarly, the mission of the Alaska Department of Fish and Game (ADF&G) is to "manage, protect, maintain, and improve the fish, game and aquatic plant resources of Alaska ... to ensure that Alaska's renewable fish and wildlife resources and their habitats are conserved and managed on the sustained-yield principle, and the use and development of these resources are in the best interest of the economy and well-being of the people of the state." The mission statements for the NPS and ADF&G indicate that both agencies have a duty to not only conserve wildlife but also provide for the "enjoyment" or "benefit" of the public through use of the resource. Such statements indicate that an EBM approach has the potential to be successful within both agencies.

Goals

The goals of a state wildlife-management agency might emphasize a diversity of types of wildlife habitat, species, and recreation experiences such as hunting, trapping, and wildlife viewing throughout the state. Goals for management at site- or tract-specific settings might provide more specific direction, such as emphasizing the need for sustained resource health (general statements about no loss of certain types of critical habitat or optimal populations for certain species), a few high-quality hunting or wildlife-viewing and learning experiences, and the minimization of conflicts between those types of wildlife recreation and resource health.

Objectives (Indicators and Standards)

An example of objectives for wildlife-related recreation in Denali National Park might include statements that urge the creation of low-development, solitude- and challenge-oriented viewing opportunities in backcountry areas while urging the creation of accessible, non-challenge-oriented viewing opportunities from the park road bus system. Indicators and standards that help define meaningful measures for solitude and challenge could be developed for quality backcountry viewing in Denali National Park, just as they could be developed for accessibility and safety along the road.

Planning, however, is typically an iterative process, and planners often have to revisit their goals and objectives as actions and alternatives are debated or new information is developed concerning the link between actions and indicator variables. These elements also differ depending upon the scale of the planning effort. Consideration of the scale of the planning effort is important because planning can be done for a specific site, a tract of public land with several potential sites, a region such as a municipality or national forest with several tracts of mixed management, or even a statewide or larger regional area which involves consideration of a range of agency missions.

The following discussion outlines a planning framework for wildlife recreation and suggests how EBM concepts fit into that framework.

Mission. Mission statements provide a concise summary of the fundamental social benefits that the agency has been charged to provide. Such mission statements represent a direct expression of social values without an attempt at providing specifics related to resource management. While mission statements do not provide much specificity for management actions, it is the mission of the agency that indicates whether or not EBM approaches to wildlife-recreation management have the potential to work within the context of a particular agency. Because a philosophy of providing service to people is central to EBM approaches, agencies that do not have a mission that clearly focuses on providing benefits or enjoyment to people are unlikely to successfully adopt EBM for wildlife viewing or any other purpose.

Goals. Goals describe society's shared vision of desirable future conditions. Like mission statements, goals are qualitative, abstract statements that are a reflection of values and are rarely controversial. But goals typically provide some programmatic direction about what managers are trying to provide. Goals are the restatement of societal,

text continues on page 108

Box 4. Roles of the Planner and Stakeholders in the Planning Process

Planner: Within EBM planning the planner and planning agency ensure two primary things: (1) a competent decision, and (2) a fair planning process. The planner's responsibility in making a competent decision includes making sure that necessary scientific and technical information is collected and adequately considered in the planning decision. This includes biophysical data concerning the resource and social science data concerning demand for experiences and factors affecting such experiences. A decision that is not based on solid scientific information is incompetent, even if it is reached through a consensus of stakeholders. The EBM philosophy of planning strongly adheres to guidance of the planning process via scientific information. The planner and planning agency must also work to develop a collaborative partnership with interested stakeholders to ensure a fair decision-making process that engenders support and trust for the agency and the resulting plan. This requires direct, two-way communication with key representatives of all interested stakeholders. Such collaboration takes place throughout the planning process, but it is primarily focused on defining a vision and goals for desired future conditions.

Stakeholder: The background of stakeholders will vary greatly from process to process and within each EBM planning process. The primary role of the stakeholder is not as technical expert. Stakeholders are participating in the planning process to ensure that a diversity of interests are represented during consideration of desired future conditions and the acceptability of different means for obtaining those conditions. Direct participation of stakeholder interests is vital to building trust among all parties and with the planning agency. Such participation can also help generate innovative approaches to management. However, input from a limited number of stakeholders neither overrides scientific information nor substitutes for social scientific-based information concerning demand for EBM wildlife-viewing opportunities.

Box 5. Collaborative Planning Activities

Stage	Activities
Goal setting (Partnership building)	Creation of stakeholder planning committee representing agencies, local government, and varied local interests
	Creation of a technical advisory committee to serve the needs of the stakeholder planning committee
	Community visioning sessions with citizens
	Public forums and dialogues focused on the planning issues
	Community surveys
Objectives (standards and indicators)	Broad-based user surveys and focus groups
	Facilitated sessions of the stakeholder planning committee
	Interactive public workshops
Alternatives development	Stakeholder planning committee facilitated group sessions
	Interactive public workshops
Preferred alternative	Stakeholder planning committee facilitated group sessions
	Interactive public workshops
Monitoring and evaluation	Creation of volunteer citizen monitoring groups
	Community events focused on collecting monitoring information
	Community surveys
	Stakeholder planning committee facilitated group sessions
	Interactive public workshops

community, and personal values that exist among the public being served by the management agencies and for whom the resource is being managed to produce benefits (Davidoff and Reiner 1962).

For this reason, meaningful goals cannot simply be developed by agency personnel and applied to management. To have broad-based support, management goals must resonate with the values of the public. The best way to assure that they do is to have the public directly define these goals through face-to-face interaction between the agency and other stakeholders and through mechanisms such as public surveys that can help determine the range of values and the intensity with which they are held in different segments of the public. Open collaboration with all interested stakeholders is essential at the goal-setting stage because it is during this stage that broad decisions about the future are being made. Direct interaction of agency planners with local communities and other stakeholders not only engenders trust among stakeholders that the decision-making process is fair, but it also affords the opportunity for more in-depth, detailed discussion about stakeholder interests than can be obtained through methods such as mail surveys.

Experience-based Management concepts are part of the goal-setting process because the wildlife-recreation "product" is expressed in terms of experiences and benefits, albeit at a broad qualitative level. It is at this goal-setting stage that agencies should be initially considering what wildlife-related experiences and benefits will be produced (Driver 1985). Development of recreation opportunity classes related to wildlife is essential to ensuring that EBM is effectively incorporated into subsequent planning and management activities (see Chapter 5 for detail on how to develop recreation opportunity classes).

Recreation opportunity classes are categorical descriptions of the types of wildlife-viewing experiences, settings, and activities people desire (see Chapter 5), and provide a tool for specifying the types of wildlife-related experiences that are possible given the mix of biophysical, social, and managerial attributes in an area. It is at the goal-setting stage that the range of compatible and incompatible activities and experiences can be identified and defined within a recreation opportunity class—for example, whether or not wildlife-related recreation will be limited to non-consumptive activities. Such decisions can be made on a regional scale (e.g., prohibition of hunting in most national parks) or they can be site specific (e.g., hunting in some state wildlife-management areas but not in others). While the definition of opportunity classes involves an interplay of desired experiences, activities, and settings, the decision about the range of compatible recreation opportunities is a function of established legal mandates and policy, resource constraints, and public demand.

Defining appropriate recreation opportunity classes involves collecting information from users via focus groups, surveys, and other social science research methods. However, it also requires the validation of the defined opportunity classes with the stakeholders themselves. While these classes may be initially defined via cluster analysis of survey data, wildlife viewers and other stakeholders must accept the definitions if the classes are going to be useful as a management tool. For this reason, the stakeholder partners and other public representatives must be involved in the research to define opportunity classes and in shaping the definitions that are ultimately derived from this research. Opportunity classes that are accepted only by the researchers and managers who generated them will be of no value if the public does do accept them as meaningful classifications.

Objectives. Objectives are more concrete statements that specify the intentions of goals in clear terms. To assure clarity in providing future direction, objectives should be (Schomaker 1984; Manning 1999): (1) quantifiable in discrete terms (e.g., not simply more or less of this, but 25 percent more or 30 percent less); (2) bounded in space and time (i.e., should clearly specify when and where the quantifiable objective is to be reached); (3) realistic (i.e., objectives must be plausibly attainable based on known information and technology, but they must also be somewhat challenging to obtain); (4) outcome oriented (i.e., objectives should focus on what is being produced through management and not what resources are used in the management process).

Objectives serve two primary functions. First, they form the basis for a system of quantitative indicators and standards that define the desired conditions of the resources being managed and quality of the experiences being produced from these resources for specific recreation opportunities. Second, they provide guidance regarding the allocation of resources (usually geographically, but temporal, financial, or administrative allocations are also possible) among the different types of opportunities.

Specification of different recreation opportunity classes provides a basis for defining an array of objectives that are each compatible with and reflect the goals and mission of the agency or organization. Objectives should be defined with appropriate recreation opportunity classes in mind, and they must also be developed with the direct involvement of stakeholders. Objectives are not developed simply as guidelines for resource managers; rather they represent in concrete terms the consensus agreement among stakeholders (including managers) about what social outcomes (experiences and benefits) are desirable.

Indicators and standards. To be meaningful, objectives must be communicated in terms that are readily observable or measurable. This step involves choosing measurable variables to define and give meaning to the objectives, and is at the center of every major recreation-planning system in use today (Stankey et al. 1985; Shelby and Heberlein 1986; Driver et al. 1987; Graefe et al. 1990; National Park Service 1997). Briefly, indicators are measurable social or biophysical variables that are closely linked to a recreation opportunity. Standards on an indicator define a range of conditions under which a particular wildlife viewing opportunity is produced. A thorough discussion of developing and applying standards and indicators is provided in Chapter 8.

Constructing Allocation Alternatives. The focal point in planning for recreation experiences and benefits is the allocation decision, which is typically initiated concurrent with defining objectives in the planning process. As planning moves into the development and evaluation of alternatives, the key is to clarify how resources (e.g., space, time, personnel, funds, and natural resources) will be allocated to meet the objectives. Stated another way, it is a declaration of who gets what, when they get it, and how they get it. (See Box 6 for a discussion of criteria for evaluating the outcomes of allocation decisions). For example, if the objective is to provide wildlife-viewing opportunities along a scenic highway, alternatives could include developing a visitor center and four roadside pullouts of a certain size and design. The basis of this kind of allocation decision, and one principal task of the planner, involves two key activities: (1) assessing the demand for specific experience opportunities and the supply of resources and settings that produce such opportunities, and (2) comparing the level of demand and supply of such opportunities.

Because the allocation decision is all about how the decision will impact the various stakeholders, the public must be involved in defining different allocation alternatives and in the process of assessing demand and supply of the resources. *How* that decision is made can be just as important as what the decision is in determining whether or not it will be viewed as a just, or fair, decision (Thibaut and Walker 1975; Lind and Tyler 1988; Tyler et al. 1997). Exclusion of interested and impacted stakeholders from the process will likely cause it to be seen as unfair and lead to lack of support for the allocation decision (see Box 6 for a discussion of *procedural justice*).

Box 6. What's Fair When Allocating Resources?

Distributive Justice

The process of selecting alternatives is a decision about a fair allocation of resources; however, the planner can assist deliberations with development of alternatives that can be evaluated against clear allocation philosophies. Based on Morton Deutsch's (1975) ideas, Shelby and Heberlein (1986) have discussed four key concepts that can guide allocation of recreation resources. The most simple and straightforward criterion is equality, which suggests that each person should have an equal chance at obtaining benefits (in this case, the opportunity for a particular type of viewing opportunity). But this approach may be perceived by some as unfair because it does not recognize that some users have more invested in an opportunity than others. The second concept, equity, is an allocation criterion that explicitly addresses this issue. With equity, allocation is weighted based on a ratio of inputs to outputs: if inputs are unequal, then outputs (allocations) should be unequal. A classic example is the difference in license fees charged to residents versus nonresidents. It can be argued that residents provide greater inputs than nonresidents to creating hunting opportunities in the state (e.g. taxes, land ownership/ habitat) and consequently should be given preference in allocation. A third criterion is efficiency, by which resources are allocated to their most highly valued use(s). More accurately, in the case of wildlife viewing this would mean allocating a mix of opportunities, which would maximize the benefits obtained by viewers. Given this approach, higher-value opportunities (which might be identified by a willingness to pay entrance fees) gain precedence over those with lower value to their potential users. A final allocation criterion is need. This criterion suggests giving priority to certain potential users who may be disadvantaged or disabled from receiving opportunities via another system of allocation. In some cases, such as hunting, allocation is guided by a combination of these concepts. For example, in Colorado, a variety of hunting seasons have been created which differ in their allocation philosophies. Some are based on based on equality (a simple lottery system for moose hunting); some on equity (differential pricing for residents

box continues

versus nonresidents);others on efficiency (differential pricing or waiting for specific hunts via a preference-point system); and yet others on need (youth hunts and hunts for individuals with disabilities).

Procedural Justice
Fairness is not just about the result of an allocation decision (distributive justice), but also involves perception of fairness concerning the process of making decisions. This concept is known as procedural justice (Thibaut and Walker 1975; Lind and Tyler 1988) and is increasingly being recognized by decisionmakers and planners in making decisions concerning natural resources and the environment (Lawrence et al 1997; Lauber and Knuth 1999). It is central to the ideals of environmental dispute resolution and management (Wondolleck and Yaffee 2000; Crowfoot and Wondolleck 1990; Susskind and Cruikshank 1988) and the involvement of stakeholders in making natural-resource decisions (Decker et al. 1996). While there is not complete agreement in the literature, procedural justice appears to depend largely on factors associated with the relationship of parties to decision authorities and other members of the community (Tyler 1994; Tyler and Degoey 1995). Key relational variables include a decision authority who is neutral and trustworthy, and respects the interests of stakeholders. Fair processes are also defined by the opportunity for stakeholders to participate, voice their concerns, and influence decision making.

Assessment of Demand. One of the primary information needs required to construct viable allocation alternatives is a demand assessment that quantifies opportunities desired by the public. Using EBM, this exercise involves identifying public preferences for the specific types of opportunities identified in the recreation opportunity typology. For this reason, traditional approaches to demand assessment are not very useful in EBM.

Typically, demand assessments either use historical data to anticipate likely trends in participation in recreation activities or develop predictive models incorporating information such as price of activity participation and socio-demographic variables such as age, gender, education, and ethnicity (Hof and Kaiser 1983; Walsh 1986; Flather and Cordell 1995). Such a broad-based assessment of activity demand does not provide the level of specificity for desired recreation experiences that is needed for EBM. Under the EBM approach, demand assessments focus not on demand for activities, or even facilities, *per se*, but on the biophysical, social, and managerial conditions that create a particular opportunity type (as defined via indicators and standards).

A demand assessment using EBM must still address the two primary questions essential to any assessment of demand: (1) How is the population that is being served defined? and (2) What measures of participation and opportunity preference are needed? The definition of the service population is a function of agency mission, goals of the planning area, scale of the planning decision, and public interest in the resource. The service population may be as broad as the public who live in or have visited a certain geographic region. For example, given the mission of the National Park Service, the service population of locations such Rocky Mountain National Park or Denali National Park will likely include all residents of the region as well as national and international visitors to the parks. On the other hand, the service population for locations of more local or regional interest may be residents of a particular geographic locale who participate in or are interested in participating in wildlife viewing. Assuring that the definition of the service population is congruent with the goals of a planning area is important to defining the appropriate range of public preferences to be considered.

The second question in a demand assessment asks what measures of participation and value of participation should be obtained. Using EBM, measures of participation should relate closely back to the social and psychological outcomes associated with wildlife viewing that are the primary motivation for recreation participation, as well as the physical, social, and managerial settings that facilitate production of these key experiences (see Chapter 5).

The determination of how the value of wildlife-viewing participation will be measured is extremely important because such values will greatly influence how allocation will proceed. One approach is to use an economic analysis that allows for an explicit comparison of the benefits and costs that result from implementing specific allocation strategies, in a common monetary metric. If this approach is taken, measures of value will most likely be obtained using a travel cost method (Walsh 1986) or a form of contingent-valuation method such as willingness-to-pay assessments (Mitchell and Carson 1989; Loomis 1993). In practice, the economic approach is used infrequently in recreation planning. Instead, an informal and subjective approach to evaluating alternatives is often used (Driver et al. 1991a). With this approach, a wide variety of attitudinal assessment techniques can be used to measure value, including

---

Box 7. Methods for Demand Assessment

Data-collection methods for assessing demand vary, but the most frequently used approach is a survey of the service population. When coupled with probability sampling techniques, surveys are particularly useful in determining the distribution of characteristics in a given population of interest, which is the essential task in demand assessment. In addition, surveys are advantageous because (1) they are conducive to quantification and easy summarization, and (2) they can be implemented in a relatively short period of time and are cost efficient. Despite these advantages, surveys can be disadvantageous because they do not allow for an in-depth assessment of a given topic. Consequently, demand surveys are typically supplemented with stakeholder processes or other forms of public involvement. Techniques that offer promise for future testing would combine the advantages of stakeholder processes and the broadly representative survey methods. For example, a survey might identify key groups within a population from which a small sample could be obtained as part of a planning advisory group that could provide a more thorough understanding of experience preferences. (See Chapter 7 for a detailed discussion).

---

ratings such as degree of perceived benefits, importance, likelihood of taking trips, and degree of desirability (see Box 7).

Assessment of Supply. Assessing resource supply when using EBM for wildlife viewing involves answering two questions: (1) What is the current availability of wildlife-viewing opportunities defined within the recreation opportunity typology? and (2) What is the potential for expanding available opportunities? Supply assessment using EBM requires an inventory of not simply sites but of sites in the context of all the resources that go into producing specific wildlife-viewing experiences, including the specific type of activity and the biophysical, social, and managerial setting attributes associated with a site.

For example, experiences produced by bear viewing may vary not only with the motivations of the visitors but also from site to site depending on the concentration of bears, proximity of bears, behavior of bears toward other bears and toward humans, number of other people at the site, level of visitor autonomy, remoteness of

Providing information about wildlife resources is the key to ensuring that visitors achieve the experiences they desire at a viewing location. (Photo by David Fulton)

the site, and other factors such as facilities and level of development. It is not enough to say there are *x* number of locations to view brown bears in Alaska. An EBM supply inventory for brown-bear viewing in Alaska would also focus on the attributes of a site that facilitate specific wildlife-viewing experiences. Additionally, an inventory needs to identify and categorize not just sites currently within the management system but also sites that are not currently managed by agencies but provide viewing opportunities and other sites that have the potential to produce different wildlife-viewing opportunities than currently available in the management system.

Designing Specific Allocation Alternatives. From a comparison of supply and demand, and through collaborative efforts with stakeholders, a number of different allocation scenarios are crafted along with their likely consequences. An allocation alternative comprises two important elements: a description of the ends to be accomplished and a description of the specific means by which they will occur. The ends are described using terminology derived from the recreation classification system. For example, one alternative might pose a strategy of increasing emphasis on viewing opportunities for recreationists who are highly involved with wildlife viewing and interested in creative expression (painting, photography, poetry) involving wildlife. This strategy might be warranted because, in an evaluation of supply and demand, it is found that the demand for and value of these opportunities is relatively high, yet most of the current opportunity and management effort is targeted toward other, less-specialized opportunities.

The second critical component of creating an alternative is a discussion of the means by which opportunities will be provided.

Creation of opportunities does not necessarily mean development of facilities or even the designation of specific areas for a particular type of opportunity, although either of these actions is possible. Opportunities can be created by a wide variety of actions (or combinations of them), such as introducing interpretive programs, providing information on how to participate or where to go, creating limited-entry permit systems, limiting conflicting uses (such as hunting), timing of uses, site development (trails, pullouts, visitor centers, signing), marketing, or "distance participation" through video connections to the site presented at central locations or via the World Wide Web. The package of actions proposed within a strategy should be guided by an overriding concern about the type of opportunity to be provided.

Collaboration in Demand and Supply Assessment. The stakeholder planning partners are thoroughly involved through all stages of the demand and supply assessments. While technical expertise may reside with agencies or university scientists and planners, the assessment is useful only if the entire decision-making community believes and accepts the data. Actual participation of stakeholders during assessment of demand and supply will likely take the form of providing oversight and direction to technical experts through a stakeholder planning committee.

Determine Consequences (Benefits/Costs) of Alternatives. After allocation alternatives have been defined, the next planning step is an analysis of the benefits and costs of each potential management alternative. In a traditional planning model, this analysis still dominates the decision-making process. It has a presumption of objectivity, particularly when phrased in terms of benefits and costs (the economic model). In fact, however, evaluation remains an essential and unavoidable part of the process because values must be placed on the various benefits and costs, and such values are intrinsically subjective. Valuation is at its core a social process and requires broad agreement among key stakeholders and interests. To assure that there is not a hidden, but fundamental, disagreement concerning the consequences of various alternatives, valuation assumptions must be clearly communicated and acceptable to all affected parties.

Chapters 7 and 12 provide further information about these analyses and the evaluative component. A point to remember, however, is that changing evaluations could require changes in indicators/standards/objectives. For example, after you determine that providing solitude-oriented recreation in a high-demand area may mean significant changes in the managerial setting (e.g., a

Individuals seek a range of viewing experiences. They are not always focusing on the experiences we would expect. (Photo by David Fulton)

permit system), the agency and users may wish to opt for less-stringent encounter standards for defining solitude.

Selecting an Alternative. Finally, the planning process comes down to choosing among the alternatives (each with their associated objectives, indicators, and standards). To guide selection of an alternative, each is described according to the likely consequences it will produce. These consequences might include likely participation levels (by opportunity class), revenue generated, costs incurred, economic impact, effects on target wildlife, as well as other consequences. Final selection comes from several sources, but in keeping with the collaborative notion, the selection decision should arise from negotiated positions of stakeholders and information from users as well as legal mandates and scientific information. Collaborative involvement of stakeholders is important, because, although based on scientific information, the actual selection decision is intrinsically a political decision, or a decision about what is a just, or fair, allocation of resources as well as cost effective.

Implementation. Regardless of the specific allocation decision, the alternative selected must be a readable plan that provides a clear blueprint of explicitly what experiences or benefits will be produced, where and when they will be produced, the quality and quantity that will be produced, and for whom they will be produced. This plan must also specify the means of production, or what actions will be specifically taken to achieve the objectives defined by the indicators and standards (see Chapter 9). Once allocation decisions are made, the next phase of EBM planning directs implementation through the development of site, project, and program "action plans." These action plans provide detail on things such as how

Even while visiting the same site and engaged in similar activities,
different individuals may desire very different outcomes from their
experience. Here a professional photographer shares viewing space with
a father and son at McNeil River Falls State Game Sanctuary, Alaska.
(Photo by David Fulton)

permit systems will be implemented, how sites will be laid out, the types of structures that will be developed, how to time use, what type of information to be presented and by what means it will be conveyed, the scheduling of programs, and other specific decisions. Successful application of EBM depends on the degree to which action plans adhere to and incorporate (1) the overarching intent specified in objectives, and (2) the descriptions of recreation opportunity types. The objectives and classification typologies developed by research and agreed to by the involved stakeholders are the key to assuring action plans that achieve the desired outcomes of the public.

Monitoring and Evaluation. The final step in a planning process involves implementing the actions and, then, monitoring and evaluating consequences of the plan. This step recognizes that planning must involve some mechanism for evaluating consequences and revisiting past decisions to make appropriate changes in strategies and actions. This step is a "feedback loop" that allows changes in actions or standards, as needed. Such a feedback loop alters the linear, rational-comprehensive plan, to an adaptive planning process that is cyclical and responsive to changes in the system under consideration.

Like any planning effort, the EBM approach has little value unless it is used in an ongoing fashion. Once allocation decisions are made and area/site/program plans are implemented, how can we be sure that the plans were successful? Such assurances are made through the evaluative process. Monitoring and evaluation is directed by the specifically defined management objectives that describe the specific quantitative outcomes desired through management and the specific actions that will be taken to achieve those outcomes. These desired outcomes are quantified through the use of indicators and standards, and it is these specific, quantitatively expressed standards that are used as the basis for monitoring and evaluation. At a minimum, evaluation should address the following questions:

• To what extent were desired opportunities for wildlife viewing realized? (i.e., How much use occurred via programs/sites developed for specific types of opportunities?)

• How did visitors evaluate the quality of these opportunities?

• Did site conditions stay within the bounds of standards that were established?

• What cost was expended in providing opportunities and what was its ratio to actual participation?

To answer these questions, an array of evaluative systems must be developed and implemented throughout a visitor-use season. This might include actions such as visitor registration, observation of use or wildlife movements, regular inspection of site facilities, and post-visit evaluations of users.

Stakeholders are a central part of monitoring and evaluation efforts. Use of volunteers from stakeholder groups to help design and implement monitoring projects is an invaluable way to retain the interest and energy of the community of stakeholders. Evaluation of the plan also includes all parties who helped develop it and continue to have a stake in management and decision making. Through such efforts, evaluation becomes the foundation for a recurring cycle of "fine-tuning" action plans and for periodic revision of allocation planning involving all stakeholders.

Conclusion

EBM is focused on defining the specific recreation experiences desired by the public. To ensure public agencies develop wildlife-viewing opportunities that match these desired experiences, planning processes that combine elements of comprehensive and collaborative planning are used. While EBM planning decisions are founded on solid social science information, the public is directly involved in these decisions at all stages of planning.

Summary Points

• Good planning for wildlife-viewing recreation actively involves stakeholders in developing cost-effective decisions that protect resources and are expressed in specific experience-based outcomes chosen with careful consideration of their consequences.
• EBM relies on a planning process that adheres to collecting and using valid and reliable scientific information. Decisions made in EBM have a rational basis derived from a comprehensive consideration of available information.
• Success of EBM also relies on a collaborative partnership among management agencies and other interested stakeholders. Agencies have a responsibility for guiding the planning process based on scientific information and expert judgment, but consensus building requires direct dialogues with interested stakeholders to ensure concerns relating to procedural justice are addressed (see Chapter 7).
• The elements of the EBM approach are implemented through a rational-comprehensive planning process that relies on sustained, collaborative partnerships among the agencies, communities, and users interested in and impacted by the wildlife-viewing opportunities (Figures 6-1 and 6-2).
• Stages in the planning include defining the missions and goals; clarifying objectives; identifying, evaluating, and selecting alternatives; implementation; action; and monitoring.
• Through the planning process, wildlife-viewing opportunity classes are defined; demand and supply assessments of these opportunities are completed; allocation alternatives are defined and evaluated; and actions to develop wildlife-viewing opportunities are implemented and monitored.

Literature Cited

Bailey, J. A. (1982). "Implications of muddling through for wildlife management." *Wildlife Society Bulletin,* 10, 363-69.

Braybrooke, D., and C. Lindblom (1963). *A Strategy of Decision.* London, England: The Free Press of Glencoe.

Bruns, D. (1997). Benefits-based management: an expanded recreation-tourism management paradigm. Presentation at the 1997 International Symposium on Human Dimensions of Natural Resource Management in the Americas. February 25- March 1, 1997. Belize City, Belize Central America.

Buckley, W. (1968). *Modern Systems Research for the Behavioral Scientist.* Chicago, IL: Aldine.

Cortner, H. J., and M. A. Moote (1999). *The Politics of Ecosystem Management.* Washington, DC: Island Press.

Crowfoot, J. E., and J. M. Wondolleck (1990). *Environmental Disputes: Community Involvement in Conflict Resolution.*Washington DC: Island Press.Washington, DC

Culhane, P.J., and H.P. Friesma (1979). "Land use planning for public lands." *Natural Resources Journal,* 19, 43-74.

Daniels, S., and G. Walker (1996). "Collaborative learning: Improving public deliberation in ecosystem-based management." *Environmental Impact Assessment Review,* 16, 71-102.

Davidoff , P., and T. A., Reiner (1962). "A choice theory of planning". *Journal of the American Institute of Planning.* 28, 103-15.

Decker, D. J., C. C. Krueger, R. A. Baer, B. A. Knuth, and M. E. Richmond (1996). "From clients to stakeholders: a philosophical shift for fish and wildlife management". *Human Dimensions of Wildlife.* 1(1):70-82.

Decker, D. J., and L. C.Chase 1997. "Human dimensions of living with wildlife—a management challenge for the 21st century." *Wildlife Society Bulletin,* 25, 788-95.

Deutsch, M. (1975). "Equity, equality, and need: What determines which value will be used as the basis for distributive justice?" *Journal of Social Issues,* 31, 137-39.

Driver, B. L. (1985). "Specifying what is produced by management of wildlife by public agencies." *Leisure Sciences, 7*(3), 281-95.

Driver, B. L. (1999). "Management of public outdoor recreation and related amenity resources for the benefits they provide."In H. Cordell (Ed.), *Outdoor Recreation in American life: A National Assessment of Demand and Supply Trends.* Champaign, IL: Sagamore Publishing.

Driver, B. L., P. J.Brown, T. Gregoire, and G. H. Stankey (1987). "The ROS planning system: Evolution and basic concepts." *Leisure Sciences,* 9, 203-14.

Driver, B. L., P. Brown, and G. Peterson (Eds.) (1991a). *The Benefits of Leisure.* State College, PA: Venture Publishing, Inc.

Driver, B. L., H. E. A. Tinsley, and M. J. Manfredo (1991b). "The paragraphs about leisure and recreation experience preference scales: Results from two inventories designed to assess the breadth of the perceived psychological benefits of leisure." In B.L. Driver, P. Brown, and G. Peterson (Eds.), *The Benefits of Leisure.* State College, PA: Venture Publishing, Inc.

Driver, B. L. and D. Bruns (1999). "Concepts and uses of the benefits approach to leisure." In T. Burton and E. Jackson (Eds.), *Leisure Studies: Prospects for the Twenty-first Century.* State College, PA: Venture Publishing, Inc.

Etzioni, A. (1967). "A mixed-scanning: a third approach to decision-making." *Public Administration Review*, 27, 385-92.

Etzioni, A. (1968). *The Active Society*. New York: Free Press.

Flather, C. H. and H. K. Cordell (1995). "Outdoor recreation: Historical and anticipated trends." In R.L. Knight and K.J. Gutzwiller (Eds.), *Wildlife and Recreationists: Coexistence through Management and Research*. Washington, DC: Island Press.

Friedmann, J. (1973). *Retracking America:A Theory of Transactive Planning*: New York: Schoken.

Friedmann, J. (1987). *Planning in the Public Domain: From Knowledge to Action*. Princeton, NJ: Princeton University Press.

Graefe, A.R., F.R. Kuss, and J. J. Vaske (1990). *Visitor Impact Management: The Planning Framework*. Washington, DC: National Parks and Conservation Association.

Henning, D. H., and W. R. Mangun (1989). *Managing the Environmental Crisis: Incorporating Competing Values in Natural Resource Administration*. Durham, NC: Duke University Press.

Hof, J. G., and H. F. Kaiser (1983). "Long-term outdoor recreation participation projections for public land management agencies." *Journal of Leisure Research*, 15, 1-14.

Hudson, B. M. 1979. "Comparison of current planning theories: Counterparts and contradictions." *American Planning Association Journal*, 45, 387-98.

Kartez, J. D. (1989). "Rational argumensts and irrational audiences: psychology, planning and public judgment." *Journal of the American Planning Association*, 55, 445-56.

Lauber, T. B., and B. A. Knuth (1999). "Measuring fairness in citizen participation: A case study of moose management." *Society and Natural Resources*, 11, 19-37.

Lawrence, R. K., S. E. Daniels, and G. E. Stankey (1997). "Procedural justice and public involvement in natural resource decision making." *Society and Natural Resources*, 10, 577-89.

Lee, K. N. (1993). *The Compass and the Gyroscope: Integrating Science and Politics for the Environment*. Washington, DC: Island Press.

Lind, E. A., and T. R. Tyler (1988). *The Social Psychology of Procedural Justice*. New York: Plenum Press.

Lindblom, C. E. (1959). "The science of muddling through." *Public Administration Review*. 19:79-88.

Loomis, J. B. (1993). *Integrated Public Lands Management: Principles and Applications to National Forests, Parks, Wildlife Refuges, and BLM Lands*. New York: Columbia University Press.

Manfredo, M. J., J. J., Vaske, and L. Sikorowski (1996). "Human dimensions of wildlife management." In A. W. Ewert, Ed., *Natural Resource Management: The Human Dimension*. New York: Westview Press.

Manning, R. E. (1999). *Studies in Outdoor Recreation: Search and Research for Satisfaction* (2nd Edition). Corvallis, OR: Oregon State University Press.

Mitchell, R. C., and R. T. Carson (1989). *Using Surveys to Value Public Goods: The Contingent Valuation Method.* Washington, DC.: Resources for the Future.

Shelby, B., and T. A. Heberlein (1986). *Carrying Capacity in Recreation Settings.* Corvallis, OR: Oregon State University Press.

Schomaker, J. (1984). Writing quantifiable river recreation management objectives. *Proceedings of the 1984 National River Recreation Symposium,* 249-53.

Stankey, G. H., D. N. Cole, R. C. Lucas, M. E. Petersen, and S. S. Frissell (1985). *The Limits of Acceptable Change (LAC) System for Wilderness Planning.* Report INT-176. Ogden, UT: U.S. Department of Agriculture, Forest Service.

Susskind, L., and J. Cruikshank (1988). *Breaking the Impasse: Consensual Approaches to Resolving Public Disputes.* New York: Basic Books.

Thibaut, J., and L. Walker (1975). *Procedural Justice: A Psychological Analysis.* Hillsdale, NJ: Lawrence Erlbaum.

Tyler, T. R. (1994). "Psychological models of the justice motive: antecedents of distributive and procedural justice." *Journal of Personality and Social Psychology,* 67, 850-63.

Tyler, T. R., and P. Degoey (1995). "Collective restraint in social dilemmas: procedural justice and social identification effects on support for authorities." *Journal of Personality and Social Psychology,* 69, 482-97.

Tyler, T. R., R. J. Boeckmann, H. J. Smith, and Y. J. Huo (1997). *Social Justice in a Diverse Society.* Boulder, CO: Westview Press.

United States Department of the Interior, National Park Service (1997). *VERP: The Visitor Experience and Resource Protection Framework—A Handbook for Planners and Managers.* Denver, CO: Denver Service Center.

United States Department of the Interior, National Park Service Web site (2002). National Park Service Organic Act, 16 U.S.C.1. http://www.nps.gov/legacy/mission.html (accessed June 2002).

Walsh, R. (1986). *Recreation Economic Decisions.* State College, PA: Venture Press.

Wondolleck, J.M. and S.L. Yaffee (2000). *Making Collaboration Work: Lessons from Innovation in Natural Resources Management.* Washington, D.C.: Island Press.

Informing the Planning Process Through Citizen Participation

T. Bruce Lauber, Lisa C. Chase, and Daniel J. Decker

Introduction

CITIZEN PARTICIPATION PLAYS an important role in wildlife management (Decker et al. 1996). Public wildlife managers have used a variety of approaches to get citizens involved in wildlife management. For example:

• the Colorado Division of Wildlife held a series of public meetings and conducted a mail survey to help them understand citizens' attitudes about elk management in the suburbs west of Denver where elk populations were growing larger (Chase and Decker 1998);

• a citizen task force made up of stakeholders with different interests was organized in Irondequoit, New York, to suggest deer-management strategies that would both give people the opportunity to view deer and protect people from deer-related problems (Stout et al. 1997); and

• the Texas Parks and Wildlife Department organized a birding competition, the Great Texas Birding Classic, to help recreational birders raise money to protect bird habitat (Scott et al. 1999).

Because citizen participation is becoming a regular feature of wildlife management, it is important to consider how it can enhance an experience-based approach to planning wildlife-viewing recreation.

What is Citizen Participation?

When people say "citizen participation," they can mean a lot of different things. So, what exactly do *we* mean when we use the phrase?

Citizen participation is the involvement of stakeholders in informing, making, understanding, implementing, or evaluating wildlife-management decisions.

According to this definition, citizen participation may include such diverse activities as: conducting a survey of suburban residents to find out whether they would like to be able to see deer and other wildlife in their community; empowering a representative group of citizens to serve as a task force to generate recommendations

124

Having citizens monitor local populations of birds and other wildlife is one type of citizen participation. (Photo by Paul Curtis)

about site development for unique wildlife-viewing opportunities (e.g., heron rookery, eagle wintering areas); and asking citizen volunteers to monitor the populations of different birds near their homes.

Many things that wildlife managers do could be called citizen participation, making it hard to understand the concept, let alone plan for its use. Not surprisingly, then, one of the biggest obstacles to using citizen participation effectively is the failure of agencies to define fully what they hope to get out of it. To help demonstrate how citizen participation may benefit wildlife-viewing management, we will discuss four basic ways that citizen participation is used: to improve information about important stakeholders that managers and advisory committees use to make wildlife-management decisions; to improve the quality of judgment in management decision making; to improve the implementation of management actions; and to improve the social climate in which management occurs.

Improving Information

One of the most common reasons for citizen participation is to provide managers with better information about stakeholders—their needs, desires, beliefs, values, and/or behaviors. This kind of information can help to make better management decisions in lots of different ways. For example, Martin (1997) showed how asking people about the kinds of wildlife-viewing experiences they wanted could lead to better management. He surveyed nonresident visitors to Montana and found that people interested in seeing wildlife could be split into three groups, which he called specialists, intermediates, and novices. People in these groups wanted different kinds of wildlife-viewing experiences. For instance, specialists liked to avoid crowds at well-developed viewing facilities when watching wildlife. Martin concluded that if managers are to keep most wildlife viewers happy, they need to understand what the different segments of wildlife viewers want from their recreation experiences and provide them with the types of opportunities that will satisfy them.

Improving Judgment

Unfortunately, sometimes having more information about what stakeholders want does not make decision making any easier for managers; in fact, the added information can show just how complicated a situation really is. Stout et al. (1994) described how hard it is to manage white-tailed deer when people are interested in deer for very different reasons: some want chances to view deer, others want to hunt deer, and still others are concerned about problems that deer can cause. Conflict between stakeholders' interests in wildlife is common. Even when a manager understands just what stakeholders want, making a final wildlife-management decision is not easy. Managers are faced with the unenviable task of choosing which stakeholders' needs, wants, and desires will be satisfied, and which will not. When faced with such choices, it is likely that *no* decisions will be acceptable to *all* stakeholder groups.

Stout et al. (1994) argued that in situations of high conflict it can help to have the stakeholders themselves suggest a decision that balances the needs and concerns of all who are interested. They discuss one way, the citizen task force, that stakeholders can become involved in making a judgment about what management decision is best. On a citizen task force, stakeholders with different interests work directly with each other to reach a decision that balances legitimate stakeholder needs. Citizen task forces have been found to lead to reasonable decisions and to make these decisions more acceptable to different stakeholders (Stout and Knuth 1994; Stout et al. 1996).

Improving Implementation

Making a good decision is only one necessary component of sound management. Decisions also need to be implemented effectively and efficiently. Sometimes implementation requires stakeholder participation as well. For example, the community of Irondequoit, New York, struggled with the question of how to manage a large suburban deer herd for many years. Eventually, as noted above, with the help of a citizen task force, a decision was reached to (1) reduce the size of the deer herd through a selective culling program; and (2) conduct research on contraception as a way to manage the

Managing suburban deer herds to meet wildlife-viewing and other objectives poses unique challenges for manager. (Photo by Paul Curtis)

size of the deer herd in the future. However, the New York State Department of Environmental Conservation was not capable of implementing these recommendations on its own. It had neither the funding nor the staff available to carry out these actions, and various local government laws and regulations needed to be modified before these actions could take place. Consequently, an interagency task force consisting of representatives of various state and local government agencies was organized to oversee deer management in Irondequoit. The task force ensured that the steps necessary to carry out the deer-management program were completed and assigned responsibility for completing each step. Stakeholder involvement in the implementation of the program was evidenced by participation of (1) police officers from the town of Irondequoit to implement the selective culling program; (2) citizen volunteers to monitor deer involved in the contraception research; (3) university researchers to carry out the contraception research; and (4) local government to fund implementation of the program.

Improving the Social Climate

Although citizen participation can be used to improve management decisions and actions, it also contributes to wildlife management in less direct ways. Sound wildlife management depends on a citizenry that supports and contributes to management decisions and actions. Citizen participation, therefore, is commonly implemented to improve the social climate in which wildlife management occurs by influencing people and the way they relate to each other. This can happen in four basic ways: (1) by transforming beliefs or attitudes; (2) by changing behaviors; (3) by improving relationships between stakeholders; and (4) by increasing the capacity of people to contribute to policy making and management. Attempts to improve the social climate of management would include such diverse activities as: gathering and considering public input before developing a management plan in order to build community ownership of that management plan (e.g., Mitchell et al. 1997); developing programs that will encourage stakeholders to support habitat-conservation efforts financially (e.g., Scott et al. 1999); engaging diverse stakeholders in team-building efforts to improve their relationships and their ability to work together to develop mutually acceptable management plans (e.g., Guynn 1997); and helping citizens develop the skills and knowledge they need to contribute more effectively to wildlife-management processes in the future (e.g., Landre and Knuth 1993b).

Citizen participation may not contribute directly to decision making in every situation to which it is applied, but it can have considerable overall impact on management.

Range of Approaches to Citizen Participation

In any given situation, citizen participation may be intended to meet all of the purposes described above, or only one or two. The characteristics of each situation determine the choice of appropriate objectives of citizen participation. These objectives, in turn, influence which approach to citizen participation is taken. Approaches to citizen participation differ according to the degree of control that stakeholders have compared to the agency, the particular techniques of citizen participation that are used, and the participants included in the process.

In this section, we discuss five approaches to citizen participation: expert authority, passive-receptive, inquisitive, transactional, and co-managerial (Decker and Chase 1997). On one end of the spectrum, the authoritative approach keeps the control of decision making squarely (and nearly exclusively) within the realm of the management agency. The passive-receptive and inquisitive approaches also keep decision making within the management agency; however, the agency accepts or even seeks stakeholder input to be considered as decisions are made. In contrast, the decision making is shared by stakeholders and managers in the transactional approach. In the co-managerial approach, stakeholders and managers share influence over both decisions *and* actions.

For the approaches in which stakeholders have little control, the objectives of citizen participation are relatively simple. As stakeholders play a larger and larger role in the management process, however, the objectives of citizen participation necessarily become more numerous and complicated. We will review each of the five approaches in turn.

Expert Authority Approach

In situations where the expert authority approach is followed, managers assume the role of experts, making decisions and taking actions unilaterally. They do not intentionally rely on input from stakeholders and do not share control of the management decision-making process. Control remains with the wildlife-management agency. Nonetheless, citizen participation can still be important because stakeholders can have a substantial influence over whether management decisions are successfully implemented. Consequently, the objective of citizen participation under this approach—if there is any—is to improve the social context or climate for management by building stakeholder support for decisions or actions.

Stakeholder support may be built through education and communication under the authoritative approach, but the usefulness of education and communication is not limited to this approach. Indeed, they are instrumental to the success of all the approaches described, though other objectives and forms of participation also become important.

Passive-receptive Approach

Under the passive-receptive approach, managers recognize that stakeholders' beliefs, attitudes, values, behaviors, and experiences need to be considered in making good wildlife-management decisions. Although managers do not actively seek citizen input or involvement, they are open to that input when it is offered. Stakeholders, therefore, have a greater influence in management than under the expert authority approach, but only those stakeholders who take the initiative to reach out to managers. Although these stakeholders may influence decisions, control over those decisions remains with the agency.

The objectives of citizen participation under the passive-receptive approach are not only to build support for management decisions and actions, but also to improve the quality and quantity of information used to make decisions. Understanding vocal stakeholders' beliefs, attitudes, and values, therefore, becomes an important objective in citizen participation in addition to education and communication processes.

The passive-receptive approach is a process where "the squeaky wheel gets the grease." The "squeaky wheels" in this case are the stakeholders who take the initiative to contact managers and communicate their concerns and desires, which often means organized interest groups who do not represent the full range of interests in wildlife. The following approaches to citizen participation are better at involving a more representative sample of citizens.

Inquisitive Approach

As under the passive-receptive approach, managers taking the inquisitive approach recognize the need to consider stakeholders' perspectives if they are to make sound wildlife-management decisions. And seeking this information, rather than passively receiving it, can have the dual objectives of both improving decisions and improving public acceptance of decisions. Using this approach, wildlife managers take an active role in seeking out this information, recognizing the potential biases involved in considering the perspectives of only those stakeholders who take the initiative to contact the agency. But like the expert authority and passive-receptive approaches, control remains with the agency; it is the agency that decides whether and how to reflect different perspectives in its final management decisions.

Surveys and public meetings are two of the most common techniques used in the inquisitive approach. Surveys include mail-back questionnaires, telephone interviews and in-person interviews. Depending on the type of information sought, the survey will target different participants. The survey may be a census (where everyone is contacted); more often, a survey targets a randomly selected sample because the population is so large that reaching everyone would be too expensive and time consuming. When a randomly selected sample is contacted, inferential statistics can be used to generalize to the entire population. Alternatively, a study may target certain segments of the population, such as particular stakeholder groups like homeowners, hunters or those with extreme opinions. Many more techniques exist under the inquisitive approach; the ones we described are those most commonly adopted in wildlife management.

Transactional Approach

Even when considering systematically collected information about stakeholders' perspectives, managers often have a daunting task in making management decisions. Stakeholders frequently have conflicting perspectives and deciding how those perspectives are to be balanced in management decisions is difficult. Managers' personal biases may be revealed in how they weigh the importance of different perspectives and may result in a decision that is unacceptable to many stakeholders.

Citizen task force members in Irondequoit, New York, explore the effects of deer on their community. A citizen task force is an example of a transactional approach. (Photo by Paul Curtis)

This type of outcome is of particular concern in politically charged issues involving diverse perspectives where stakeholders do not have experience with managers, or where mutual trust between stakeholders and wildlife managers has not been established. Managers in such settings often rely on transactional approaches to involve stakeholders in decision making. In these approaches, stakeholders are allowed to deliberate among themselves and judge the implications of all available information (and perhaps call for additional information to be generated) for management decisions, while managers primarily serve to manage the process and provide technical advice. Managers give some control over decisions to stakeholders, although they may retain the power to reject or approve stakeholders' final recommendations. Thus, the control is shared in this approach.

Allowing stakeholders to deliberate with each other when making recommendations for wildlife management can improve decisions. Rather than having a management agency decide which needs, desires, and values should be reflected in a decision and the weight they should be given, this approach brings a more diverse range of perspectives to bear on the problem. As stakeholders deliberate with each other in an effort to reach consensus, they must negotiate with each other about how to balance different perspectives. A more thorough analysis of management problems and a more balanced solution to those problems can result. The transactional approach can also fulfill some of the objectives that have been discussed under the other approaches, including improving the social information base of decisions and improving the social climate of management by building ownership in and support for agency decisions and actions. However, managers must make deliberate efforts to extend their efforts toward citizen participation beyond those few citizens involved in transactional approaches if they want these benefits of citizen participation to extend to the entire community.

Co-managerial Approach

The co-managerial approach is still being defined and refined in the context of wildlife management in North America (Schusler 1999). This approach recognizes that as demands for attention to local wildlife issues continue to increase in the future, it may become strategically important for agencies to move beyond involving stakeholders in decision making and begin to share responsibility for other elements in the management process, too—such as the funding and implementation of management actions. Given the broad role that stakeholders could have in management, all of the objectives for citizen-participation programs described above would be important, requiring complex and multi-faceted citizen-participation programs.

Co-management combines many techniques from other approaches in an attempt to go beyond traditional citizen-participation practices. In addition to the various techniques described in the four approaches above, governing boards of citizens and managers may oversee agency decisions and activities. Community members rather than wildlife managers may be responsible for education and communication. Participants will be involved in several levels and aspects of management, perhaps taking on the responsibility of financing, implementing, and evaluating management actions. Some participants may thus play large roles with much influence. Other participants may only want to be informed on occasion. As experiments with co-management continue throughout North America, the list of techniques and roles for participants will expand.

Citizen Participation and Experience-based Management

The Nature of Experience-based Management

The fundamental presupposition of an Experience-based Management (EBM) approach is that it is important to integrate protection and service in the management of wildlife-viewing recreation (see Chapter 5). Managers must protect the wildlife resource while striving to serve people by providing them with the opportunities that they want for wildlife-viewing recreation. In providing people with the recreational opportunities that they want, managers should consider three elements: the psychological outcomes that people desire; the recreational activities associated with those outcomes; and the setting in which recreation is sought, including its physical, social, and managerial attributes.

In order to utilize an EBM approach Manfredo et al. (Chapter 5) and Fulton et al. (Chapter 6) identified a number of basic steps that need to be incorporated in the planning process:

• Decisionmakers must determine the types of recreational opportunities they will consider providing. This choice will be influenced by existing policy, resource constraints, and public demand.

• Standards and criteria must be developed that will serve to guide management decisions. Providing quality recreation experiences is a standard that is fundamental to an EBM approach. Other standards and criteria will vary from context to context.

• Demand for different types of recreation experiences must be assessed. This step involves both defining the population that managers wish to serve and evaluating the types of experiences members of this population participate in and desire.

• The ability to supply various recreation experiences must be assessed. Both current recreational opportunities and the ability to expand these opportunities should be considered.

• Different alternatives for providing wildlife-viewing-recreation experiences need to be defined.

• The various consequences of each allocation alternative should be identified.

• An allocation alternative must be selected.

• The alternative selected must be implemented.

• Monitoring and evaluation of the impacts of the implementation should be conducted.

Why is Citizen Participation Critical to Experience-based Management?

In the first section of this chapter, we listed four basic reasons why citizen participation is used: improving management decisions by improving information; improving management decisions by improving judgment; improving the implementation of management decisions; and improving the management climate, which consists of (1) beliefs and attitudes, (2) behaviors, (3) relationships, and (4) capacity.

Each step in an EBM planning process can benefit from citizen participation implemented for one or more of these reasons.

• Gathering information from diverse stakeholders can lead to: (1) a more comprehensive set of criteria and standards for judging management plans; (2) a more thorough evaluation of existing participation and desires related to wildlife recreation; and (3) a more exhaustive set of allocation alternatives (in addition to a better understanding of the different consequences of those alternatives).

• Involving diverse stakeholders in choosing or recommending a particular allocation alternative can improve judgment by ensuring that each allocation alternative has been considered from a variety of different perspectives.

• Involving stakeholders in the implementation of management plans can increase the range of recreation experiences that managers may be able to supply because it expands the knowledge, funding, and labor available for management.

• Encouraging stakeholder participation in these and other stages of the planning process can improve the management climate by building support for management plans. For example, stakeholder involvement in decision making often can increase the appreciation that stakeholders have for the trade-offs that managers face when they must balance providing recreation opportunities with meeting other important needs (such as protecting the wildlife resource or adhering to existing policy). Indeed, active involvement makes stakeholders more likely to accept decisions that do not meet all of their needs (Lauber and Knuth 1999).

Transactional or co-management approaches to citizen participation are probably necessary to supply the full range of benefits that stakeholder involvement can provide. The passive-receptive and inquisitive approaches focus primarily on information gathering. These approaches can indeed improve the information base available for EBM planning through the use of techniques such as surveys. Having diverse stakeholders interact with managers and each other in a transactional approach, however, is superior if managers are hoping to ensure that a decision has been thoroughly considered from multiple perspectives. Stakeholders are not just involved in providing information but in judging the implications of all available information for management decisions. If an agreement on a management recommendation between diverse stakeholders is reached, managers can be assured that it has been evaluated from a variety of viewpoints. A transactional approach also can be very successful in building stakeholder ownership of and support for management plans, because stakeholders tend to be more vested in a plan if they have helped to craft it. Indeed, transactional approaches have been shown to educate participating stakeholders about the substantive aspects of management issues (improving their ability to contribute to a wise decision) and improve stakeholders' awareness of the interests and concerns that others bring to the table (Landre and Knuth 1993b; Stout et al. 1996).

Many of the benefits of a transactional approach arise from the give-and-take of deliberation over an issue between stakeholders with different perspectives. Such opportunities for deliberation simply are not provided in approaches to citizen participation designed only to gather information. Indeed, even when the primary purpose of citizen participation is providing information (e.g., helping to develop a list of indicators that stakeholders think are important in evaluating management plans), interaction between stakeholders with different perspectives can be helpful. Such interaction can lead to a brainstorming effect in which the ideas suggested by one participant can help to generate creative ideas among others (Fisher and Ury 1981).

A co-management approach has the potential to extend the benefits of stakeholder involvement even further. Although involving stakeholders in management decision making may be very beneficial, sometimes more is needed. Situations may arise in which cooperation between different stakeholders is necessary to ensure that the range of recreational benefits sought can be provided. For example, under some circumstances, jurisdictional complexity will prevent state wildlife-management agencies from being able to provide the desired recreational benefits on their own. Multiple agencies in state and/or local government have responsibilities in some circumstances that may have an impact on managing wildlife recreation, and local laws and regulations may constrain the kinds of recreation opportunities that state agencies can provide. Under other circumstances, state agencies may simply not have the capacity to provide the types of recreational opportunities sought without additional funds and labor. Assistance and participation from local government and citizens' groups may be necessary in such situations, and stakeholder participation in decision making is not sufficient. A more comprehensive co-management approach in which stakeholder involvement is incorporated throughout the management process is needed.

Planning Citizen Participation in an
Experience-based Management Approach

Beyond these general recommendations about the best approaches to citizen participation in EBM planning, a number of specific guidelines for successful citizen participation can be gleaned from the literature.

(1) Set clear objectives for citizen participation. Citizen participation is very complex in that it can be used for so many different purposes. It may be tempting to plunge into citizen-participation planning without clearly defining the important objectives in a given scenario. This may be a recipe for disaster, however, because it may result in important implicit objectives not being met. Recommendations to clearly define all objectives prior to planning citizen-participation efforts are common in the literature (e.g., Enck and Brown 1996; Lauber and Knuth 1996). The purposes of citizen participation that we have listed in this chapter can serve as an initial guide in planning, but objectives must be tailored to the specific characteristics and needs of each management context.

(2) Choose techniques that will help you meet your objectives. After objectives have been defined clearly, it is more straightforward to choose the techniques that can help to meet them. We have argued for the benefits of a transactional approach to EBM planning, in which diverse stakeholders may interact directly with each other in deliberative bodies, such as citizen task forces, to develop management objectives and strategies, and a co-management approach, in which various stakeholders from local government and organizations may work closely with agencies throughout the management process. The techniques we have described under other approaches to citizen participation, such as the inquisitive approach, may also be useful in planning. The use of various information-gathering techniques, such as surveys, public meetings, and focus groups, for example, may be able to provide a citizen task force with the information it needs about a community to consider thoroughly all aspects of a management issue (Stout et al. 1994).

Furthermore, if one of the objectives of citizen participation is to build support for the implementation of a management plan, it will not be enough for an agency to work with the small number of stakeholders that can serve on a citizen task force. Rather, the agency will need to reach out to other stakeholders through strategies designed to work with larger audiences, such as public meetings, surveys, and other techniques (Pelstring 1998).

(3) Provide the resources for citizen participation to work. Citizen participation can be complex and expensive. Considerable staff time will be needed to plan and implement citizen-participation strategies. Additional resources may be required for meeting facilities, outside facilitators, transportation, communication, and refreshments. Nevertheless, if citizen participation is carefully conceived and executed, it is worth the time and monetary cost. It is far more worthwhile to invest the necessary resources initially than to have a citizen-participation effort that fails to meet its objectives because the necessary investment was not made (Siemer and Decker 1990; Stout and Knuth 1994).

(4) Involve diverse stakeholders. When planning citizen participation, it is important to think broadly about the stakeholders that may be interested in and affected by a management program. Given that one purpose of citizen participation is to improve management decisions, the more relevant diverse perspectives that are brought to bear on a problem, the more thoroughly that decision will be considered (Enck and Brown 1996; Pelstring 1998). Furthermore, if all important stakeholders are not included in a decision-making process, there will be a danger that the resulting management plan will not meet their interests or even that some disaffected stakeholders may oppose it. Despite its importance, involving relevant stakeholders is one of the most challenging aspects of citizen participation (Lauber and Knuth 1996; Chase and Decker 1998; Chase et al. 1999), and involving nontraditional stakeholders is particularly difficult (Enck and Brown 1996; Lauber and Knuth 1996), so managers should plan on spending significant time and energy on this step.

(5) Demonstrate receptivity to stakeholder input. Research has shown that stakeholder acceptance of management decisions depends in part on whether they truly believe their input has been considered by agency decision makers (Siemer and Decker 1990; Stout et al. 1994; Lauber and Knuth 1999). Under some circumstances, stakeholders may suspect that agencies are involving them in management as part of a token effort to appease the public without truly considering their input. If citizen-participation efforts are perceived this way, they may do more harm than good. On the other hand, if stakeholders perceive that an agency has genuinely considered their input, this can enhance perceptions of the fairness of management. Consequently, agencies need to develop clear guidelines for how citizen input will be used and communicate those guidelines to interested stakeholders.

(6) Pay attention to relationships. Good working relationships between stakeholders are particularly important in transactional and co-management approaches to citizen participation. In some management scenarios, stakeholders with conflicting interests may be working closely together on management issues. These stakeholders may even have clashed with each other in the past. The productivity of citizen-participation efforts will depend in part on the capacity of these stakeholders to work together productively. Under such circumstances, fostering good working relationships between stakeholders, although it may not seem directly related to the task at hand, can be critical to the success of citizen participation (Landre and Knuth 1993a; Stout and Knuth 1994; Loker 1996).

(7) **Educate your participants.** One typical goal of citizen participation is to improve management decisions. This goal can only be achieved if management decisions are based on good information and sound reasoning. If stakeholders are going to influence a management decision, therefore, managers should usually assume that they will need to be educated about the management topic at hand, including such components as (1) education about the biology of a species under consideration; (2) education about the political constraints on management; and (3) education about the interests and perspectives of other stakeholders. For stakeholders involved in transactional and co-management approaches, some effective education will occur through interaction with other stakeholders with different knowledge and perspectives (Landre and Knuth 1993b; Lauber and Knuth 1996; Pelstring 1998). Educating participating stakeholders is too important simply to assume that it will occur, however. Agencies should explicitly consider (1) in what ways stakeholders need to be educated; and (2) how that education will occur.

(8) **Communicate about your management processes.** One benefit of citizen participation is increased public acceptance and awareness of management plans and activities. People are more likely to accept decisions and actions if they believe that diverse stakeholders have been fairly involved in their development. This benefit can only occur, however, if the public at large is aware of citizen-participation processes. It is surprisingly easy to overlook communicating about citizen-participation efforts beyond those stakeholders who are directly involved in them. Nevertheless, such communication can influence widespread public acceptance of management plans and actions. Agencies should plan, therefore, on communicating explicitly with the public about the planning process and how stakeholders have been involved in it (Siemer and Decker 1990; Lauber and Knuth 1997).

(9) **Plan for implementation in advance.** If an agency hopes to achieve a better decision by involving the public, these benefits will never accrue if the decision cannot be implemented. Sometimes the ability to implement decisions is not rooted solely in an agency but depends on the approval of other state- and local-government-

agency stakeholders. If these other stakeholders can influence the management process, they need to be involved in it from the beginning. Under such circumstances, a co-management approach to citizen participation may be most appropriate, with other citizen-participation strategies incorporated under its umbrella. Involving additional stakeholders early in the planning process may be cumbersome, but it can help to avoid situations in which a decision is reached but cannot be implemented.

Summary Points

Citizen participation has a lot to offer to an EBM approach to managing wildlife viewing. To make good use of citizen participation, managers need to understand a few key points:

• The term "citizen participation" is applied to many different things. Stakeholders may be involved in informing, making, understanding, implementing, or evaluating management decisions.

• Citizen participation is most likely to be successful if managers have a clear sense of what they want to get out of it. It is typically implemented for one of four basic purposes: (1) improving the quality of information in decision making; (2) improving the quality of judgment in decision making; (3) improving the implementation of decisions; or (4) improving the social climate of management by influencing stakeholders' beliefs and attitudes, changing stakeholders' behaviors, improving relationships between stakeholders, or increasing the capacity of stakeholders to contribute to management.

• Depending on one's purposes, any one of several approaches to citizen participation could be appropriate. These approaches differ according to the degree of control held by agencies relative to stakeholders, the particular citizen-participation techniques used, and which stakeholders are included in the process. The approaches range from the expert authority approach, in which management agencies retain complete control over management, to co-management, in which control over and responsibility for all aspects of management is broadly shared between agencies and stakeholders.

• Citizen participation can benefit an EBM approach to wildlife-viewing recreation in many ways. It has the potential to play an important role in each step of the EBM planning process.

• Several guidelines can help to ensure the success of a citizen-participation effort. These include: (1) set clear citizen-participation objectives; (2) choose techniques that will help you meet your objectives; (3) provide the resources for citizen participation to work;

(4) involve diverse stakeholders; (5) demonstrate receptivity to stakeholder input; (6) pay attention to relationships; (7) educate your participants; (8) communicate about your management processes; and (9) plan for implementation in advance.

Literature Cited

Chase, L. C., and D. J. Decker (1998). "Citizen attitudes toward elk and participation in elk management: a case study in Evergreen, Colorado." *Human Dimensions of Wildlife* 3(4):55-56.

Chase, L. C., W. F. Siemer, and D. J. Decker (1999). *Deer Management in the Village of Cayuga Heights, New York: Preliminary Situation Analysis from a Survey of Residents.* Human Dimensions Research Unit Publication 99-1. Ithaca, NY: Department of Natural Resources, New York State College of Agriculture and Life Sciences, Cornell University.

Decker, D. J., C. C. Krueger, R. A. Baer, Jr., B. A. Knuth, and M. E. Richmond (1996). "From clients to stakeholders: a philosophical shift for fish and wildlife management." *Human Dimensions of Wildlife* 1:70-82.

Decker, D. J., and L. C. Chase (1997). "Human dimensions of living with wildlife—a management challenge for the 21st century. *Wildlife Society Bulletin* 25(4):788-795.

Enck, J. W., and T. L. Brown (1996). *Citizen Participation Approaches to Decision-making in a Beaver Management Context.* Human Dimensions Research Unit Publication 96-2. Ithaca, NY: Department of Natural Resources, New York State College of Agriculture and Life Sciences, Cornell University.

Fisher, R., and W. Ury (1981). *Getting to Yes.* New York: Penguin Books.

Guynn, D. E. 1997. "Miracle in Montana—managing conflicts over private lands and public wildlife issues." *Transactions of the North American Wildlife and Natural Resources Conference* 62:146-54.

Landre, B. K., and B. A. Knuth (1993a). "The role of agency goals and local context in Great Lakes water resources public involvement programs." *Environmental Management* 17(2):153-66.

Landre, B. K., and B. A. Knuth (1993b). "Success of citizen advisory committees in consensus-based water resources planning in the Great Lakes Basin." *Society and Natural Resources* 6(3):229-57.

Lauber, T. B., and B. A. Knuth (1996). *Citizens' and Agency Staff Members' Evaluation of Decision-making Procedures: A Case Study of the New York State Moose Reintroduction Issue.* Human Dimensions Research Unit Publication 96-1. Ithaca, NY: Department of Natural Resources, New York State College of Agriculture and Life Sciences, Cornell University.

Lauber, T. B., and B. A. Knuth (1997). "Fairness and moose management decision-making: the citizens' perspective." *Wildlife Society Bulletin* 25(4):776-87.

Lauber, T. B., and B. A. Knuth (1999). "Measuring fairness in citizen participation: a case study of moose management." *Society and Natural Resources* 12:19-37.

Loker, C.A. (1996). *Human Dimensions of Suburban Wildlife Management: Insights from Three Areas of New York State*. Human Dimensions Research Unit Publication 96-6. Ithaca, NY: Department of Natural Resources, New York State College of Agriculture and Life Sciences, Cornell University.

Schusler, T. M. (1999). *Co-management of Fish and Wildlife in North America: A Review of Literature*. Human Dimensions Research Unit Publication 99-2. Ithaca, NY: Department of Natural Resources, New York State College of Agriculture and Life Sciences, Cornell University.

Martin, S. R. (1997). "Specialization and differences in setting preferences among wildlife viewers." *Human Dimensions of Wildlife* 2(1):1-18.

Mitchell, J. M., G. J. Pagac, and G. R. Parker (1997). "Informed consent: gaining support for removal of overabundant white-tailed deer on an Indiana state park." *Wildlife Society Bulletin* 25(2):447-50.

Pelstring, L. M. 1998. Stakeholder outreach and citizen task forces: an examination of the DEC's public participation efforts relating to deer management. M.S. thesis. Ithaca, NY : Department of Communication, Cornell University.

Schusler, T. M. (1999). *Co-management of Fish and Wildlife in North America: A Review of Literature*. Human Dimensions Research Unit Publication 99-2. Ithaca, NY: Department of Natural Resources, New York State College of Agriculture and Life Sciences, Cornell University.

Scott, D., S. M. Baker, and C. Kim (1999). "Motivations and commitments among participants in the Great Texas Birding Classic." *Human Dimensions of Wildlife* 4(1):50-67.

Siemer, W. F., and D.J. Decker (1990). *An Evaluation of Public Meetings Held by the DEC Bureau of Wildlife (October 1989)*. Ithaca, NY: Human Dimensions Research Unit, Department of Natural Resources, New York State College of Agriculture and Life Sciences, Cornell University.

Stout, R. J., D. J. Decker, and B. A. Knuth (1994). *Public Involvement in Deer Management Decision-making: Comparison of Three Approaches for Setting Deer Population Objectives*. Human Dimensions Research

Unit Publication 94-2. Ithaca, NY: Department of Natural Resources, New York State College of Agriculture and Life Sciences, Cornell University.

Stout, R. J., and B. A. Knuth (1994). *Evaluation of a Citizen Task Force Approach to Resolve Suburban Deer Management Issues.* Human Dimensions Research Unit Publication 94-3. Ithaca, NY: Department of Natural Resources, New York State College of Agriculture and Life Sciences, Cornell University.

Stout, R .J., D. J. Decker, B. A. Knuth, J. C. Proud, and D. H. Nelson (1996). "Comparison of three public-involvement approaches for stakeholder input into deer management decisions: a case study." *Wildlife Society Bulletin* 24(2):312-17.

Stout, R. J., B. A. Knuth, and P. D. Curtis (1997). "Preferences of suburban landowners for deer management techniques: a step towards better communication." *Wildlife Society Bulletin* 25(2):348-59.

Indicators and Standards: Developing Definitions of Quality

Jerry J. Vaske, Doug Whittaker, Bo Shelby,
and Michael J. Manfredo

Introduction

MANAGERS DO NOT SET OUT with a goal of providing "bad" wildlife-viewing experiences. Similarly, recreationists want "good" experiences. Even members of the general public who seldom visit places where wildlife are present are likely to think high-quality viewing opportunities should be available. Everyone's goals are thus the same. At issue is what differentiates "good" from "bad" wildlife viewing. Consider the following scenarios:

Scenario 1. Mt. Evans (70 miles west of Denver): As the car climbed the mountain road toward the peak of Mt. Evans, the young family of four from the east coast watched in awe as the small herd of bighorn sheep crossed in front of them and grazed quietly along the roadside. The children, armed with their Kodak Instamatics, excitedly snapped pictures of the sheep that would for sure impress their New York friends. A few miles away on a remote mountainside, another photographer equipped with her Nikon had just captured on film the most magnificent bighorn she had ever seen in twenty-five years of observing wildlife.

Scenario 2. Brooks River in Katmai National Park: After traveling thousands of miles along the Alaska highway, followed by a float-plane trip to Katmai National Park, and a half-hour hike to the Brooks River Falls viewing platform, the two college students (along with forty other people) were able to watch several large grizzlies fish for sockeye salmon. It was a sight that they would not soon forget.

Scenario 3. Rocky Mountain National Park (RMNP): The elderly couple had moved to Colorado after the husband had retired. Their neighbors had told them about the large numbers of elk that could be seen and heard bugling during fall evenings in RMNP. Thinking this might be a fun thing to do, the couple packed a lunch and drove up to the park. Although they had to wait twenty minutes to get through the entrance gate due the line of cars in front of them, they immediately saw one bull and two cows attempt to cross the road by dodging the hundred or so cars parked along the shoulder. Over the course of the next few hours, the couple watched intensely as several large bulls warned the young spikes away from their respective harems.

Three different scenarios, yet should one experience be considered higher quality than the others? If yes, which experience and why? Does the answer lie in the types of animals (bighorn sheep, grizzly bears, elk) or numbers of animals (few versus many) observed? Do the numbers of other people (none versus many) influence the quality of a viewing opportunity? Is it important to be alone, or can one be alone with others? Are the characteristics of the visitors themselves (young versus old, party size, novice versus experienced viewers) or the characteristics of the visit (chance encounter with wildlife along the roadside or a strenuous hike to a remote viewing area) important considerations when defining quality wildlife viewing?

The simple answer is that all of these considerations *might* be important when defining appropriate conditions for a given type of experience. Unfortunately, such a conclusion does not provide much guidance for wildlife managers who struggle with these issues on a daily basis. A better answer begins with the experience-based planning efforts (see Chapters 5 and 6) that define desired conditions relative to distinct opportunities.

In developing these opportunity definitions, it is important to recognize that neither the roadside nor the solitude-oriented wildlife-viewing experience is inherently of higher quality; they are simply different. While the experienced wildlife photographer interested in the "perfect" shot of a bighorn might find roadside picture taking tame and unchallenging, the family who simply wants to see and learn about wildlife does not. Similarly, the family may not be comfortable hiking into unknown remote locations, or viewing animals without the safety provided by their vehicle. In all cases, the issue is defining the conditions that contribute to high quality for that type of opportunity. There are high-quality and low-quality wildlife-viewing experiences in the backcountry, just as there are high-quality and low-quality roadside viewing experiences.

Qualitative definitions, however, are not enough. Virtually all recent planning frameworks recommend identifying and establishing *quantitative* impact indicators and standards (e.g., the Limits of Acceptable Change [LAC], Stankey et al. 1985; Carrying Capacity Assessment Process [C-CAP], Shelby and Heberlein 1986; Visitor Impact Management [VIM], Graefe et al. 1990; Visitor

Experience and Resource Protection [VERP], National Park Service 1997). (See Box 1.)

Indicators are the biophysical, social, managerial, or other conditions that managers and visitors care about for a given experience. *Standards* restate management objectives in quantitative terms and specify the appropriate levels or acceptable limits for the impact indicators (i.e., how much impact is too much for a given indicator). Standards identify conditions that are desirable (e.g., availability of and proximity to animals), as well as the conditions that managers don't want to exceed (e.g., encounters with other people, wildlife flight reactions, incidents of wildlife-human conflict). (See Box 2.)

Despite the prominence of these definitions, relatively little information exists for how good standards should be developed, the criteria for choosing indicator variables for which standards are set, the characteristics of good standards, or the sources of good standards (Whittaker and Shelby 1992; Belnap 1998). This chapter examines each of these issues.

Box 1. Common Features of Planning Frameworks

Recent planning frameworks (EBM, LAC, C-CAP, VIM, VERP) recommend using indicators and standards through a sequential process. By attending to the following basic steps, wildlife managers have a "game plan" for dealing with human impacts.

(1) developing goals that categorize the benefits to be achieved;

(2) defining appropriate experience opportunities for specific management objectives;

(3) identifying key impact indicators;

(4) setting quantitative standards for the selected impact indicators;

(5) inventorying and monitoring existing conditions against the standards; and

(6) linking management actions to standards when impacts exceed standards.

The frameworks provide a model within which standards develop from a process. They are a means to an end, not an end in themselves (Shelby et al. 1992). Setting standards is a key step in these planning frameworks, but it is only a necessary, not sufficient, condition for successful management.

Box 2. Why Indicators and Standards Are Important

Quantitative indicators and standards serve several important functions:

(1) Standards articulate in unambiguous terms what outputs management is trying to provide. Wildlife-viewing experiences are created through the interaction of social, biological, and physical conditions, and the visitors' expectations and preferences for those conditions (Manfredo et al. 1996a). While managers do not create *experiences*, they are responsible for creating *opportunities* for experiences by manipulating social, environmental, and managerial conditions. Quantitative standards help shape those opportunities (i.e., a demand function) and signal whether or not that opportunity is possible given existing conditions (i.e., a supply function). For example, it may not be possible to offer a remote viewing experience in an urban setting.

(2) Standards help establish priorities for management, focus on future conditions, and allow managers to be proactive. There is a need to look ahead to what actions might be employed to meet standards, as well as a need to look back at the opportunity definitions management is trying to provide (Vaske et al. 2000). Standards define minimum or optimal conditions and allow managers to note when impacts are approaching defined levels, rather than waiting for problems to occur and then reacting to them (Whittaker and Shelby 1992). From a larger societal perspective, standards help clarify which stakeholders might benefit from a proposed management action.

(3) Standards focus attention on specific conditions and problems or benefits. In contrast to traditional wildlife management where the number of visitors (quantity) in an area was emphasized (Vaske et al. 1995), standards return the managers' attention to the quality of recreation opportunities. By concentrating on the conditions that create experiences, the probable causes of unacceptable impacts as well as the potential benefits to different stakeholders can be identified (Graefe et al. 1990).

(4) Indicators and standards provide a base for measuring the rate and magnitude of change and for evaluating the

acceptability of that change. The literature sometimes confuses the concepts of impact change and evaluation (Shelby and Heberlein 1986). The confusion can be illustrated by the term "wildlife harassment." Harassment refers to both a change (an objective impact—e.g., the birds flew away when the people approached) and a value judgment that the impact exceeds some standard. While most people would agree that management actions are necessary when wildlife harassment occurs, there is less consensus about what constitutes harassment. All human use has some impact. Whether the impact is harassment depends on management objectives (e.g., protect the migratory birds), standards (e.g., migratory birds should never be flushed from their nesting areas because of the presence of humans), expert opinion, and public values. Breaking concepts like harassment or crowding into two parts—the impact component (change in wildlife behavior or experiential change) and the evaluative component (the acceptability of the change)—provides a foundation for thinking about potential problem situations.

(5) Standards help the public clearly understand what management is trying to accomplish. The public is often confused by wildlife planning because plans fail to articulate management objectives in specific and explicit terms. Standards reduce this ambiguity and help clarify for the public why planning is important (Vaske et al. 2001).

Standards link concrete, on-the-ground conditions with more intangible, qualitative experiences. While experiences are social psychological entities, standards are tangible and specific. With the development of quantitative standards, a more rational discussion of the area's objectives can occur with the different stakeholders (Whittaker and Shelby 1992). For example, comparing existing conditions against the standards provides a quantitative estimate of whether any experiential changes are within the limits specified by standards, and whether the benefits suggested to accrue to stakeholders have been realized.

Walrus viewing on Round Island, Alaska. The number of viewable wildlife may be a good biological and wildlife-experience indicator, but it may lack specificity and responsiveness because the number of animals at a location is often influenced by more than just viewing-impact variables and may not respond to some management prescriptions. (Photo by Julie Whittaker)

Criteria for Choosing Indicators

Before standards can be developed, appropriate impact indicators must be selected. As used in other sciences (e.g., medicine, agriculture, forestry), indicators are variables that reflect the "health" of something (Ott 1978). Indicators identify what conditions will be monitored (e.g., a person's blood pressure), while the standards define when those conditions are acceptable or unacceptable. For example, the American Heart Association defines high blood pressure (an indicator) as greater than or equal to 140 mm HG systolic pressure (a standard) or greater than or equal to 90 mm Hg diastolic pressure (a standard).

Although any number of variables could be monitored, it is important to identify those indicators that are most linked to issues of concern (Graefe et al. 1990). Thus, while a physician could monitor a stroke victim's kidney functions, it is more efficient to focus on the individual's blood pressure. The same logic applies to selecting indicators for wildlife-viewing opportunities. A manager could count the number of vehicles at trailhead parking lots, but past research suggests that monitoring how individuals distribute themselves in time and space throughout the park or refuge, or how they interact with wildlife and other visitors, are better indicators of recreation-opportunity differences (Shelby and Heberlein 1986; Kuss et al. 1990).

It is also important to recognize that there is no single "best" indicator or set of indicators. The choice of indicators and standards depends on the particular impact under consideration and the specific characteristics of the site. In other words, indicators and standards should be specific to the resource and opportunities provided at the site. The key is to select those impact indicators that matter the most for a given experience (Whittaker 1992). Although indicators and standards are site specific, it is possible to identify criteria for choosing indicators.

Criteria for Choosing Indicators
- Specificity and responsiveness
- Sensitivity
- Measurability
- Integration with management objectives
- Impact importance

Specificity and Responsiveness

Indicators are only useful if they refer to specific conditions created by human use. As noted above, an overall measure of human density in an area is too vague unless it is linked to the impact conditions associated with that level of use (e.g., encounters with others, loss of solitude-oriented wildlife-viewing opportunities). Specific indicators might focus on the abundance of selected wildlife species, the frequency of wildlife sightings, or wildlife reproductive success.

Indicators should reflect impact changes related to impacts caused by human activity rather than those caused by natural events (see Box 3). Unfortunately, disentangling human from natural impacts is complex. Wall and Wright (1977) suggest four factors that limit ecological studies and introduce difficulties in identifying human impacts: (1) there are often no baseline data for comparison to natural conditions; (2) it is difficult to disentangle the roles of humans and nature; (3) there are spatial and temporal discontinuities between cause and effect; and (4) in light of complex ecosystem interactions, it is difficult to isolate individual components. Some impacts take the form of naturally occurring processes that have been speeded up by human interference. Even without human activity, however, severe impacts can occur due to natural causes that render the impacts associated with recreational use insignificant (Schreyer 1976).

Box 3. Natural vs. Human Impacts

The Trustees of Reservations' (TTOR) program of biological and social research at three barrier beaches in Massachusetts (Cape Poge Wildlife Refuge and Wasque Reservation, Edgartown; Coskata-Coatue Wildlife Refuge, Nantucket; and Crane Beach, Ipswich) illustrates this distinction between human and natural impacts (see Vaske et al. 1992, for a summary). Between 1986 and 1989, a program for research and protection of rare shorebirds was instituted to assist in the development of TTOR management plans. Biological impacts to nesting piping plovers were measured through direct observation and surveys of predator populations (Rimmer and Deblinger 1990), and a study of human disturbance on migratory shorebirds at Crane Beach (Deblinger et al. 1989). The social research consisted of a series of on-site visitor surveys at each beach (Vaske et al. 1992).

Findings from the research at Cape Poge-Wasque and Crane Beach indicated that predators (e.g., skunks, raccoons, and red fox) were causing greater impact than humans. Two types of protection were applied to mitigate these impacts. Small wire-mesh fences were installed around nests to protect piping plovers from predators (see Vaske et al. 1994, for a review). Outside of these exclosures, symbolic fencing composed of a single strand of twine was erected to eliminate disturbance by visitors.

The Crane Beach study also indicated a natural zoning existed between visitors and the birds. Most boaters were attracted to a portion of the beach which, due to habitat considerations, the birds did not use for either feeding or resting. By designating restricted boat-landing areas outside of the feeding flats and restricting visitors from bird-resting areas in the dunes, managers could ensure that migratory shorebirds could stop over at Crane Beach undisturbed. Moreover, by restricting human use from nesting areas during critical seasons, as opposed to prohibiting use altogether, they could permit the coexistence of both plovers and humans.

Sensitivity

The indicator needs to be sensitive to changes in conditions during relatively short time periods; Merigliano (1989) suggests within one year. Such changes may be reflected in biological conditions (e.g., the presence or absence of particular wildlife species) or the human experience (e.g., the frequency of encounters with others at a viewing site). If the indicator only changes after impacts are substantial or never changes, the variable lacks the early warning signs that allow managers to be proactive (Whittaker and Shelby 1992).

Visitor satisfaction, for example, is often a major management objective and has been one of the most commonly used indicators of recreation quality (Shelby and Heberlein 1986). If, as traditionally assumed, enjoyment from a recreation experience is inversely correlated with the number of people present, reported satisfaction ratings should provide the basis for setting standards. Studies in a variety of settings, however, have consistently found that recreationists are generally satisfied with their experience independent of the use intensities they experienced (Kuss et al. 1990).

A variety of explanations have been offered to account for these findings. For example, to cope with the negative consequences of increasing numbers of visitors (e.g., loss of solitude), some individuals modify their standards for what is acceptable. The end result is a "product shift" or change in the character of the wildlife-viewing experience at a given area (Shelby et al. 1986). Other people who are more sensitive to user densities may stop visiting an area altogether if adjustments, either attitudinal (product shift) or behavioral (e.g., visiting during off peak times, visiting less frequently), fail to bring about the desired experience. With all of these explanations, the current visitors to a heavily used area may be as satisfied as visitors five or ten years ago when use levels were much lower, but are receiving a different type of experience.

While overall satisfaction measures are not sensitive to changing use conditions, other measures of recreation quality (e.g., perceived crowding and normative tolerance limits) do show the requisite variation. Perceived crowding, for example, combines the descriptive information (the density or encounter level experienced by the individual) with evaluative information (the individual's negative evaluation of that density or encounter level). When people evaluate an area as crowded, they have at least implicitly compared the impact they experienced with their perception of a standard. Findings from a comparative analysis of 35 crowding studies and

59 different settings and activities indicated that crowding varies across recreational settings and activities, time or season of use, resource availability, accessibility, or convenience, and management strategies designed to limit visitor numbers (Shelby et al. 1989). This variability has allowed recreation researchers and managers to use crowding as a useful indicator.

Measurability

Indicators should be easily and reliably measurable in the field. When choosing impact indicators, it is important to specify the level of detail at which selected indicators will be measured and evaluated. The scale of measurement may range from sophisticated indices using quantitative measurements to subjective visual rating schemes. The choice of an appropriate level of measurement will depend on such factors as the availability of funding and personnel, number of sites that must be evaluated, and frequency of measurement and site evaluation.

To illustrate, early crowding studies employed multiple-item scales (Shelby et al. 1989). While such scales consider a concept from different points of view and provide the data necessary for estimating reliability coefficients, the mathematical calculations involved in combining survey items into a single scale score sometimes make it difficult to compare results and can render the findings less understandable to managers (Shelby et al. 1989). To overcome these problems, Heberlein and Vaske (1977) developed a single item that asks people to indicate how crowded the area was at the time of their visit (see Box 4).

The crowding measure alone is not a perfect substitute for information about use levels, impacts, and evaluative standards that a more complete study can provide. Nevertheless, one can easily collect data with a single crowding item, thereby providing considerable insight about a study site. The single-item crowding measure is easy to interpret and compare across studies, and has been widely used in outdoor-recreation research (Shelby et al. 1989). The consistency of these findings makes the crowding measure a good indicator for addressing social impacts.

Integration with Management Objectives

Indicators need to be linked to the management objectives that specify the type of experience to be provided. For example, if a management objective is to provide a low-density backcountry wildlife-viewing experience, the indicators should focus on the number of encounters between visitors, perceptions of crowding, and encounter norm tolerances (see Box 5). Alternatively, if a

Box 4. Crowding Indicator and Standards

Most theorists recognize a difference between density and crowding, but even scientists sometimes use the word "crowding" inappropriately when referring to high density. Density is a descriptive term that refers to the number of people per unit area. It is measured by counting the number of people and measuring the space they occupy, and it can be determined objectively. Crowding, on the other hand, is a negative evaluation of density; it involves a value judgment that the specified number is too many. The term *perceived crowding* is often used to emphasize the subjective or evaluative nature of the concept.

Researchers have developed a relatively simple measure of perceived crowding (Heberlein and Vaske 1977). The question asks people to indicate how crowded the area was at the time of their visit. Responses were given on the scale below:

1	2	3	4	5	6	7	8	9
Not at all Crowded		Slightly Crowded			Moderately Crowded		Extremely Crowded	

In this item, two of the nine scale points label the situation as uncrowded, and the remaining seven points label it as crowded to some degree. The rationale is that people may be reluctant to say an area was crowded because crowding is an undesirable characteristic in a recreation setting. An item that asked "Did you feel crowded?" might lead most people to say "No." The scale is sensitive enough to pick up even slight degrees of perceived crowding, just as measures of undesirable chemicals (e.g., pollutants or carcinogens) are sensitive to even low levels of these substances.

Shelby et al. (1989) developed crowding standards based on this indicator. Their comparative analysis of 59 different settings and activities suggested that when ≤ 35 percent of the visitors feel crowded, density levels in the area were not a problem. For locations where between 50 and 60 percent of visitors felt crowded, the setting was approaching its carrying capacity, and visitors started to experience access and displacement problems. Locations and activities where over 65 percent of the visitors felt crowded were considered over carrying capacity.

management objective involves roadside viewing opportunities, the indicators might be linked to visitor safety and the probability of seeing wildlife.

Useful impact indicators are those that can be treated by management prescriptions. A seemingly eloquent solution to a human-caused impact that cannot be addressed by management actions does not resolve the problem condition. The most useful indicators reflect multiple impact conditions. Because wildlife managers typically have small monitoring budgets, indicators that can be used to represent several different impacts allow managers to focus their attention and efforts while being reasonably assured that the overall quality of a given experience is maintained (Whittaker and Shelby 1992). Crowding or norm tolerances are examples that often reflect several other interaction-type indicators such as encounters with others or competition for space on wildlife-viewing platforms.

Impact Importance

Finally, and most importantly, indicators should represent important impacts (Whittaker 1992). For example, if managers, stakeholders and visitors are not concerned about a social impact or researchers are not able to show how an impact negatively influences wildlife populations, developing standards is difficult

Viewer approaching Dall sheep in Denali National Park, Alaska. Standards such as "80 percent proability of viewing sheep from a distance as close as 50 feet" help managers identify acceptable approach distances and help visitors understand the type of viewing experience they are likely to receive in an area. (Photo by Doug Whittaker)

Box 5. Normative Indicators and Standards

The concept of norms provides a theoretical framework for collecting and organizing information about users' evaluations of conditions and has proven to be sensitive to changing use conditions. As defined by one research tradition, norms are standards that people use to evaluate behavior or the conditions created by behavior as acceptable or unacceptable (see Vaske et al. 1986; Vaske et al. 1993; Shelby et al. 1996; for reviews). Norms thus define what behavior or conditions should be, and can apply to individuals, collective behavior, or management actions designed to constrain collective behavior. This normative approach allows researchers to define social norms, describe a range of acceptable behavior or conditions, explore agreement about the norm, and characterize the type of norm (e.g., no tolerance, single tolerance, or multiple tolerance norms; Whittaker and Shelby 1988).

Normative concepts in natural-resource settings were initially applied to encounter impacts in backcountry settings (encounter norms measure tolerances for the number of contacts with other users). The focus on encounters in backcountry worked because encounter levels were generally low, survey respondents could count and remember them, and encounters have important effects on the quality of experiences when solitude is a feature. Most studies showed that encounter norms across these backcountry settings were stable and strongly agreed upon, usually averaging about four encounters per day (Vaske et al. 1986).

More recently, norm concepts and methods have been applied to a greater diversity of impacts and settings (Shelby and Vaske 1991; Shelby et al. 1996). Research on encounter norms in higher-density frontcountry settings, for example, has demonstrated more variation in visitors' tolerances for others as well as lower levels of agreement (Manning et al. 1996; Vaske et al. 1996; Vaske and Donnelly 1998; Donnelly et al. 2000). This led some researchers to examine norms for interaction impacts other than encounters (Shelby et al. 1987; Martinson and Shelby 1992; Whittaker 1992; Whittaker and Shelby 1993). Norms for recreationist proximity, percentage of time within sight of others, incidents of discourteous behavior, competition

box continues

for specific resources, and waiting times at access areas have all been examined. These alternative interaction impacts are often more salient than encounters in higher-use settings (Basman et al. 1996; Whittaker and Shelby 1996). Taken together, this work suggests that normative data are sensitive to changing use conditions, can facilitate understanding visitors' evaluations of social and environmental conditions, and have proven helpful to managers.

Normative standards may also provide a gauge for estimating benefits to society. If, for example, a management objective is to enhance the flow of dollars into a community's economy by creating more wildlife-viewing opportunities, one indicator might be the occupancy rate at local motels. The standard in this situation might be 50 percent occupancy.

to justify. If wildlife viewers are more interested in photographing elk than the number of people standing next to them, frequency of seeing elk becomes a better indicator of quality experiences than social-interaction variables. Alternatively, if visitors consider solitude in viewing experiences as more important than number of animals seen, encounters with other visitors becomes an important quality indicator.

Characteristics of Good Standards

Specific standards are established for each impact indicator and define an acceptable level of impact for each indicator. Just as impact indicators reflect management goals and objectives, standards are quantifiable value judgments concerning what the agency is attempting to achieve. Based on previous work (Schomaker 1984; Graefe et al. 1990; Whittaker and Shelby 1992; Vande Kamp 1998), the following discusses several important characteristics of good standards.

Characteristics of Good Standards
- Quantifiable
- Time Bounded
- Attainable
- Output Oriented

Quantifiable

Standards restate management objectives in quantitative terms. A good standard unequivocally states the level of acceptable impact. Such statements define how much is acceptable in quantitative terms. For example, a good standard might specify that visitors should be able to watch wildlife with fewer than ten other people present. Specifying that there should only be "a few other people present" is not a good standard because it does not define how many constitutes "a few."

Time Bounded

"Time-boundedness" complements the quantifiable component of a good standard (Whittaker and Shelby 1992). Quantifiable standards only state "how much" is appropriate. Time-bounded standards specify "how much, how often" or "how much by when." This is especially important for wildlife-viewing impacts because such opportunities often have a seasonal component. Seeing five hundred elk in Rocky Mountain National Park is a common occurrence for a fall evening, but a rare event during the summer when the elk are at higher elevations. Such seasonal differences in viewable wildlife often correlate with fluctuations in visitor numbers. The number of day visitors to RMNP who are explicitly interested in viewing and photographing elk, for example, is substantially greater in the fall than other seasons. Time-bounded standards recognize such variation.

Attainable

Management standards need to be reasonably attainable. When standards are too easy, little is accomplished. If they are too difficult to achieve, both managers and visitors are likely to become frustrated. Good objectives and standards should "moderately challenge" the manager and staff (Schomaker 1984).

For each important indicator, standards should be set at levels that reflect management's intent for resource or experiential outcomes in the area (Whittaker and Shelby 1992). While standards that are difficult to attain are generally undesirable, they may still be necessary. A "no litter" standard, for example, may not be attainable, but is still correct. The cynical excuse for not setting appropriate standards is that managing for some conditions is "too hard." On the other hand, management strategies designed to meet a standard may produce sufficient positive change to warrant the effort. Without standards, it is too easy to do nothing (management by default).

**Box 6. Different Experiences—
Different Indicators and Standards**

Table 8-1 (see page 160) describes three different elk-viewing experiences: (1) highly specialized backcountry, (2) specialized frontcountry, and (3) general interest roadside elk viewing. For each experience, potential impact indicators and associated standards are presented. Although the scenarios are hypothetical, the indicators represent the types of impact conditions managers must typically address. In addition, while some of the standards are based on considerable previous research (e.g., encounter norms for the backcountry, percent feeling crowded), other standards (e.g., wildlife flight distance) require empirical validation. They are presented here to highlight the link between specific indicators and standards and to demonstrate how the standards might change for different viewing experiences.

Five different categories of impact indicators (e.g., development, crowding/norm tolerances) are shown in Table 8-1. Consistent with the principles of Experience-based Management, minimal development, opportunities for solitude, and few regulations characterize a highly specialized backcountry viewing experience. Standards associated with this experience type suggest that there should be no physical barriers separating humans and wildlife at prime viewing areas, and that only small ($\leq$ 50 sq. feet) temporary viewing blinds constructed from natural-colored fabric would be allowed. To enhance the opportunity for solitude, the standard for encounters with other groups on the trails is $\leq$ 4 groups per day, with zero other people at one time at prime viewing areas, and $\leq$ 35 percent of visitors feeling crowded while viewing wildlife. Groups of more than four individuals would be restricted from prime viewing areas and visitors would have the opportunity to roam beyond designated trails. The standard for wildlife flight distance for the backcountry experience was set at $\leq$ 200 feet to emphasize that wildlife in this scenario have not become habituated to the presence of people. To obtain a closer view necessitates more skill on the part of the viewer, thereby adding to the challenge of the experience.

In contrast, the standards for the general interest roadside elk-viewing experience emphasize more development (e.g., barriers to separate humans and wildlife, relatively large [$\leq$ 600 sq. feet], single-level, permanent blinds, with paved trails, ≤ 8 feet in width). Because the presence of other visitors does not detract from the quality of the experience, the crowding/norm tolerance standards for the general interest experience are not as stringent as those described for the other two experience opportunities.

In general, the standards for the specialized frontcountry experience fall in between the standards specified for the other two experiences. It is important to note, however, that the standards do not always vary among different experience types. With all three experiences, for example, managers are likely to have a standard of zero incidents of humans being injured by wildlife and zero incidents of wildlife being disturbed by humans.

While the specifics shown in Table 8-1 may vary somewhat depending on resource constraints and the range of experience opportunities available, they do illustrate the linkage between indicators and standards and highlight how standards may change depending on management objectives. Planners and managers should think of Table 1 as a starting point for selecting indicators and setting standards.

Output Oriented

Standards should be "output" rather than "input" oriented (Schomaker 1984). This distinction suggests that managers should focus on the conditions to be achieved rather than the way the standard is met. For example, a standard that specifies "150 people per day in a wildlife-viewing area" is not a good standard because it refers to an action (use limits) rather than an acceptable impact. "Less than ten encounters per day" or "no more that 35 percent of the visitors feeling some level of crowding" are better standards because they emphasize the acceptability of different impact conditions.

Sources for Selecting Indicators/Developing Standards

Identifying characteristics of good standards is a useful exercise, but it does not provide much information about what standards should be (see Box 6), or where they should come from. Many different management and research efforts have developed or recommended various standards, utilizing a variety of techniques

text continues on page 162

Table 8-1. Hypothetical impact indicators and standards for three elk viewing experience opportunities

Impact Indicator	Highly Specialized Backcountry Elk Viewing	Specialized Frontcountry Elk Viewing	General Interest Roadside Elk Viewing
Development			
Physical barriers separating humans and wildlife at prime viewing areas	0 Barriers	≤ 10% areas with barriers	Barriers installed wherever needed to increase safety
Viewing blinds / hides			
Type	Temporary blinds only	Temporary or permanent	Permanent blinds
Material	Natural colored fabric	Wood clad indigenous structure	Wood clad equipped with binoculars
Size	≤ 50 sq. feet	≤ 200 sq. feet	≤ 600 sq. feet
Design / Shape	DNA	Multi-level. Irregular design to enhance privacy between viewers	Single level. Square or rectangular design. Multiple access points
Trail surface	Indigenous mulch	Unpaved	Paved
Trail treadway width	≤ 1 foot	≤ 4 feet	≤ 8 feet
Crowding / Norm Tolerances			
Encounters with other groups on the trails	≤ 4 groups per day	≤ 4 groups per hour	≤ 25 people per hour
Number of people in sight at one time at prime viewing areas	0	≤ 10 people	≤ 50 people
Percent of viewers feeling crowded at prime viewing areas	≤ 35%	≤ 35%	≤ 50%

Impact Indicator	Highly Specialized Backcountry Elk Viewing	Specialized Frontcountry Elk Viewing	General Interest Roadside Elk Viewing
Viewing distances to concentrations of elk	80% probability of viewing distance within 50 feet	50% probability of viewing distance within 50 feet	80% probability of viewing distance within 200 feet
Regimentation			
Group size limits	≤ 4 people per party	≤ 8 people per party	≤ 25 people per party
Ranger escort (No / Yes)	No	Yes	Yes
Freedom to roam beyond designated trails (No / Yes)	Yes	No	No
Human-Wildlife Interaction Number of human-wildlife incidents involving:			
Incidents with injuries to humans	0	0	0
Disturbance to elk	0	0	0
Wildlife flight distance	≤ 200 feet	≤ 50 feet	≤ 50 feet

Roadside wildlife viewers near Eielson Visitor Center in Denali National Park, Alaska. Standards for social interaction levels may be a key issue for many viewing opportunities. (Photo by Julie Whittaker)

or sources of information. A review of the most common sources and techniques follows.

Sources for Selecting Indicators/Developing Standards
- Laws and policy mandates
- Manager's professional judgment
- Biological research
- Public involvement
- Wildlife viewer or population surveys

Laws and Policy Mandates

Laws and policy mandates may provide guidelines for selecting specific impact indicators and developing appropriate standards for desirable wildlife-viewing experiences. Most laws, however, are written in broad and often vague language. Directives such as "provide high-quality viewing experiences" or "minimize human-wildlife conflict" lack the specificity necessary to set quantitative standards.

Manager's Professional Judgment

Wildlife managers often develop standards based on their interpretation of laws and policy mandates, their knowledge of the area, their understanding of the recreation opportunities, and their knowledge of conditions that define those opportunities (Whittaker and Shelby 1992). By imposing their idea of what is appropriate, or even their own personal values, in the decision-making process, managers have implicitly been setting standards for years (Manfredo et al. 1995). An argument can be made, however, for setting standards more explicitly. First, although wildlife-management standards have traditionally been based solely on professional judgment and biological expertise, the increasingly political nature of all natural-resource actions implies that decisions made in isolation are likely to generate considerable public scrutiny (Shepherd 1990; Bright and Manfredo 1993). Second, although it has been assumed that managers understand the acceptability of different resource and experiential conditions, a growing body of empirical evidence suggests considerable differences between the views of managers, visitors, and organized interest groups (Magill 1988; Shelby and Shindler 1990; Gill 1996). By formalizing the process for developing standards and including different points of view, managers gain a greater understanding of their objectives, have more justification for their actions, and are able to be more proactive when potential problem situations arise (Whittaker and Shelby 1992).

Biological Research

Science-based wildlife research has been and always will be an important component in developing standards. Wildlife data help clarify what management goals are biologically possible and describe how management actions affect wildlife impacts (Vaske et al. 2001). Biological research by itself, however, cannot predict which alternatives are more or less desirable. For example, scientists are often assumed to be the most appropriate individuals to set standards for acceptable air- and water-pollution levels. When viewed from the larger societal perspective, however, this assumption is invalid. The scientific data describe the consequences of allowing a certain number of pollutants per volume of air or water (e.g., X number of people will die at contamination level Y). Whether this risk level is considered acceptable depends on legislation or other government functions. Even at extremely low levels of water pollution, some people are likely to become ill. It is impossible to set a standard until the acceptability of various risk levels has been identified (Whittaker and Shelby 1992).

Public Involvement

Traditional public involvement (e.g., focus groups, public meetings) represent another important strategy for developing standards, especially for social-impact indicators and standards. Wildlife viewers are experts in identifying the characteristics of an experience they find most important. When given the opportunity to communicate their preferences, individuals are typically willing to express their views. Small focus-group meetings with different interest groups, for example, provide a useful starting point for identifying which impacts matter more (Whittaker 1992). Standards can be developed from input provided by participants at larger public meetings, but it is often difficult to focus discussion on specific issues at these meetings. Moreover, individuals who attend public hearings and voice the loudest concerns may not represent all constituents.

Although these traditional techniques for soliciting citizen participation provide useful information (Creighton 1980; Bleiker and Bleiker 1990), wildlife managers are increasingly adopting a stakeholder approach to involving public interests (Decker et al. 1996; Chapter 7). Approaches such as transactive planning and co-management bring diverse interests and stakeholders in direct communication with one another and with agency decision makers to fashion collaborative solutions to wildlife-management challenges. For example, agencies now routinely form citizen task forces, roundtables, advisory councils, and stakeholder planning teams to assist agency personnel with planning tasks and decisions (Beatley et al. 1995; Stout et al. 1996; Curtis and Hauber 1997). When multiple stakeholders have a voice in developing standards, polarized views about acceptable conditions and experiences are likely to emerge. Under these conditions, some negotiation and compromise must occur to develop standards that will be supported by the different publics and interest groups.

Wildlife Viewer or Population Surveys

Perhaps the most useful source for developing standards involves user or population surveys. Even the best public-involvement efforts tend to neglect the occasionalist or generalist wildlife viewers or the "general public" in favor of special-interest groups who voice strong opinions on a topic. When surveys adhere to scientific principles (e.g., reliability, validity, representativeness, generalizability), the approach is especially useful for developing standards for social indicators (Manfredo et al. 1996b). Past research highlights a few key issues to consider when using surveys to develop standards (see Shelby and Heberlein 1986; Graefe et al. 1990; for reviews).

First, the survey should include a range of impact conditions and gauge which of those impacts are more important (Whittaker 1992). Managers may ultimately establish standards for only a few key impact indicators. However, because surveys are usually conducted before this decision is made, asking about several different types of impact (e.g., human-interaction impacts, wildlife impacts) allows some flexibility in choosing different indicators. If respondents are asked to consider the relative importance of different impacts, the survey can facilitate the indicator selection process.

Second, questions about users' personal standards should be direct, involve quantitative response categories, and be easy to understand (Whittaker and Shelby 1992). As noted previously, extensive research has failed to demonstrate a consistent relationship between impact variables (e.g., encounters with others) and general evaluative measures (e.g., satisfaction). Most researchers recommend focusing on the evaluation of impacts themselves (Shelby and Heberlein 1986). For example, surveys might ask respondents to report the number of encounters they are willing to have per day or to rate acceptable encounter levels for different experiences. An effective technique used in several studies involves parallel questions about the amount of impact individuals experienced and the amount of impact they are willing to tolerate. Statistical comparisons of such results provide data about where to set standards and allow definition of an impact problem (Vaske et al. 1986; Shelby et al. 1987).

Third, when asking about quantitative estimates of acceptable impact levels, respondents should be allowed to specify that "this impact does not matter to me" or that "the impact matters but I cannot give a number" (Roggenbuck et al. 1991; Hall et al. 1996). Some wildlife viewers, especially those with little experience, may not have opinions about acceptable impact levels or may not even be aware of the impact situation (Donnelly et al. 2000).

Finally, analysis of survey data should go beyond simple frequencies or measures of central tendency (Vaske et al. 1986; Shelby and Vaske 1991; Shelby et al. 1996). Such measures are useful starting points, but closer examination of the response distributions reported by different groups or the level of group agreement are also important for developing standards.

Summary Points

- Indicators and standards represent explicit statements about what conditions managers are trying to provide, and are crux decisions in planning and management processes. Although indicators are site specific, it is important to consider the criteria outlined here when selecting among the range of possible choices (e.g., ensuring that they are sensitive to changing conditions, measurable and integrated with management objectives). Similarly, several general criteria were outlined for developing standards (e.g., that they should be quantifiable, time bounded, attainable, and output oriented).

• Knowing the characteristics of good and bad standards represents an initial step in the planning process. At some point, managers need to actually develop standards for the impact indicators of concern. Five techniques were outlined in this chapter: relying on laws and policies, professional judgment, biological research, public involvement, and user or population surveys. Each of these sources of information has advantages and disadvantages; none should be viewed as the "best" single technique and all are probably necessary if management is to be successful.

• Although much has been learned over the past few decades, it is important to remember the lessons learned and the gaps in our knowledge. First, while crowding and norms for acceptable behaviors and conditions in backcountry areas are well understood, research in frontcountry areas is still in its infancy. More research exploring the range of norms frontcountry visitors use when making evaluative judgments is clearly needed. Second, because not all stakeholders share the same norms, managers need to adopt a collaborative planning process where all interested parties can express their views on acceptable management actions. Third, the task of managing human-wildlife impacts is not over when a management action has been implemented. Monitoring of key impact indicators is critically important to determine whether the actions are producing the desired outcomes without altering other characteristics of the experience or the resource.

• A final word of caution. Selecting key impact indicators and setting standards do not guarantee good management. Standards are only as good as the information upon which they are based. Perfect knowledge of all impacts, however, is impossible. The classic error is failure to set a standard early (because of a lack of information), and assume that no decision has been made. Letting impacts incrementally increase is also a decision—management by default instead of by design. Standards are a key to management by design and lie at the heart of good wildlife management.

Literature Cited

Basman, C. M., M. J.Manfredo, S. C. Barro, J. J. Vaske, and A. Watson (1996). "Norm accessibility: An exploratory study of backcountry and frontcountry recreation norms." *Leisure Sciences*, 18, 177-91.

Beatley, T., T. J. Fries, and D. Braun (1995). "The Balcones Canyonlands conservation plan: A regional multi-species approach." In D. R. Porter and D. A. Salvesen (Eds.), *Collaborative Planning for Wetlands and Wildlife*. Washington, DC: Island Press.

Belnap, J. (1998). "Choosing indicators of natural resource condition: A case study in Arches National Park, Utah, USA." *Environmental Management*, 22, 635-42.

Bleiker, H., and A. Bleiker (1990). *Citizen Participation Handbook for Public Officials and Others Serving the Public*. (5th edition). Monterey, CA: Institute of Participatory Management and Planning.

Bright, A. D., and M. J. Manfredo (1993). "An overview of recent advances in persuasion theory and their relevance to natural resource management." In A. W. Ewert, D. J. Chavez, and A. W. Magill (Eds.), *Culture, Conflict and Communication in the Wildland–urban Interface*. Boulder, CO: Westview Press.

Creighton, J. L. (1980). *Public Involvement Manual: Involving the Public in Water and Power Resource Discussions*. Washington, DC: United States Department of Interior, Government Printing Office.

Curtis, P. D., and J. R. Hauber (1997). "Public involvement in deer management decisions: Consensus versus consent." *Wildlife Society Bulletin*, 25, 399-403.

Deblinger, R. D., J. J. Vaske, and M. P. Donnelly (1989). "Integrating ecological and social impacts into barrier beach management." In *Proceedings of the Northeast Recreation Research Symposium*. (Technical Report NE-132). Burlington, VT: U.S. Department of Agriculture, Forest Service, Northeast Forest Experiment Station.

Decker, D. J., C. C. Krueger, .R. A. Baer, B. A. Knuth, and M. E. Richmond (1996). "From clients to stakeholders: a philosophical shift for fish and wildlife management." *Human Dimensions of Wildlife*, 1, 70-82.

Donnelly, M. P., J. J. Vaske, D. Whittaker, and B. Shelby (2000). "Toward an understanding of norm prevalence: A comparative-analysis." *Environmental Management*, 25, 403-14.

Gill, R. B. (1996). "The wildlife professional subculture: The case of the crazy aunt." *Human Dimensions of Wildlife*, 1, 60-69.

Graefe, A. R., F. R. Kuss, and J. J. Vaske (1990). *Visitor Impact Management: The Planning Framework*. Washington, DC: National Parks and Conservation Association.

Hall, T., B. Shelby, and D. Rolloff (1996). "Effect of varied question format on boaters' norms." *Leisure Sciences*, 18, 193-204.

Heberlein, T. A., and J. J. Vaske (1977). *Crowding and Visitor Conflict on the Bois Brule River*. (Technical Report WIS WRC 77-04). Madison, WI: University of Wisconsin, Water Resources Center.

Kuss, F. R., A. R. Graefe, and J. J. Vaske (1990). *Recreation Impacts and Carrying Capacity: A Review and Synthesis of Ecological and Social Research*. Washington, DC: National Parks and Conservation Association.

Magill, A. (1988). "Natural resource professionals: The reluctant public servants." *The Environmental Professional*, 10, 295-303.

Manfredo, M. J., J. J. Vaske, and D. J. Decker (1995). "Human dimensions of wildlife management: Basic concepts." In R. L. Knight and K. J. Gutzwiller (Eds.), *Wildlife and Recreationists: Coexistence through Management and Research*. Washington, DC: Island Press.

Manfredo, M. J., B. L. Driver, and M. A. Tarrant (1996a). "Measuring leisure motivation: A meta-analysis of the recreation experience preference scales." *Journal of Leisure Research*, 28, 188-213.

Manfredo, M. J., J. J. Vaske, and L. Sikorowski (1996b). "Human dimensions of wildlife management." In A. W. Ewert (Ed), *NaturalResource Management: The Human Dimension*. New York: Westview Press.

Manning, R. E., D. W. Lime, W. A. Freimund, and D. G. Pitt (1996). "Crowding norms at frontcountry sites: A visual approach to setting standards of quality." *Leisure Sciences*, 18, 39-59.

Martinson, K. S., and B. Shelby (1992). "Encounter and proximity norms for salmon anglers in California and New Zealand." *North American Journal of Fisheries Management*, 12, 559-67.

Merigliano, L. (1989). "Indicators to monitor the wilderness recreation experience." In D. Lime (Ed.), *Proceedings, Managing America's Enduring Wilderness Resources Symposium*. Minneapolis, MN: U.S. Department of Agriculture, Forest Service.

Ott, W. R. (1978). *Environmental Indices: Theory and Practice*. Ann Arbor, MI: Ann Arbor Science Publications, Inc.

Rimmer, D. W., and R. D. Deblinger (1990). "Use of predator exclosures to protect Piping Plover nests." *Journal of Field Ornithology*, 53, 263-68.

Roggenbuck, J. W., D. R. Williams, S. P. Bange, and D. J. Dean (1991). "River float trip encounter norms: Questioning the use of the social norms concept." *Journal of Leisure Research*, 23, 133-53.

Schomaker, J. H. (1984). "Writing quantifiable river recreation management objectives." In J. S. Popadic, D. Butterfield, D. Anderson, and M. R. Popadic (Eds.), *Proceedings of the 1984*

National River Recreation Symposium. Baton Rouge, LA: University of Louisiana.

Schreyer, R. (1976). "Sociological and political factors in carrying capacity decision making." In *Proceedings of the Third Resources Management Conference*. Ft. Worth, TX: U.S. Department of the Interior, National Park Service.

Shelby, B., and T. A. Heberlein (1986). *Carrying Capacity in Recreation Settings*. Corvallis, OR: Oregon State University Press.

Shelby, B., N. S. Bregenzer, and R. Johnson (1986). "Product shift as a result of increased density: Empirical evidence from a longitudinal study." Paper presented at the first national symposium on Social Science in Resource Management. Corvallis, OR.

Shelby, B., D. Whittaker, R. Speaker, and E. E. Starkey (1987). *Social and ecological impacts of recreation use on the Deschutes River Scenic Waterway*. Report to the Oregon Legislature. Corvallis, OR: Oregon State University.

Shelby, B., J. J. Vaske, and T. A. Heberlein (1989). "Comparative analysis of crowding in multiple locations: Results from fifteen years of research." *Leisure Sciences*, 11, 269-91.

Shelby, B., and B. Shindler (1990). "Evaluating group norms for ecological impacts at wilderness campsites." In *Proceedings of the Third Symposium on Social Sciences in Resource Management*. College Station, TX: Texas A&M University.

Shelby, B., and J. J. Vaske (1991). "Using normative data to develop evaluative standards for resource management: A comment on three recent papers." *Journal of Leisure Research*, 23, 173-87.

Shelby, B., G. Stankey, and B. Shindler (Eds.) (1992). *Defining Wilderness Quality: The Role of Standards in Wilderness Management —A Workshop Proceedings*. (General Technical Report PNW-GTR-305). Portland OR: U.S. Department of Agriculture, Forest Service, Pacific Northwest Research Station.

Shelby, B., J. J. Vaske, and M. P. Donnelly (1996). "Norms, standards and natural resources." *Leisure Sciences*, 18, 103-23.

Shepherd, W. B. (1990). "Seeing the forest for the trees: 'New perspectives' in the Forest Service." *Renewable Resources Journal*, Summer, 8-11.

Stankey, G. H., D. N. Cole, R. C. Lucas, M. E. Petersen, and S. S. Frissell (1985). *The Limits of Acceptable Change (LAC) System for Wilderness Planning* (Report INT-176). Ogden, UT: U.S. Department of Agriculture, Forest Service, Intermountain Forest and Range Experiment Station.

Stout, R. J., D. J. Decker, B. A. Knuth, J. C. Proud, and D. H. Nelson (1996). "Comparison of three public-involvement approaches for stakeholder input into deer management decisions: A case study." *Wildlife Society Bulletin*, 24, 312-17.

U.S. Department of the Interior, National Park Service (1997). *VERP: The Visitor Experience and Resource Protection (VERP) Framework, A Handbook for Planners and Managers*. Denver, CO: U.S. Department of the Interior, National Park Service, Denver Service Center.

Vande Kamp, M. E. (1998). *The Use of Existing Information in the Process of Setting Social Standards for Proposed Wilderness Area of Zion National Park*. (Technical Report NPS/CCSOUW/NRTR-98-07, NPS D-127). Seattle, WA: University of Washington.

Vaske, J. J., B. Shelby, B., A. R. Graefe, and T. A. Heberlein (1986). "Backcountry encounter norms: Theory, method and empirical evidence." *Journal of Leisure Research,* 18, 137-53.

Vaske, J. J., R. D. Deblinger, and M. P. Donnelly (1992). "Barrier beach impact management planning: Findings from three locations in Massachusetts." *Canadian Water Resources Association Journal,* 17, 278-90.

Vaske, J. J., M. P. Donnelly, and B. Shelby (1993). "Establishing management standards: Selected examples of the normative approach." *Environmental Management,* 17, 629-43.

Vaske, J. J., D. W. Rimmer, and R. D. Deblinger (1994). "The impact of different predator exclosures on Piping Plover nest abandonment." *Journal of Field Ornithology,* 65, 201-9.

Vaske, J. J., D. J. Decker, and M. J. Manfredo (1995). "Human dimensions of wildlife management: An integrated framework for coexistence." In R. Knight and K. Gutzwiller (Eds.), *Wildlife and Recreationists: Coexistence through Management and Research*. Washington, DC: Island Press.

Vaske, J. J., M. P. Donnelly, and J. P. Petruzzi (1996). "Country of origin, encounter norms and crowding in a frontcountry setting." *Leisure Sciences,* 18, 161-76.

Vaske, J. J., and M. P. Donnelly (1998). *An Evaluation of the Glacier Gallery in the Columbia Icefield Visitor Centre*. Human Dimensions Research Unit Publication 37. Fort Collins, CO:Colorado State University, Human Dimensions in Natural Resources Unit.

Vaske, J. J., D. Whittaker, and M. P. Donnelly (2000). "Tourist impact management in North American national parks." In R. Butler and S. Boyd (Eds.), *Tourism and National parks: Issues and Implications*. New York: John Wiley and Sons.

Vaske, J. J., D. C. Fulton, and M. J. Manfredo (2001). "Human dimensions considerations in wildlife management planning." In D. Decker, T. Brown, W. F. Siemer (Eds.), *Human Dimensions of Wildlife Management in North America*. Bethesda, MD: The Wildlife Society.

Wall, G., and C. Wright 1977. *The Environmental Impact of Outdoor Recreation.* (Publication Series No. 11). Waterloo, Ontario: University of Waterloo, Department of Geography.

Whittaker, D. (1992). "Selecting indicators: Which impacts matter more?" In B. Shelby, G. Stankey, and B. Shindler (Eds.), *Defining Wilderness Quality: The Role of Standards in Wilderness Management—A Workshop Proceedings.* (General Technical Report PNW-GTR-305). Portland, OR: U.S. Department of Agriculture, Forest Service, Pacific Northwest Research Station.

Whittaker, D., and B. Shelby (1988). "Types of norms for recreation impacts: Extending the social norms concept." *Journal of Leisure Research,* 20, 261-73.

Whittaker, D., and B. Shelby (1992). "Developing good standards: Criteria, characteristics and sources." In B. Shelby, G. Stankey, and B. Shindler (Eds.), *Defining Wilderness Quality: The Role of Standards in Wilderness Management—A Workshop Proceedings.* (General Technical Report PNW-GTR-305). Portland, OR: U.S. Department of Agriculture, Forest Service, Pacific Northwest Research Station.

Whittaker, D., and B. Shelby (1993). *Kenai River Carrying Capacity Study: Important Conclusions and Implications.* Report to Alaska State Parks. Anchorage, AK: U.S. Department of the Interior, National Park Service, RTCA project report.

Whittaker, D., and B. Shelby (1996). "Norms in high-density settings: Results from several Alaskan rivers." Paper presented at the 6th International Symposium on Society and Resource Management. Pennsylvania State University, May.

Choosing Actions: Problem Definition, Identifying Strategies, and Evaluation Criteria

Doug Whittaker, Jerry J. Vaske, and Michael J. Manfredo

Introduction

DEFINING EXPERIENCES THROUGH OBJECTIVES, indicators, and standards is the heart of Experience-based Management (EBM), but choosing and implementing management actions are the steps that carry the process to completion. Opportunities for high-quality experiences are created when concrete actions produce biophysical, social, or managerial conditions that people can enjoy. Because actions are tangible and achievement-oriented, however, they can sometimes become a premature focus during planning. When stakeholders and managers approach planning efforts with specific actions in mind, it can be challenging to avoid advocating potential solutions before there is agreement about the problems they are trying to solve.

EBM urges a more systematic and deliberative process. The fundamental prerequisites to choosing actions are (1) agreement about the mix of opportunities to be provided, and (2) definitions of the experiential and resource conditions (standards) that define high-quality versions of those opportunities. Actions are then judged for their ability to meet those standards while minimizing other unwanted consequences.

This chapter reviews some of the issues involved in developing and choosing among actions. In general, one can think of three stages in this process. The first stage focuses on **problem definition**, which is fundamentally addressed by adopting the EBM planning framework (not a focus of this chapter; see Chapters 5 and 6). EBM urges managers to explicitly decide what opportunities they are trying to provide, establish standards that define high-quality experiences and natural-resource health, and then compare existing conditions with those standards to determine if actions are needed.

Under this system, problems exist and actions are needed when conditions are incongruent with standards. For many existing opportunities, planners focus on actions that will help sustain conditions at certain levels, usually in the face of increased impacts from higher use levels or changing visitor behavior. When developing new opportunities, however, the "problem" is how to create novel target experiences, usually through enhanced access and development, while ensuring that those experiences will be provided into perpetuity or without displacing a valued existing opportunity.

Problem definition is the most important step of the action selection process. It gives clarity and cause to the expenditure of scarce managerial resources. In the EBM approach, managers are encouraged to express problems in context of the ends to be achieved (e.g., opportunities provided, resource conditions to be maintained, benefits to local communities). The discussion of action selection thus begins with agreement about what is occurring at a resource. More specifically, it clarifies the type of recreation opportunity to be provided, asserts what is to be accomplished, and thus provides ultimate guidance as to whether the "right" action has been selected.

In many cases, debate or controversy about the "right" action is, at root, a debate over the appropriate type of recreation opportunity to be provided. Further, when conflicts or impacts are present, the most effective solution to the immediate problem may undesirably alter the opportunity available. For example, managers may consider responding to increased vegetation impacts with site hardening, which is likely to solve the short-term problem. However, hardening may also change perceptions of the area's "naturalness." If it is chosen, the managers might solve one problem, but only by altering the "end-state" they are trying to achieve. Visitors seeking a more natural experience might be "displaced" by conditions at the less natural area, while those insensitive to a more developed experience "invade" the site (creating a "product shift"). If the target opportunity features a primitive environment without human modification, site hardening would not be the best choice; in this case, limits on visitation, site resting, or site restoration might be more appropriate. These action choices become obvious only after one has been clear about the type of opportunity to be provided.

Having defined the problem, the second stage focuses on **identifying strategies**—generating lists of potential actions that could be used to address those problems. Planners, stakeholders, and the public often embrace this creative "brainstorming" task, and can usually develop long lists of potential actions. These efforts, however, can be made more efficient and comprehensive if ideas for actions are developed systematically. The first half of this chapter reviews one system for identifying actions, providing examples of common wildlife-viewing actions within each category.

The third and final stage of the process focuses on **applying evaluation criteria to choose actions** from the list of alternatives developed from the problem definition effort. This is the end-point of the process, and it requires evaluative judgments about which consequences are better or worse in four major areas: (1) effectiveness, (2) appropriateness, (3) public acceptability, and (4) administrative, financial, or legal feasibility. The second part of this chapter discusses each of these criteria, providing examples where each may be decisive when making action choices.

Identifying Potential Management Actions

Identifying actions is an interactive task that encourages planners, stakeholders, and the interested public to share their knowledge and experience to generate an exhaustive list of actions that could be used to address problems. Because this task avoids judging the merits of any action, and final decisions are further away in the process, these sessions tend to be non-controversial and highly productive.

Brainstorming sessions, however, can also be undisciplined, with ideas coming from multiple sources and addressing multiple issues or problems. If efficiency and comprehensiveness are important, a more systematic approach may be useful. Not only will such a system help ensure an exhaustive list of actions, but it will group similar actions together and may subsequently help planners identify some of the consequences of those groups.

One such system organizes actions by three general approaches: (1) capital-development actions that build things or modify the environment; (2) education actions that provide learning opportunities or help modify human behavior through persuasion efforts; and (3) regulation actions that modify human behavior through more coercive efforts, including use limits that ration visitation. The following section reviews these approaches, providing examples of common wildlife-viewing actions, and briefly discussing how those actions might be used to affect categories of biophysical, social, and managerial conditions.

Development Actions

Development actions typically employ a "technical fix" approach and usually refer to human-built structures or other capital improvements that modify the environment. Development actions are important when creating new opportunities and often have decisive roles when maintaining or enhancing existing opportunities. Applied to wildlife viewing, these types of actions tend to be employed in four major ways:

(1) to minimize human impacts to biophyical resources or social experiences;

(2) to enhance wildlife population numbers or concentrate wildlife in viewable areas;

(3) to improve viewing quality and provide opportunities for education or interpretation; and
(4) to accommodate the sheer volume of use (provide expected facilities) or attract greater numbers of users, thus increasing the numbers of viewing experiences.

Development Actions to Minimize Human Impacts
Development actions can be designed to reduce human impacts to biophysical resources. For example, one can minimize vegetation trampling in alpine-tundra settings with boardwalks that keep people off sensitive plants (Cole 1979; 1987). In general, these actions are designed to focus use away from sensitive sites and harden heavily used sites such as campsites and viewing areas (van der Zande et al. 1984; Hammitt and Cole 1987).

Wildlife-viewing actions in this category often help separate animals from concentrated human use through **barriers, platforms, or other designated viewing areas** (Larson 1995). These structures minimize disturbance impacts on wildlife by maintaining distance between viewers and wildlife (Altman 1958), are often crucial for human safety, and help prevent human-wildlife conflicts. For example, Kenyan safari camps sometimes have human enclosures ("wildlife free zones") that wildlife are discouraged from entering.

Lower River bear-viewing platform at Brooks River in Katmai National Park, Alaska. Platforms help minimize wildlife disturbance while allowing close-proximity viewing, but may also change development levels. (Photo by Doug Whittaker)

Similarly, passive barriers (downed logs and other vegetation) have been placed across trails to discourage brown bears from using trails that access lodge and campground areas at Alaska's Brooks River in Katmai National Park. These "developments" help prevent surprise encounters between humans and bears, and minimize the opportunity for bears to develop an attraction for human food sources that exist around lodges and campgrounds (Squibb 1991).

Provision of **platforms and hides** is a related action, and they are notable for helping protect wildlife and improve viewing. Platforms improve some experiences by providing viewing positions above vegetation, establishing safety margins from dangerous wildlife, or minimizing the sights and sounds created by humans so that people can be closer without startling wildlife.

Platforms also concentrate human use to particular locations and encourage predictable patterns of human behavior, allowing wildlife to become habituated to viewers. Habituation, which is often confused with attraction, is a waning of response to repeated neutral stimuli (Thorpe 1956; Eibl-Eibelsfeldt 1970); by contrast, attraction involves reinforcing stimuli that cause a change in "natural" behavior (Whittaker and Knight 1998). Habituation can thus help protect wildlife species from viewing disturbances, because the human use does not distract them from their daily activities, which may be crucial for survival (Edington and Edington 1986; Erwin 1989).

In addition, one could argue that habituated animals (those that do not react to human presence) allow people to see how animals behave "naturally"—when humans provide only neutral stimuli. An effective example of this includes the glass-enclosed underwater hides at Mzima Springs in Kenya, which allow close observation of potentially dangerous hippopotami while they feed and bathe in the water without disturbance by viewers.

One could argue that highly involved viewers are less likely than occasionalists or generalists to appreciate habituated wildlife, because the skill and challenge involved in viewing habituated animals is less. However, all types of viewers are likely to appreciate the opportunity to see wildlife that does not react to human disturbances. Habituation to viewers may have other negative consequences for wildlife, but for viewers it is almost always positive (Whittaker and Knight 1998).

Development Actions to Enhance or Concentrate Wildlife
While most development alternatives focus on minimizing human impacts, it is sometimes possible to "develop" improved habitat or create attractions that increase or concentrate species and provide

more viewing opportunities. Similar to other "technical fixes," this approach includes **large-scale habitat manipulations** (e.g., prescribed burns that increase ungulate browse; increasing woody debris or building concrete structures that alter depths and velocities to provide cover and rearing habitat for game fisheries), as well as smaller-scale attraction efforts such as providing **supplementary food sources** during certain times of the year. In these cases, the actions manipulate the environment to enhance wildlife concentrations for human use.

Wildlife-viewing examples of this type include birdbaths and feeders used to attract songbirds to suburban neighborhoods, or artificial nesting islands used to attract geese or other waterfowl to existing wetlands. Supplementary artificial food sources provided to certain species during periods of harsh weather (e.g., hay for wintering elk; fish cannery scraps for bald eagles) are other examples with obvious viewing implications, although these programs are typically initiated to enhance a species' survival rates.

Development Actions to Enhance Education and Interpretation
A third type of development focuses on facilities that allow for education and interpretive opportunities that enhance wildlife-viewing opportunities (Manfredo et al. 1996) and link with education and interpretation actions discussed below. Highly involved viewers are particularly interested in detailed information about wildlife, and are likely to be avid supporters and users of **visitor centers** that help them plan their trips (Manfredo et al. 1992; Manfredo and Larson 1993). These facilities are also crucial for shaping the quality of opportunities for occasionalist and generalist viewers (Manfredo and Larson 1993), who may not be willing or able to gain access to backcountry areas where some species are found.

An example in this latter category is found at the Pratt Museum in Homer, Alaska, where visitors can manipulate remote video cameras placed at a bear viewing area (McNeil River) and a bird rookery (Gull Island in Kachemak Bay), both many miles away. Visitors can view wildlife (albeit on a TV screen) in real time, as well as direct what they want to observe (controls manipulate the cameras' angle, direction, and zoom, as well as operate lens wipers to clear off rain, bird guano, or sea spray). The system provides viewing opportunities for many more people than is possible on-site, with much less risk of disturbance.

Although observing wildlife on a TV screen is not equal to viewing animals in person, the approach does create viewing opportunities for many individuals who would not otherwise have the experience.

For many people, wildlife viewing is a valued vicarious experience made possible through books, TV, videos, and museum/visitor-center programs. Similar to wilderness areas that few people visit, "existence value" may outweigh "use value" for many kinds of wildlife; developed facilities that focus on these vicarious experiences and help people learn about wildlife honor that non-use value.

Development Actions to Attract or Accommodate Use

Finally, **basic development facilities** such as parking lots, toilets, and campsites can all be used to attract or accommodate use, which is perhaps the most traditional focus of management efforts to create a new viewing opportunity or cope with an evolving one. Many viewing opportunities are discovered by highly involved pioneer users who require few facilities beyond provisions for access. As more people learn about an opportunity, however, a "wide spot on the side of the highway" or a primitive trail may not be sufficient, particularly for the less-involved users who demand greater convenience or amenities. Managers typically encourage this demand with incremental facility development, but sometimes fail to consider how this may change the type of experience offered over time.

Educational Actions

Educational actions typically refer to measures that employ a "cognitive fix" approach and represent systematic persuasion efforts by managers to modify human behaviors that are causing unacceptable biophysical or social impacts. Educational actions, however, are also used to enhance the quality of opportunities by providing information about available viewing options or interpreting the natural resources people want to understand. Wildlife education is particularly relevant to highly involved viewers, but can also enhance the experience for less specialized viewers (Manfredo et al. 1992; Manfredo and Larson 1993).

Educational Actions to Minimize Impacts

Managers sometimes view educational actions as a panacea for human-caused impact problems (Roggenbuck 1992). "If people only understood what impacts they cause," proponents declare, "we could get them to behave differently." Compared to regulatory approaches, education is also preferred by many managers because it is less obtrusive (Fish and Bury 1981).

In wildlife-viewing settings, educational actions often focus on **teaching viewing etiquette** (toward both wildlife and other viewers) and **minimum-impact practices** (e.g., no-trace camping, human-waste disposal). Attempts to establish norms for ethical behavior around wildlife and other wildlife viewers are evident in agency viewing literature and are common in some popular media (e.g., *Outside, Backpacker, Outdoor Photographer*). Although highly involved wildlife viewers are probably aware of these ethical codes, their effect on behavior is unclear and deserves more research attention (additional discussion of this issue is presented below).

Another educational alternative works to reduce impacts in a more indirect fashion, by **dispersing use through information** about use levels or other viewing conditions (Lucas 1981; Krumpe and Brown 1982). For example, if viewing-platform densities at different times of the day or season are publicized, some people are likely to avoid times with higher use to match their experiential preferences. In contrast to persuasion efforts designed to change behavior that may not be in an individual's short-term interest (e.g., keeping distant from a subject species), information intended to disperse use helps visitors choose the type of experience they want, or allows them to cognitively prepare for the conditions they are likely to find. Research suggests this kind of information is highly valued by users, although it rarely appears to have more than a moderate influence on use patterns (Roggenbuck 1992).

Educational Actions to Provide or Enhance Learning about Wildlife
The second type of education alternative focuses on interpretation, which differs from information designed to instruct visitors on where to go or how to behave (Tilden 1957). Interpretation is about the use of information to "inspire [visitors], to help them see and experience new things, to develop greater understanding and appreciation, and to stimulate increased stewardship of our natural, cultural, and historical resource heritage" (Gallup 1997). Learning about the natural world has long been recognized as an important motivational component of many outdoor-recreation pursuits (see Chapter 4), but wildlife viewing is often particularly focused on interpretation and learning objectives.

There are many interpretation options and an extensive literature on learning related to natural resources that suggest that different media and techniques are differentially effective for different types of information and learners (Ham 1992; Knudson et al. 1995). Examples in wildlife-viewing settings include "personal interpretation" actions such as **ranger-led excursions** or **presentations at campgrounds or visitor centers**, as well as "non-personal" actions such as **multimedia interpretive programs** (e.g., slide shows, videos, computer programs), **on-site interpretive** stations (e.g., kiosks, trailhead bulletin boards), and **written materials** (e.g., brochures, maps, and guides). In many cases, these actions require infrastructure that directly link them to development actions such as visitor centers, but it is often important to keep them separate during problem definition and action identification. While some education and interpretation actions dictate certain types of development, others do not; during the generation of ideas it is often better to split rather than lump actions so planners will be aware of the comprehensive range.

Regulatory Actions

Regulatory actions refer to those which employ a "structural fix" approach; the focus here is on changing human behavior to minimize biophysical or social-experience impacts even if one cannot change people's attitudes and norms toward those behaviors first. Regulations are essentially formal norms enforced through specific external sanctions, and they become necessary when educational alternatives fall short.

In reality, educational and regulatory approaches are complementary rather than competitive (Lucas 1982). Many regulations are used as reinforcement for educational efforts, while all regulations need to be widely known through education to be effective. In some cases, regulations are as much about raising the level of awareness about problem behaviors (and the impacts they cause) as they are about enforcement.

Regulatory examples in wildlife viewing settings include **area closures** or **prohibitions on certain types of use**, and are commonly used to address human impacts on wildlife or other biophysical resources. They have been advocated by a many researchers in a wide variety of settings; some illustrative examples include:

• Fishing prohibitions on river segments near bear-viewing locations at Alaska's McNeil River to prevent bear-attraction problems.

• No-boating segments on rivers near bald eagles' nesting areas or buffer-zone spatial restrictions around feeding areas to prevent general disturbance (Anthony et al. 1995).

• No waterborne activity (boating or swimming) in coves where manatees breed (to prevent disturbance), and no-wake regulations in critical manatee habitat areas (to prevent mortality or injury from boat collisions) (O'Shea 1995).

• Leash laws or regulations prohibiting dogs on beaches with nesting-bird colonies so they are not disturbed (Burger 1995).

Humpback whale in Resurrection Bay, near Kenai Fjords National Park, Alaska. Education efforts designed to keep people from approaching wildlife too closely may require regulatory backup because close approaches bring undeniable benefits for viewers.
(Photo by Doug Whittaker)

Regulations may also address some social impacts. Regulations that **separate uses by time or space**, for example, are often central to addressing use conflicts such as motorized vs. non-motorized use (Shelby 1980). However, other types of recreation conflicts (e.g., hunting vs. viewing) may represent conflicts in values and may not be so easily defused through zoning actions (Vaske et al. 1995).

In wildlife-viewing settings, subtle variations in use conflicts may be apparent and require a regulatory approach. The viewing behavior norms of highly involved users, for example, are likely to differ from those with creative orientations or those who are more generalist or occasionalist in their perspective (Manfredo et al. 1992). On bear-viewing platforms at Alaska's Brooks River, "at large comments" (e.g., witticisms about the way a bear looks or behaves) and "grandstand effects" (e.g., clapping when a bear catches a fish jumping the falls) appear acceptable to generalist viewers but quite distracting to highly involved viewers. In this case, a kind of *de facto* temporal zoning has occurred, with more-involved viewers avoiding the platforms during the middle of the day when there are high densities of generalist users. In early morning and in the evening, the platforms tend to feature lower densities, fewer day tourists, and considerably different types of social interaction among viewers (Whittaker 1993). Managers could probably address the needs of highly involved visitors through platform etiquette regulations, but educational alternatives may also meet similar objectives. It is also possible to develop variable regulations at different times or locations in order to provide different types of experiences.

A final class of regulatory actions adopt **use limit or carrying capacity** approaches to solving impact problems. A persistent notion among recreation managers, the media, and the public is that higher use levels equate with higher impact levels, even though data suggest that links between use and impacts are complex (Shelby and Heberlein 1986; Kuss et al. 1990). Advancements in research as well as the development of several similar visitor-impact planning frameworks (e.g., C-CAP, LAC, VIM, and VERP; see Chapter 8) are essentially efforts to cope with this complexity instead of focusing on use limits as a single "magic" solution (Shelby and Heberlein 1986; Graefe et al. 1990). Nonetheless, use limits remain a powerful strategy for dealing with some impacts, including those associated with wildlife-viewing experiences where solitude and primitive conditions are featured.

Evaluation Criteria for Selecting Potential Actions

After problem definition and action identification is complete, the focus shifts to choosing actions by applying evaluation criteria. The goal here is to choose actions in an explicit manner, linking opportunities, standards, and strategies in a coherent, defensible package. Judgments about the merit of actions are made at this stage and so the level of controversy may be high. Systematic consideration of the choices, and the ability to trace decisions back to their source, therefore become important.

One system of evaluation criteria identifies four fundamental issues: (1) **effectiveness**, (2) **appropriateness**, (3) **public acceptability**, and (4) administrative, financial, or legal **feasibility**. The following section reviews each of these, provides examples of how these criteria apply to some common wildlife-viewing actions, and discusses potential trade-offs between actions that solve one set of problems but may also have other unintended or unwanted experiential consequences.

Effectiveness

Effectiveness is the first criterion and relates to the ability of an action to provide desired biophysical, social, facility or managerial conditions for a target experience. Effectiveness is about what works and what does not. In some cases, actions will obviously cause a

Education efforts can be effective when the preferred behavior brings obvious benefits to viewers; eagles at a viewing area in Homer, Alaska, will take flight en masse if they hear a car door open. (Photo by Julie Whittaker)

desired change (e.g., when facilities are designed to accommodate a certain volume of use), while in others the link between actions and results may be less clear (e.g., whether a regulatory program will eliminate certain behaviors causing unacceptable impact levels). It is beyond the scope of this paper to review the effectiveness of a full range of actions, but two examples can help suggest situations where effectiveness may be a key issue.

The first example focuses on educational efforts designed to change visitor behavior to protect the quality of recreation experiences or the biophysical resource (e.g., advocacy of minimum-impact techniques or viewing etiquette). While persuasion research applied to other natural-resource issues suggests that some behavior modifications are possible with well-developed educational efforts, production of the kind of long-term, lasting change envisioned by education proponents is often both challenging and complicated (Roggenbuck 1992). Designing effective educational campaigns requires clear understanding of persuasion theory and practice, an understanding that is often distinctly missing from many natural-resource management efforts (Manfredo et al. 1992). Without diverting into a review of these issues, the following are a few findings from this literature applicable to viewing management:

• Educational efforts addressing unintentional or uninformed behavior appear more effective for visitors with low knowledge levels, and are thus likely to be more successful for changing behavior among occasionalist, generalist, or low-involvement viewers, each of whom are more likely to fit this description (Manfredo and Bright 1991).

• Among information efforts, personal efforts (e.g., face-to-face contacts with a ranger, live presentations) appear more effective than nonpersonal efforts. Among the nonpersonal efforts, multimedia programs (e.g., slide shows, computer programs, videos) appear to be slightly more effective than static written messages (e.g., signs, brochures) at increasing knowledge about an area or appropriate behaviors, but there is sparse evidence that any of these alone can result in enduring behavior changes (Roggenbuck 1992).

• Studies of littering behavior show personal contact and role modeling by rangers appear to be more effective than nonpersonal techniques, but that environmental prompts (e.g., litter receptacles) can reduce the likelihood of the depreciative behavior (Roggenbuck 1992). This has implications for combining education efforts with passive barriers (e.g. a small post placed on a trail) to discourage viewers from approaching sensitive areas at particular times (e.g., nesting sites). A system of these posts appears to be effective at preventing approaches to walrus haul-out areas on Alaska's Round Island, although viewers at this remote, limited-access area are highly involved individuals likely to be attentive to impact issues.

The notion that people can learn how to behave appropriately—and want to—is a fundamental assumption among wildlife-advocacy groups and many managers, but some caution is warranted. Studies suggest there is a widely held behavior norm not to litter (Muth and Clark 1978), yet litter remains a problem in many recreation settings in part because a few violators may produce substantial impacts.

Education and persuasion attempts are also likely to be less effective when the behavior in question produces personal rewards in addition to causing undesirable impacts. Photographers can often create better images by getting closer to, or provoking reactions from, wildlife. For example, Stellar's sea lions often display stress through intraspecies fighting when boats approach too closely. But even if viewers recognize the stress they are causing, there is undeniably more action to watch and many people find it difficult to keep their distance. In these types of cases, supporting regulatory approaches are likely to be necessary to achieve compliance.

In contrast, educational efforts designed to minimize "signs-of-use" impacts (e.g., litter, fire rings, switchback shortcuts) or social impacts (e.g., encounters, competition for viewing sites) are likely to be effective in many viewing settings. In these cases, the behavior may be unintentional or uninformed, and being persuaded often brings rewards (e.g., a cleaner environment, less-crowded conditions), even if these are only experienced in the long term (Roggenbuck 1992).

The limitations of education do not relieve managers of the responsibility to continue persuasion efforts, but managers should not expect them to dramatically change visitor behavior either. As with many behaviors based on environmental ethics, widespread conformity depends on whether people recognize the consequences of their actions and take responsibility for them (Schwartz 1968). Mirroring ideas suggested in Hardin's "tragedy of the commons" (1968), this notion suggests that widespread adoption of wildlife-viewing ethics requires people to recognize wildlife as a commons property and understand how inappropriate behaviors degrade viewing for themselves or others.

A second example focuses on use limits. In general, use-limit alternatives are more effective when addressing social impacts such as encounter levels or competition for sites and facilities (Kuss et al. 1990). In contrast, many biophysical impacts appear less directly related to use levels because initial or relatively low levels of use may create proportionately larger impacts (Hammitt and Cole 1987; Kuss et al. 1990). For example, the first few groups to pioneer a campsite appear to have the greatest impacts on vegetation loss; subsequent groups then camp in the same areas and typically cause marginal additional impact (Cole 1987). Some wildlife disturbance impacts may fall into this pattern because some wildlife learning research suggests many animals have the capacity to adjust or habituate to human uses over time, while initial encounters may cause flight (Knight and Cole 1995; Whittaker and Knight 1998). However, other research suggests that disturbance impacts can have cumulative effects, in which case more people over a longer time period may increase disturbance problems (Anthony et al. 1995).

Considerable work has focused on relationships between use levels and social impacts in backcountry settings (particularly encounters), and this work is directly applicable to backcountry wildlife-viewing opportunities as well (Vaske et al. 1986). Solitude is likely to be an important component of certain kinds of wildlife-viewing experiences, and use limits are a direct way to address many interaction impacts that can degrade solitude. This can be particularly true in settings where wildlife is concentrated (e.g., bears fishing for salmon, sheep at a mineral lick, elk in a winter lower-elevation refuge), and viewing use levels are likely to be concentrated as well (Whittaker 1997). Educational or regulatory actions in these higher-density situations may have limitations; at some point the sheer numbers of people will alter the experience and use limits may be the only effective approach to maintaining standards.

Appropriateness

Appropriateness refers to the ability of an action to be consistent with the target experience, and is the counterpoint to effectiveness.

Effectiveness asks whether an action will fix a problem; appropriateness explores whether that fix will have other undesirable effects—this is an issue because virtually all actions have unintended consequences and trade-offs. Development that addresses biophysical impacts, for example, may change developmental settings or increase interaction impacts. Similarly, regulations that address user conflicts or minimize wildlife disturbance may change the managerial setting, or diminish visitors' sense of freedom, a key issue in primitive, backcountry viewing settings (Burke et al. 1979; Lucas 1982). Appropriateness focuses on these trade-offs, weighing their relative benefits and costs.

There is a long list of wildlife-viewing actions where appropriateness is central; three examples are provided to illustrate the concept. The first example examines supplementary feedings that increase or concentrate viewable wildlife such as eagles. This action will undeniably increase the number of eagles in an area, but it is an open question whether that is appropriate. Manipulating an ecosystem—even with the best of intentions—may have unintended biological consequences (Steinbeck and Ricketts 1941); more eagles might mean a decrease in small mammal populations in the area or modify interactions between eagles and other bird species in a scavenging guild (Knight and Knight 1984; Knight et al. 1991). Manipulating the number of eagles might also alter viewing experiences by decreasing the skills and challenge level required of viewers. More-involved viewers may prefer opportunities that are not created by human modifications or attractants. Finally, concentrations of viewable wildlife may exacerbate human-wildlife conflicts or increase social-interaction (crowding) impacts at the area.

A second example focuses on platform/hide development, which also may affect experiences in unintended ways. Designed to place viewers in close proximity to target wildlife, platforms also add a barrier between viewers and wildlife. Viewing from a human-made structure also makes the setting less primitive, and may diminish the skill and challenge component important with some viewing experiences. As with hunting, some of the enjoyment from wildlife viewing comes from learning where animals may be found or how to approach them without causing flight. Platforms eliminate the need for these skills because they identify viewing locations and limit viewer movement. Such changes are likely to be major issues for experienced wildlife viewers interested in ground-level encounters and "stalking" animals. But they may be unimportant to generalist viewers simply interested in being close to animals.

Platforms may also concentrate use, which may exacerbate interaction and competition impacts between visitors, further affecting certain types of experiences. Data from Alaska's Brooks River bear-viewing area, for example, suggest that most viewers prefer capacities lower than twenty to thirty people, even when a platform can physically accommodate over a hundred (Whittaker 1997). In addition, visitors during the fall viewing season prefer lower-density experiences than summer visitors, expressing preferences for capacities less than ten to fifteen people at one time. Managers should use caution when designing platforms to accommodate large numbers of people; social-experience considerations may dictate a smaller size than either biological or physical-facility considerations.

Finally, platforms may limit the variety of viewing positions in an area, dictating particular views that may be judged less favorably by some wildlife viewers. Bear-viewing platforms, for example, are typically built high enough to provide some safety from bears. But for experienced photographers, this "top-down" perspective may be less appealing than ground-level views (Whittaker 1993). The design of viewing platforms to accommodate the needs of serious photographers (Manfredo et al. 1992; Manfredo and Larson 1993) might also consider issues such as platform shake and space for tripods. These variables are probably irrelevant for occasionalist viewers, but can be decisive for the highly involved or creative types.

A final example focuses on the potential undesirable consequences associated with a variety of educational or regulatory actions, which might include the loss of challenge or an increased managerial "footprint." On the challenge issue, some wildlife viewers are very focused on the discovery and skill-development components of their viewing experiences, and may not want too much educational information that tells them where to go or how to behave (Lucas 1981). Some occasionalist viewers may also show a distinct lack of interest in learning information; for example, it is not hard to imagine a family of tourists in Yellowstone content to simply watch, not learn about, a herd of bison in a meadow. Wildlife managers with biology training often assume the general public has a deep interest in learning about wildlife, but interest levels are likely to be variable even within the wildlife-viewer types identified by Manfredo et al. (1992). Assessing the learning needs and information limits of different types of users is a good area for natural-resource education researchers, and may help planners choose actions with the appropriate type and level of information.

On the issue of the managerial "footprint", educational and regulatory actions may also be intrusive or change the managerial-setting even as they help prevent biophysical or social change. People commonly identify "escape" and "freedom" with wildland recreation such as wildlife viewing; admonishments, rules, or ranger contacts all have the potential to detract from the more primitive versions of those experiences (Lucas 1982). In response to these considerations, many natural-resource agencies emphasize an "education first" approach in an effort to minimize these effects.

Public Acceptability

Public acceptability refers to the willingness of stakeholders or the general public to support an action, and is often distinct from considerations of effectiveness and appropriateness. Interest groups sometimes oppose an action even if they agree with the opportunity goal or recognize that an action will help provide it. Groups may simply identify with certain management traditions that preclude consideration of other actions.

The acceptability of use limit actions, for example, often follows from resource-access traditions. On multi-day whitewater-river opportunities in the West, use limits (i.e., where users have to obtain a permit to float the river) are a commonly used and accepted management tool. On day-use fishing rivers, however, permit systems to reduce crowding or other social impacts are rare. Use limits have some tradition within viewing management (permits are required to view wildlife as diverse as Alaskan brown bears and Sumatran orangutans), but these often apply to particularly charismatic species and are usually based on biological concerns. It is less clear how use limits would be accepted in other wildlife viewing settings with less charismatic species, or to address social rather than biological impacts.

Acceptability is often a negative constraint rather than a positive guide to action choice, but it directly links with fundamental planning principles that value collaborative decision making (see Chapters 6 and 7). Planners who do not factor public acceptability into their action choices risk derailing a plan from public backlash.

Feasibility

The final evaluation criterion focuses on the general feasibility of implementing the action, addressing potential administrative, financial, and legal barriers. Similar to public acceptability, feasibility often acts as a negative constraint rather than a positive guide to action choice. It is obvious, however, that actions need to fit with agency mandates, cost structures, and legislative dictates to be considered in a plan.

Agencies are usually acutely aware of these issues during planning, and it is beyond the scope of this chapter to review them in depth. Feasibility concerns, however, can influence a variety of action choices and have further ancillary effects on wildlife-viewing experiences. When developing a visitor center and programming interpretive services, for example, financial costs are central to decisions about the size and scope of the facility. And while contributions from volunteers and user fees offer ways that managing agencies can stretch their maintenance and operations budgets, these may have other effects on experiences. Many volunteers possess skills or knowledge that can dramatically enhance educational efforts, particularly if those volunteers are recruited from the ranks of highly involved users. In other cases, however, volunteers may not be professional staff and may lack the abilities to provide needed interpretive services.

User fees are another action receiving increasing interest as a way to enhance agency revenues and offset the cost of facility or area management. While fees are sometimes considered a tool for limiting use or decreasing depreciative behavior, they are fundamentally about stretching agency budgets. Unfortunately, they can also affect the managerial footprint on viewing experiences, and may affect the type of viewer that uses an area as well.

Supplementary winter feedings in Homer, Alaska attract hundreds of viewable bald eagles, but may have other unwanted biological consequences and could diminish experiential elements such as a "sense of naturalness" or the challenge of finding wildlife to view. (Photo by Julie Whittaker)

Depending upon their magnitude and method of collection, fees may be well received and appropriate for some opportunities (particularly those that feature highly developed facilities). With other opportunities, however, fees may be less appropriate and less acceptable to users (e.g., remote, wilderness-oriented experiences that feature an absence of managerial presence and few facilities). When cost feasibility drives planning decisions, other actions such as fees often come into play as well.

Conclusion

Many different strategies can be used to enhance wildlife-viewing experiences, with each offering different trade-offs (e.g., increased managerial footprint vs. lower impact, limited access vs. lower-density experiences, lower disturbance levels vs. closer proximity to wildlife). Crafting strategies that create new opportunities, minimize impact problems at existing opportunities, or enhance the benefits of different experiences requires care because any action may have other consequences for target experiences.

In many cases, sets of management actions will need to be combined to meet standards, and this may introduce complex interaction effects. Experimentation and flexibility also may be necessary, particularly until research can supply more data about user preferences for particular actions in particular experience settings. In addition, managers should not assume that actions that have worked in the past will continue to work over time. Both humans and wildlife can learn new behaviors in response to management actions and resulting conditions (Whittaker and Knight 1998); some degree of monitoring and re-visiting action decisions is a crucial part of any effective management process. Wildlife viewing is a relatively new focus for wildland managers, so there is little information about what people want or how wildlife will respond to viewing patterns. There are considerable opportunities for managers to try new techniques, document effects, and improve our understanding of how actions affect opportunities and wildlife.

Innovation can be encouraged during planning efforts at several steps. Stakeholders and the public often think about actions from their earliest involvement in a planning effort, and it is crucial to find ways to capture their ideas. As noted in Chapter 6, successful planning processes are iterative rather than serial, so it is never too early to consider possible actions that might be used to address problems or provide opportunities. Brainstorming actions during the initial scoping phase of a planning effort allow planners to anticipate possible action strategies even as they are focusing on prerequisite goals, objectives, and indicators/standards. However, it is crucial to avoid commitment to any actions until later in the process, when stakeholders and the public deserve additional opportunities to develop and react to alternative management strategies. Determining action acceptability via public surveys can further prove useful in these latter stages of the process, when planners and stakeholders have solidified definitions of the experiences they are trying to provide but require more information about the costs they are willing to pay to provide them.

Summary Points

• EBM urges use of a systematic and deliberative process for choosing actions to meet planning objectives and standards. The three basic stages in this process are problem definition (defining opportunities and deciding what impacts or other issues need to be addressed); brainstorming (generating lists of actions that might address those issues); and evaluation (applying criteria to determine which actions are effective, appropriate, publicly acceptable, and feasible).

• Problem definition is probably the most important step in the process. The essential first step is to clarify the type of recreation opportunity to be provided, which in turn provides the context for selecting the "right" action(s). EBM provides the framework for clarifying opportunities, establishing indicators and standards that define resource health or experiential quality, and determining whether actions are needed to meet those standards.

• Identifying or brainstorming potential actions is more efficient and comprehensive when actions are conceptualized by categories, each emphasizing different approaches and types of consequences. One such system focuses on three broad categories: capital development, education, and regulation (including use limits).

• Actions can be evaluated by four fundamental criteria: effectiveness (what works and what does not?); appropriateness (will an action have other unintended and unwanted experiential or biological consequences?); public acceptability (are there traditions of support or opposition for some actions?); and administrative, financial, or legal feasibility (can the managing agency implement the action?). Some actions have substantial trade-offs, solving one set of problems but creating others or changing the type of opportunity provided.

• Development actions are capital improvements that modify the environment. They are typically used to minimize human impacts to biophysical resources; enhance wildlife population numbers or concentrate wildlife in viewable areas; improve viewing quality and provide opportunities for education or interpretation; accommodate the sheer volume of use (provide expected facilities); or attract greater numbers of users, thus increasing the numbers of viewing experiences. The trade-off with many development actions is that they decrease naturalness, or attract higher use densities, potentially changing the type of viewing experience.

• Education actions often refer to systematic persuasion efforts by managers to modify human behaviors that are causing unacceptable biophysical or social impacts, but they are also used to enhance opportunities by providing information about available viewing options or interpretation. As a tool to reduce impacts, education actions often have some limitations and should not be viewed as a panacea; designing effective education campaigns or interpretation efforts is also complex.

• Regulatory actions focus on changing human behavior to minimize biophysical or social-experience impacts when educational alternatives fall short. Regulatory examples include area closures, prohibitions on certain types of use, and use limits, each of which may help reduce some types of resource or experiential impacts, or address user conflicts. However, they also may have some trade-offs by changing the managerial setting (making it more structured) or reducing the number of people who can participate.

Literature Cited

Anthony, R. G., R. J. Steidl, and K. McGarigal (1995). "Recreation and bald eagles in the Pacific Northwest." In R. L. Knight and K. J. Gutzwiller (Eds.), *Wildlife and Recreationists: Coexistence through Management and Research*. Washington, DC: Island Press.

Altman, M. (1958). "The flight distance in free-ranging big game." *Journal of Wildlife Management* 22(2):207-9.

Burger, J. (1995). "Beach recreation and nesting birds." In R. L. Knight and K. J. Gutzwiller (Eds.), *Wildlife and Recreationists: Coexistence through Management and Research*. Washington, DC: Island Press.

Burke, J. F., R. Schreyer, and J. D. Hunt (1979). "Behavior modification." *Trends* 16(4): 33-36.

Cole, D. N. (1979). "Reducing the impact of hikers on vegetation: An application of analytical research methods." In *Proceedings: Recreational Impact on Wildlands.*Report No. R-6-0011979. Seattle, WA: U.S. Department of Agriculture,, Pacific Northwest Region.

Cole, D. N. (1987). *Research on Soil and Vegetation in Wilderness: A State of Knowledge Review.*Research Paper INT-288, Ogden UT: U.S. Department of Agriculture, Forest Service, Intermountain Research Station.

Edington, J. M., and M. A. Edington (1986). *Ecology, Recreation, and Tourism.* Cambridge, England: Cambridge University Press.

Eibl-Eibelsfeldt, I. (1970). *Ethology, the Biology of Behavior.*New York: Holt, Rinehart, and Winston.

Erwin, R. M. (1989). "Responses to human intruders by birds nesting in colonies: experimental results and management guidelines." *Colonial Waterbirds* 12:104-8.

Fish, L. B., and R.. L. Bury (1981). "Wilderness visitor management: Diversity and agency policies." *Journal of Forestry* 79(9): 608-12.

Gallup, B. (1997). *Interpretation: Foundation for Sustainable Tourism.* Denver, CO: University of Colorado Press.

Graefe, A. R., F. R. Kuss, and J. J. Vaske (1990). *Visitor Impact Management: The Planning Framework.* Washington, DC: National Parks and Conservation Association.

Ham. S. H. (1992). *Environmental Interpretation: A Practical Guide for People with Big Ideas and Small Budgets.* Golden, CO: North American Press.

Hammitt, W. E., and D. N. Cole (1987). *Wildland Recreation: Ecology and Management.* New York: Wiley and Sons.

Hardin, G. (1968). "The tragedy of the commons." *Science* 78(6): 20-27.

Knight, R. L., and S. K. Knight (1984). "Responses of wintering bald eagles to boating activity." *Journal of Wildlife Management* 48:999–1004.

Knight, R.L. , D. P. Anderson, and N. V. Marr (1991). "Responses of an avian scavenging guild to anglers." *Biological Conservation* 56: 195-205.

Knight, R. L., and D. N. Cole (1995). "Wildlife responses to recreationists." In R. L. Knight and K. J. Gutzwiller (Eds.), *Wildlife and Recreationists: Coexistence through Management and Research.* Washington, DC: Island Press.

Krumpe, E. E., and P. J. Brown (1982). "Redistributing backcountry use through information related to recreation experiences." *Journal of Forestry* 80: 360-64.

Knudson, D. M., T. T. Cable, and L. Beck (1995). *Interpretation of Cultural and Natural Resources.* State College, PA: Venture Publishing.

Kuss, F. R., A. R. Graefe, and J. J. Vaske (1990). *Recreation Impacts and Carrying Capacity: A Review and Synthesis of Ecological and Social Research.* Washington, DC: National Parks and Conservation Association.

Larson, R. A. (1995). "Balancing wildlife viewing with wildlife impacts: A case study."In R. L. Knight and K. J. Gutzwiller (Eds.), *Wildlife and Recreationists: Coexistence through Management and Research.* Washington, DC: Island Press.

Lucas, R. C. (1981). *Redistributing Wilderness Use through Information Supplied to Visitors.* Research Paper INT-27. Ogden, UT: U.S. Department of Agriculture, Forest Service, Intermountain Forest and Range Experiment Station.

Lucas, R. C. (1982). "Recreation regulations—when are they needed? *Journal of Forestry* 80(3): 148-51.

Manfredo, M. J., and A. D. Bright (1991). "A model for assessing the effects of recreation communication campaigns." *Journal of Leisure Research* 23(1):1-20.

Manfredo, M. J., M. Paulson, J. Wurtz, J., and A. Bright (1992). *Development of a Recreation and Tourism Assessment System for the Rocky Mountain Arsenal.* Final Report for the Cooperative Agreement 14-16-0009-1552, Work Order No. 27. Fort Collins, CO: Colorado Fish and Wildlife Research Co-op Unit, Colorado State University.

Manfredo, M. J., and R. A. Larson (1993). "Managing for wildlife viewing recreation experiences: An application in Colorado." *Wildlife Society Bulletin,* 21, 226-36.Manfredo, M. J., B. L. Driver, and M. A. Tarrant (1996). "Measuring leisure motivation: A meta-analysis of the recreation experience preference scales." *Journal of Leisure Research,* 28, 188-213.

Muth, R. M., and R. N. Clark (1978). *Public Participation in Wilderness Backcountry Litter Control: A Review of Research and Management Experience.*General Technical Report PNW-75.U.S. Department of Agriculture, Forest Service Pacific Northwest Forest and Range Experiment Station.

O'Shea, T. J. (1995). "Waterborne recreation and the Florida manatee." In R. L. Knight and K. J. Gutzwiller (Eds.), *Wildlife and Recreationists: Coexistence through Management and Research.* Washington, DC: Island Press.

Roggenbuck, J. W. (1992). "Use of persuasion to reduce resource impacts and visitor conflicts." In M. J. Manfredo (Ed.). *Influencing Human Behavior: Theory and Applications in Recreation Tourism, and Natural Resources Management.* Champaign, IL: Sagamore Publishing.

Schwartz, S. H. (1968). "Awareness of consequences and the influence of moral norms on interpersonal behavior." *Sociometry,* 31: 355-69.

Shelby, B. (1980). "Contrasting recreation experiences: Motors and oars in the Grand Canyon." *Journal of Soil and Water Conservation* 35(3):129-30.

Shelby, B., and T. A. Heberlein (1986). Carrying Capacity in Recreation Settings. Corvallis, OR: Oregon State University Press.

Squibb, R. (1991). Bear use of Brooks River: Summary of information relevant to management. Unpublished report on file at Katmai National Park and Preserve Headquarters, King Salmon, AK.

Steinbeck, J., and E. F. Ricketts (1941). *Sea of Cortez: A Leisurely Journal of Travel and Research.* New York : The Viking Press.

Thorpe, W. H. (1956). *Learning and Instinct in Animals.* New York :Methuen and Co.

Tilden, F. (1957). *Interpreting Our Heritage.* Chapel Hill, NC: University of North Carolina Press.

Vaske, J. J., B. Shelby, A. R. Graefe, and T. A. Heberlein (1986). "Backcountry encounter norms: Theory, method and empirical evidence." *Journal of Leisure Research,* 18, 137-53.

Vaske, J.J., M. P. Donnelly, K. Wittmann, and S. Laidlaw (1995). "Interpersonal versus social values conflict." *Leisure Sciences* 17: 205-22.

van der Zande, A. N., J. C. Berkhuizen, H. C. van Latesterijn, , W. J. ter Keurs, and A. J. Poppelaars (1984). "Impact of outdoor recreation on the density and number of breeding bird species in woods adjacent to urban residential areas." *Biological Conservation* 30:1-39.

Whittaker, D. (1993). "Carrying capacity issues at Brooks River: Support materials for the 1993 Draft Development Concept Plan/ Environmental Impact Statement." Unpublished report on file at NPS Alaska Regional Office, Anchorage.

Whittaker, D. (1997). "Capacity norms on bear viewing platforms." *Human Dimensions of Wildlife,* 2, 37-49.

Whittaker, D., and R. L. Knight (1998). "Understanding wildlife responses to humans: A need for greater clarity in research and management." *Wildlife Society Bulletin,* 26: 312-17.

International Ecotourism and Experience-based Management

Maureen P. Donnelly, Doug Whittaker, and Sandra Jonker

Introduction

INTERNATIONAL CONSERVATION OF WILDLIFE is a complex and often contentious issue. "What on the surface appears to be a simple and unassailable goal of protecting wild animals and plants from forces beyond their control quickly dissolves, on closer inspection, into a complex tangle of conflicting issues: human rights versus the protection of animals and forests; the total exclusion of humans from protected areas versus the possibility of human coexistence with wildlife; exclusive state control over protected areas versus increased local participation in protected area management" (Saberwal 1999, p. 2). Raising wildlife protection above the rights of local communities to engage in traditional activities may serve to alienate residents, some of whom may engage in "backlash" behavior such as poaching with even more deleterious effects on wildlife populations than traditional activities (Damon and Vaughan 1995).

In an effort to address this situation, conservation advocates have called for the development of alternative management strategies that will benefit both indigenous groups and wildlife populations. One such approach is commonly labeled "ecotourism," and suggests that "eco-friendly development within villages may improve the financial status of villagers, thereby reducing their dependence on resources within protected areas" (Saberwal 1999, p. 3). By providing local communities with economic benefits, ecotourism activities are thought to lead to conservation of wildlife and other natural resources.

In a broad sense, ecotourism also seems simple and unassailable: who wouldn't want to develop tourism that both conserves natural areas and provides benefits to local populations? But the issues associated with ecotourism are also complex. In addition to debate about the definition of ecotourism, there are doubts about its sustainability and there is no widely accepted framework for developing or managing it. The challenges of ecotourism are in the details of how visitors interact with the natural environment and local people, and these can be influenced by a myriad of factors— only some of which are easily controlled in most international protected area settings. Equally important, ecotourism efforts must provide sustainable high-quality visitor experiences to be successful. While advocates may view ecotourism as a means for conserving natural areas and benefiting local people, this goal will never be met without some attention on maintaining experiential benefits for ecotourists. Ultimately, visitors influence the level and impacts of ecotourism through their demand for experiences and services, and their willingness to behave in ways that minimize negative impacts while enhancing positive ones.

As the concepts and focus of ecotourism become better defined, we think some of the ideas, concepts, and traditions of Experience-based Management (EBM) will prove useful for understanding and managing this alternative type of tourism. EBM is obviously applicable to managing ecotourism experiences from the visitor's point of view. But EBM could also be extended to understand and address ecotourism issues related to broader environmental and social impacts as well. Many of the notions prominent in EBM (e.g., defining desired future conditions via "opportunity" objectives, indicators and standards, linking actions to standards) are similarly applicable to ecotourism concerns, albeit with some modification. In the same way that it is possible to establish standards that define high-quality conditions for wildlife-viewing experiences, it is possible to develop standards for natural-resource health or high-quality social conditions for local residents.

In this chapter, we explore the links between the ideas of EBM and ecotourism. The chapter begins with an overview of ecotourism definitions and explains how the application of EBM concepts may help identify the variables of importance for researching and managing for ecotourism. Following this, we review the extent of demand for associated nature-based tourism, the importance of wildlife viewing associated with that demand, and a description of the range of experiential opportunities that roughly fit under the ecotourism umbrella. Finally, we review some of the ways that EBM concepts could be applied to ecotourism management, and suggest four areas that are likely to prove challenging, including: (1) the function of ecotourism in minimizing negative impacts to natural environments and wildlife, (2) the role and importance of guides; (3) integrating the participation of local people in EBM planning, and (4) the importance of partnerships between local communities, governmental agencies, nongovernmental organizations, and commercial interests. Throughout these discussions, we have tried to provide examples from international ecotourism destinations to help illustrate the issues.

Defining Ecotourism

For a word that did not even appear in dictionaries until the late 1990s, "ecotourism" now receives common use. On cable TV, a travel show notes that foreign expenditures from fifteen thousand ecotourists a year has saved a Costa Rican rain forest and its diversity of plants and wildlife. Tour companies in an outdoor magazine advertise ecotours to places as diverse as Antarctica and the Bahamas, offering activities from skin diving to wildlife viewing to folk dancing. An ecotourism website romanticizes traditional lodging in a northern Thailand village hut, while glossy brochures tout a lodge in Kenya as providing "rustic elegance for the ecotraveler" with the "same fastidious service you have come to expect from our chain worldwide."

So what exactly is ecotourism? Is it a segment of the tourist industry? Is it about certain kinds of activities, certain kinds of exotic locales, or certain combinations of both? And does it have anything to do with "roughing it," or can you be an ecotourist at an international-class hotel—if it is on the edge of an African game park? If ecotourism is to be a useful concept for both managers and academics, the answers to these questions should be clear. Unfortunately, the confusion over ecotourism is not just limited to the lay press or marketers who appropriate fashionable lingo (Cater 1994).

While many have claimed first use of the term, the expression was probably coined in the 1980s by Hector Ceballos-Lascurain. His definition essentially describes a segment of the tourism pie, and links ecotourism to a specific combination of settings ("relatively undisturbed or uncontaminated natural areas") and motivations ("studying, admiring, and enjoying the scenery and its wild plants and animals, as well as any existing cultural manifestations") (Ceballos-Lascurain 1988). This descriptive definition is a well-accepted starting point for understanding ecotourism as "travel to enjoy and appreciate nature" (Fillion et al. 1992), but it is unclear how this differs from other types of tourism to natural or rural areas with indigenous populations (Valentine 1990; Butler 1992; Cater 1994; Prosser 1994).

Some authors have warned that the tourism industry has wrongly appropriated the ecotourism label, arguing that ecotourism is often little more than "eco-sell" or "eco-exploitation" (Cater 1994; Wight 1994). As Cater (1994) points out, tourism that is ecologically based (i.e., dependent upon a natural-resource base) is not necessarily ecologically sound or sustainable. Researchers have long argued that tourist destinations have a life cycle that includes stages of discovery, emergence, growth, saturation, and decline (Prosser 1994). Ecotourism, in contrast, promises a kind of tourism where this life cycle is not inevitable and the characteristics of saturation, and decline never occur.

More evolved ecotourism definitions have accordingly gone beyond mere descriptions of settings and motivations, and usually include statements about minimizing adverse social and environmental impacts, while enhancing positive economic benefits (Valentine 1990; Zurick 1995). The International Ecotourism Society, for example, defines ecotourism in general as "responsible travel that conserves the environment and sustains the well-being of local people" (The International Ecotourism Society 1991). Similarly, but more concretely, Honey (1999) proposes that ecotourism is "travel to fragile, pristine, and usually protected areas that strives to be low impact and is (usually) small scale. It helps educate the traveler; provides funds for conservation; directly benefits the economic development and political empowerment of local communities; and fosters respect for different cultures and for human rights." Similar to EBM, this definition not only includes a focus on impacts, but suggests links to positive psychological outcomes for hosts and participants.

A related definitional perspective further extends the focus on hosts, tourists, their attitudes and actions. Under this approach, however, the central concern is the development of ethical codes of behavior and development that guide and enforce behavior (Hultsman 1995). The International Ecotourism Society's definition (1991) carries some of this advocacy flavor, which is mirrored in much of the ecotourism literature (Butler 1992; Shackley 1994; HaySmith and Hunt 1995; Hunter 1997). With a similar perspective, Cater (1994) argues that ecotourism may not be a "product" at all, but a "principle" of development and conduct in fragile natural and cultural environments.

It also appears that some researchers consider ecotourism, in its ideal state, to be about more than just travel to natural areas or minimizing adverse impacts; it also must involve learning experiences for both tourists and hosts, and the opportunity for genuine cultural exchange (Lindberg 1991; Van Den Berghe 1994; Zurick 1995). Yet another definition calls ecotourism "a scientific, esthetic, or philosophical approach" to travel and notes that one must "immerse himself in Nature" and move "outside one's own routines and modern life" (Zurick 1995). The presumption is that these events and experiences will be transforming (Bruner 1991) and can contribute to tourists' active participation in conservation efforts even after they have returned home (The International Ecotourism Society 1991; Honey 1999).

We think an integrated, encompassing definition of ecotourism needs to include a diversity of variables drawing from the concepts discussed above. Setting and tourist motivations are part of the story, as is a focus on the environmental, social, and cultural impacts of tourism. But ecotourism must also examine experiential variables for both hosts and tourists, linking sets of psychological outcomes to specific environmental, social, or managerial conditions that create them. These variables will then lead to ways of managing for these high-quality conditions and associated experiences. While there is room for debate about standards that define "real" ecotourism (e.g., How small must group sizes be? How rustic should facilities be? When is an area natural enough?), the idea of treating these conditions as variables seems more useful. Experience-based Management offers a framework that focuses researchers and managers on these issues, allowing an examination of important impacts, but also providing a way to understand the excitement and enthusiasm inspired by nature-based travel (Pearce 1982).

Defining the Extent and Characteristics of Wildlife-based Ecotourism

Ecotourism Participation

It is difficult to obtain accurate statistics on the extent of ecotourism participation. Data are not being systematically collected by government, the private sector, or the World Tourism Organization (WTO), partly because of ecotourism's recent emergence and partly because of the lack of a universally accepted and quantifiable definition (Fillion et al. 1992). There are some estimates, however, for nature- or wildlife-based tourism from the late 1980s and 90s, which could reflect a growing interest in the ecotourism phenomenon. In 1988, for example, it was estimated that as many as 235 million people participated in international nature-based tourism, resulting in an economic benefit of $233 billion worldwide. A significant portion of this was wildlife-related tourism—157 million travelers, resulting in benefits of $155 billion for the economies of the countries visited (Fillion et al. 1992). In 1990, the World Resources Institute found that while tourism overall had been growing at an annual rate of 4 percent, nature travel was increasing at an annual rate of between 10 and 30 percent (Reinhold 1993). In 1993, WTO estimated that 7 percent of all international tourism was nature tourism and in 1998, they estimated that nature-related forms of tourism, including ecotourism, accounted for approximately 20 percent of total international travel (WTO 1998).

Wildlife photography tour group at Brooks River in Katmai National Park, Alaska. The promise of ecotourism is that small groups of specialized users are likely to engage in sustainable, minimum-impact activities because they become more aware of their potential impacts through extensive interpretation and education experiences. (Photo by Doug Whittaker)

Even if these figures refer to more simplistic definitions of the term, it is clear the demand for ecotourism is rising, with implications that are unlikely to have escaped the attention of commercial interests (who may also be helping to drive that demand). There are also implications for managers and advocates concerned about the ways this demand may be affecting ecotourism experiences, not to mention the impacts on natural areas and host communities. A similar dramatic rise in the demand for outdoor recreation in the 1960s and 70s was probably the force that led to increased impacts in North American backcountry areas, and the resultant advent of EBM and related management frameworks [Carrying Capacity Assessment Process (CCAP), Limits of Acceptable Change (LAC), Visitor Impact Management (VIM), and Visitor Experience and Resource Protection (VERP)] to address them. By adapting the now honed concepts, frameworks, and lessons from EBM, ecotourism interests may avoid some of the problems that historically challenged EBM researchers and managers.

The Importance of Wildlife to Ecotourism

The importance of wildlife to ecotourism experiences is well documented in the research. For example, 32 percent of tourists to Australia indicated that wildlife and nature were the aspects they most enjoyed while in the country (Fillion et al. 1992). From 1983 to 1993, visitor arrivals to Kenya grew by 45 percent (372,000 to

826,000). The Kenya Wildlife Service estimates from 1995 indicate that 80 percent of Kenya's tourist market is drawn by wildlife, generating approximately one-third of the country's foreign-exchange earnings (The International Ecotourism Society, 1998). Revenue from Kenya's wildlife parks was estimated at $11.9 million in 1995.

Similarly, surveys of Europeans and Japanese traveling to North America indicated that 69 to 88 percent felt that wildlife and birds were important factors in choosing North America as a travel destination (Tourism Canada 1989). A study of North American ecotourists asked respondents about the important elements of their trips and found that the top two responses were (1) the wilderness setting, and (2) wildlife viewing (Wight 1996). Similarly, tourists to Mexico, Belize, Dominica, Costa Rica, and Ecuador were attracted to these areas for bird-watching and to observe other wildlife (Boo 1990). Informal interviews with tourists coming to see the natural and cultural sites of Peru indicated that, of those planning to visit the Amazon, about 70 percent were primarily interested in seeing animals and were less interested in indigenous cultures or Amazonian panoramas (Groom et al. 1991).

Taken together, these statistics suggest that wildlife is an essential component of an ecotourism experience and provides a significant draw for travelers. In fact, it has been estimated that, depending on the region, wildlife-related tourism accounts for some 20 to 40 percent of international tourism (Fillion et al. 1992). This fact alone, however, offers little information about how to manage for wildlife viewing to continue to attract tourism, nor to minimize adverse impacts from that tourism on natural environments, tourism experiences, or host communities. EBM, in contrast, focuses directly on these issues, urging more specific examination of ecotourism wildlife-viewing experiences.

The Wildlife-based Ecotourism Experience

A variety of experiences have been marketed as ecotourism, appealing to a broad range of wildlife enthusiasts. Some of these offerings appeal more to the generalist or occasional wildlife viewer—people who enjoy viewing and learning about wildlife in developed or high-density settings without having to endure hardship or spend considerable time to find the animals. For example, Treetops Lodge in Kenya's Aberdare National Park provides essentially sedentary animal viewing. This fifty-room facility has a roof terrace bar overlooking two waterholes that are lit by floodlights at night. In a waterhole-dependent ecological system, animals are attracted to the area despite the presence of human development—

especially when the waterhole is topped off each day and salt licks are deposited near lodge windows to lure animals closer. Guests accordingly watch the animals feed while enjoying their evening meal in comfort. They can also choose to be awakened in the middle of the night by hotel staff every time the "Big Five" (i.e., leopard, lion, elephant, rhino, and buffalo) are sighted.

Other opportunities that have been termed wildlife ecotourism are more attractive to highly involved wildlife viewers with more exacting standards for primitive, low-density, and actively participatory experiences. For example, the Parc National des Volcans in Rwanda provides a "wildlife watcher's ultimate experience" (Shackley 1996, p. 62). Small groups of tourists climb for up to eight hours through thick rainforest to spend an hour with groups of wild gorillas. Although contact with animals is forbidden, their close proximity gives tourists the temporary feeling of being part of a gorilla family. This "ultimate" wildlife experience is achieved without luxurious facilities and requires considerable physical exertion on the part of tourists. The experience also features intensive educational opportunities with guides and researchers, which help minimize safety issues and impacts to the gorillas, and meet the demand among tourists for learning experiences.

While these examples illustrate polar extremes on a wildlife-viewing ecotourism continuum, they are also useful for illustrating how EBM concepts can help separate their various qualities. The key variables include (1) social interaction (larger group sizes and tourist numbers vs. low group sizes and fewer groups); (2) the level of facility development (luxurious vs. primitive); (3) the type of human-wildlife interaction (wildlife attracted by human-modified environmental features vs. more natural settings without evident attempts to attract wildlife); and (4) the level of learning opportunity (minimal, generalist interpretation that focuses on wildlife identification vs. intensive interaction with guides and researchers). Instead of simply arguing how one opportunity is a better representation of ecotourism than the other, researchers and managers explicitly recognize differences in characteristics, and can apply a systematic process for evaluating which management actions will affect them and how.

Missing in this comparison, however, are other potential ecotourism variables, including specific economic or social impacts on local people and communities, or larger biophysical impacts on wildlife or ecosystems. But EBM principles can be extended to cover these issues as well. The fundamental steps of defining a type of opportunity with associated desired future conditions, choosing indicator variables and establishing quantitative standards that link with those conditions, and then choosing actions that help you meet your standards remain the same. In these cases, however, they are viewed from the perspective of the host communities.

Applying EBM to Wildlife-based Ecotourism: Four Considerations

There are at least two implications from the previous discussion: (1) by definition, the ecotourism experience is different from other types of tourism (e.g., mass tourism), and (2) the concepts of Experience-based Management may be applied in order to manage for those differences. There are at least four areas where EBM principles may be applied to develop indicators and standards for ecotourism experiences:

(1) Unlike mass tourism, wildlife ecotourism is supposed to limit impacts from its development. Ecotourism promotional campaigns, however, often promise viewers an intensive interaction with the natural environment, including a large quantity of close-up encounters with a diversity of wildlife species. The challenge to managers, then, is to provide high-quality wildlife-viewing experiences, while minimizing negative impacts to local environments and wildlife populations. Experience-based Management approaches can assist managers to develop indicators and standards for impacts related to very sensitive and often remote natural environments.

(2) The role and influence of guides may be more pervasive in international ecotourism settings, suggesting the need for greater research on and management attention to their characteristics and effects.

(3) While many mass-tourism efforts are developed and managed by outside interests, ecotourism development should involve a high degree of participation on the part of local communities. This involvement not only provides economic benefits to indigenous populations, but also enhances the visitor's cultural and environmental experience. The extent and type of local involvement can also be guided by EBM principles.

(4) International ecotourism settings often feature complex management partnerships among agencies, nongovernmental organizations (NGOs), and the commercial sector, suggesting the need for increased understanding of the institutional arrangements needed to implement management actions.

Ecotourism advocates integration of bioconservation efforts with high-quality tourism experiences. Tourist fees in Bukit Lawang, Sumatra, help support efforts to return once-captive orangutans to the wild, while tourists are able to view these rare primates during specific feeding times required during the "rehabilitation" process. (Photo by Doug Whittaker)

Minimizing Impacts from Ecotourism

There is a general consensus that ecotourism should involve only those activities that minimize impacts to wildlife, vegetation, soils, water, and air quality. "Efforts are made to be less consumptive, travel lighter, produce less waste, and be conscious of one's effect on the environment and on the lives of those living nearby" (Wallace 1996). In ecotourism settings, both visitors and locals are encouraged to adopt "eco-code" forms of behavior (Gauthier 1993). Bear viewing along the McNeil River in Alaska provides an excellent example of an ecotourism experience that emphasizes minimal impacts on wildlife populations. Each year more than one thousand wildlife enthusiasts compete for a chance to spend four days observing the McNeil bears. A few hundred permit winners are selected by lottery to participate in the experience. Each morning groups of no more than ten tourists pack their cameras and gear for the two-mile hike across inlets and mudflats to the falls for six hours of bear watching. By setting standards for total number of visitors and group sizes, and carefully monitoring impacts on bear populations, managers have been able to minimize negative impacts on wildlife, while providing a unique and interactive experience for participants.

But even small numbers of tourists can disturb wildlife directly through noise, feeding of animals, and overuse of critical areas (e.g. nesting areas, watering holes) and indirectly through destruction of wildlife habitat (e.g., from pollution, alteration through trail cutting). Reductions in wildlife populations can also result from the selling of wildlife-based souvenirs to ecotourists. In Peru, for example, arrows made of macaw feathers and necklaces of monkey hands are sold by lodges to tourists (Groom et al. 1991).

In order to minimize the negative consequences of ecotourism on wildlife, general ecotourism guidelines should be developed, along with more site-specific norms for appropriate behavior. Wallace (1996) has recommended a series of indicators to determine acceptable levels of visitor impacts in ecotourism settings. Some indicators that could influence impacts on wildlife include group size, mode of transportation, appropriate methods of waste disposal to reduce unacceptable wildlife behaviors (e.g., monkeys feeding on waste heaps outside of lodges in Africa), and measures of biophysical change, such as water quality and wildlife behavior.

The Role and Influence of Guides

Although most people have been sensitized to general environmental issues such as air pollution, global warming, and sanitation problems, some researchers have argued the need to increase tourists' awareness of their responsibility to protect and enhance the environmental quality of the destinations they visit (Hawkins 1995). Ecotourism draws attention to this issue, promising visitors an improved understanding of local wildlife and other natural resources, while offering a specific component of modern life where people can demonstrate pro-environmental action. Hawkins (1995, p. 267) notes "ecotourism is one of the major manifestations of environmental concern by the general public, particularly in developing countries."

Meeting this educational goal, however, depends upon integrated programs offered by ecotourism operators (usually via pre-trip information and follow-up materials) and implemented on-the-ground by guides, who offer the direct, on-site information that is likely to be most powerful and illuminating. While guides are an important component of many wildlife-viewing opportunities in North America, they may play an even larger role in international settings, particularly in remote areas. Ecotourism principles urge that guides come from local populations who can benefit both economically and culturally from the work, but they also focus on the variety of ways guides can affect the quality and impacts of ecotourism activities.

First, high-quality viewing opportunities often depend upon the skill and experience of guides to locate viewable animals. While ecotourists often have specialized viewing experience and equipment, they also typically require direct information and assistance to help them find wildlife in local environments. Guides often possess information about recent sightings, have developed efficient itineraries that maximize potential sightings, and may have developed local techniques for approaching wildlife without disturbance to provide closer-proximity viewing. While ecotourists willing to invest the time and effort can learn similar information and techniques, guides disperse it more efficiently to a greater number of people.

Second, guides can decrease safety risks in viewing settings where dangerous animals are present. While some highly involved viewers may possess the skill and experience to recognize and respond to high-risk situations, guides may have additional knowledge, training, or equipment for addressing safety concerns. Even trained biologists often employ local guides when conducting research, because local people may know individual animals or groups of animals and their behavioral tendencies. Individual wildlife can learn different responses to human behavior (Whittaker and Knight 1998), so there may be important variation within the same species in different locations. For example, the habituated brown bears at several Alaskan viewing areas (e.g., Brooks River, Pack Creek) are likely to react very differently to close-proximity viewers than bears in nearby but more remote, low-use areas. Guides are likely to be aware of these differences and can help prevent problems.

Third, guides are a central conduit for providing wildlife education, a distinguishing feature of ecotourism. Guides can disperse information not only to enhance tourist experiences (e.g., information about wildlife characteristics or behavior, habitat features, ecosystem links between different species, the history of human-wildlife interactions in the area), but to help shape visitor behavior to minimize adverse impacts on the environment or the wildlife people have come to see. Guides influence where viewers go, how they travel, how close they approach wildlife, and how they respond to wildlife behaviors. They also help to establish and enforce non-viewing environmental norms (e.g., no littering, camping in appropriate areas, proper sanitation).

Not all guides, of course, manage to produce this array of benefits. Honey (1999) notes that "good ecotourism" requires "well-trained, multilingual naturalist guides with skills in natural and cultural history, environmental interpretation, ethical principles, and effective communication." Developing these characteristics among guides, however, can be challenging in many international wildlife-viewing settings. While guiding may be relatively lucrative and desirable employment in some areas, it is seasonal and low paying in others. Attracting qualified individuals who will remain long enough to develop the above skills, abilities, and knowledge is not always possible, and institutionalized training programs that encourage development of these characteristics are not always available.

The formalized guide program developed in the Galapagos Islands provides an example of an initiative that has successfully trained and retained qualified guides. The Galapagos consist of a cluster of 120 volcanic islands and are home to a diversity of wildlife species, including sea lions, marine iguanas, giant tortoises, and a variety of rare birds. Over the past thirty years, increasing numbers of tourists have been attracted to the Galapagos to view these unique species, resulting in a growing concern about impacts. To minimize these potential impacts and improve the conservation efforts generated from ecotourism, a guide program has been developed. Under this program, naturalist guides must accompany ecotourists visiting any of the 54 designated land sites and 62 marine sites in the Galapagos. Although the guides work for tour companies, they must attend special training courses given by the park service and the Charles Darwin Biological Research Center. Most are not from the Galapagos and many have degrees in the biological or natural sciences and speak several languages. The guides serve two functions, as both guards and educators. In addition to interpreting the wildlife and other natural features for visitors, they ensure that people stay on the narrow trails and don't touch or take anything, take food onto the islands, litter, or disturb the animals (Honey 1999).

The tremendous increases in the number of visitors to the Islands, however, has forced the rapid expansion of the guide pool. This has resulted in the certification of a new class of auxiliary guides. These mostly Spanish-speaking natives of the Galapagos are very familiar with the Islands but often lack the formal scientific training characteristic of the other guides. Some feel that the introduction of these auxiliary guides has led to a decline in the level of ecological interpretation offered tourists and a decrease in the control exerted over groups, due to either a lack of conservation understanding or commitment, or a fear of losing an end-of-cruise tip (Honey 1999, p. 113). While the auxiliary guide program has raised some concerns, the program as a whole has generally been successful in minimizing negative wildlife impacts and improving the quality of the wildlife-ecotourism experience.

Although guide programs can help to protect the ecotourism attraction, there may also be some incentives for guides to provide inappropriate information or discourage appropriate tourist behavior. The quality of many viewing experiences is directly tied to the frequency and close proximity of wildlife sightings. Recognizing this, some guides will be tempted to satisfy those goals (and thus maximize their tips or status with clients), even if this comes at the expense of altered wildlife behavior or disturbance to wildlife and the ecosystem. For example, guides who present a tethered goat as a food attraction for Komodo dragons in Indonesia are trading natural behavior for improved viewing of this rare reptile. Such behavior may not be inherently negative, but it does change the type of experience that is offered and may have other undesirable consequences for the ecosystem or the long-term viability of the species.

Another example is offered by guides in Royal Chitwan National Park in Nepal, who offer elephant-back safaris to view rhinoceros and other wildlife. In this case, the trade-off appears to be between short-term and long-term viewing quality. The elephants do provide superlative viewing opportunities that are better, safer, and may have lower environmental impacts than alternative jeep-based viewing. But the travel mode also allows some excesses, as guides can use the larger elephants to chase down or corner rhinos to provide better photographic opportunities for their clients. Over the long term, this may diminish viewing opportunities in or around

Elephant-back safaris offer an exotic and safe way to view rhinoceros in Chitwan National Park in the Terai region of Nepal, and they arguably have lower impacts on the environment than alternative jeep-based tours. Tourist demand for close approaches, however, entices some guides to pursue and harass rhinos. (Photo by Doug Whittaker)

the developed areas of the park as the rhinos learn to avoid areas where elephant guiding occurs.

A final example from Manu Biosphere Reserve in Peru offers a more extreme example, where "incidents have been reported of independent guides bow hunting, digging up turtle nests, disturbing beach-nesting birds, and chasing giant otters, swimming jaguars, and tapir" to provide viewing opportunities for tour groups (Groom et al. 1991). In all of these cases, the profit motive is evident. In addition, there appears to be a complicity between guides and tourists, because, in many situations, guides only provide what ecotourists demand.

The challenge in applying EBM to these situations is the limited research on guide characteristics and behavior in ecotourism settings, the impact of guides on tourist behavior, or tourist preferences for various types of viewing experiences that guides help create. Unlike the extensive research effort into wildland recreation opportunities in North America, there is simply much less known about what ecotourists want and how guides may affect them.

Sirakaya and McLellan (1998), however, offer an initial examination of some ecotourism guide issues, relating reported behavior of ecotour companies with (1) characteristics of the company; (2) beliefs and attitudes of company employees toward ecotourism principles; and (3) perceived sanctions from agencies or clients for not meeting them. Results suggest that compliance with ecotourism principles (pro-environmental actions) is higher when operators know these guidelines, believe the guidelines provide long-term economic benefits, have previous experience with legal agencies, and have a clear picture of the costs and benefits associated with the guidelines. Concluding that "carrots work better than sticks," this research implies that ecotourism management may need to focus on improved educational efforts rather than regulations. Additional work addressing these and related questions, however, is sorely needed.

The Importance of Local Participation

A third area for applying EBM principles focuses on the extent and type of local involvement in wildlife-ecotourism efforts. International conservation of wildlife is a major challenge because of the delicate balance that needs to be established between fragile natural resources and an often-impoverished rural population. Wildlife is important to indigenous groups for the market and consumption benefits provided by hunting and fishing. Exploitation can occur, however, when the economic benefits from game are no longer sufficient to meet local needs. Implementing sustainable wildlife-management programs, therefore, requires integrating information on "the biology of game species and the economics of sustainable use with the desires of local communities" (Bodmer 1994, p. 121).

Research has clearly demonstrated that the success of ecotourism-based wildlife programs depends heavily on the involvement of local populations in both the planning and management phases (Bogdonov and Henry 1995; Shackley 1996; Wallace 1996). In the past, when ecotourism developments were introduced into an area, local populations were often displaced from the parks and restricted from using wildlife in traditional ways. Most national parks and reserves in Africa, for example, were originally established for hunters, scientists, and tourists, with little or no regard for local residents. Park management emphasized policing that forcibly evicted and kept out community members. These locals received few benefits from either the parks or tourism, and "deeply resented being excluded from lands of religious and economic value and being restricted to increasingly unsustainable areas around the park" (Honey 1999, p. 12). Management problems such as poaching, degradation of resources, and increasing hostility of locals toward parks and tourism resulted.

To combat this increasingly negative sentiment and the resulting damage to wildlife populations, the Kenyan government, in the early 1970s, put several reserves (e.g. Maasai Mara Game Reserve, Amboseli National Park) under the control of local county councils to ensure that residents received revenue from both park entrance fees and other tourism facilities. This approach was based on stakeholder and economic-development concepts suggesting that locals would protect resources that were of economic value to them, and that the answer to poverty must begin at, not simply trickle down to, local communities (Honey 1999). The success of this program suggests that, for ecotourism to enhance wildlife conservation, managers need to ensure local communities economic benefits from wildlife resources.

The howler monkey reserve in Belize offers another example of a grassroots ecotourism effort that provides benefits to local communities, conserves wildlife populations, and gives tourists an extraordinary wildlife-viewing experience. In 1985, a community sanctuary was established through the cooperative effort of several Belizean communities to protect one of the few healthy populations of black howler monkeys in Central America. The eighteen-square-mile sanctuary is managed voluntarily by landowners in unison with their farming practices. Local guides take tourists on hikes to view the monkeys in the forest canopy and to hear their rasping howl. The landowners also provide accommodation in their homes and traditional meals. The economic benefits from fees, lodging, and meals provide landowners with incentives to protect the howler monkey population, and the guiding programs help minimize negative impacts on both the land and the monkeys while maximizing the experiential benefits for visitors. This bottom-up approach to ecotourism planning and management provides excellent opportunities for sustainable wildlife development, as well as widely disseminated benefits to both locals and visitors.

Although more oriented toward consumptive wildlife use, Zimbabwe's CAMPFIRE (Communal Areas Management Program for Indigenous Resources) program provides another example of local involvement in ecotourism development. This program encourages economic growth through hunting tourism, and has resulted in an increased conservation ethic and a decrease in the amount of illegal poaching in the region. Participant communities have quotas for animal harvesting and can either sell the meat or use their quotas to develop a trophy-hunting form of ecotourism. In addition to providing economic benefits for those involved in ecotourism activities, a portion of the revenue is also distributed to subsistence farmers and small-scale industries in the area. Thus, this program provides tangible benefits to all locals in return for promoting the conservation of wildlife resources. The CAMPFIRE program provides an excellent model for local involvement that could easily be adopted for wildlife-viewing ecotourism experiences.

One of the keys to successful wildlife-ecotourism development, then, is the early establishment of partnerships and mechanisms for local input to public and private interests that operate in the area. Several indicators are recommended by Wallace (1996) for understanding and measuring local involvement, including strength and duration of local planning and advisory groups, development of local ecotourism ventures, development of tour itineraries that conform to local needs and schedules, presence of staff dedicated to community relations, and attitudes of locals toward ecotourism initiatives (p. 124).

The Importance of Partnerships

A fourth challenge in applying EBM to ecotourism settings focuses on the complex and sometimes undeveloped institutional arrangements surrounding international ecotourism destinations. Unlike most protected areas in North America, many natural and rural areas in developing countries are protected by national governments in part because of international pressure and only with considerable NGO assistance. In other cases, there may not even be a governmental institution with responsibilities to protect wildlife or experiential quality in a natural area. In either case, there may not be a structure for making proactive management decisions, or these decisions may be strongly influenced by other interests, most obviously those from conservation-oriented NGOs (e.g., World Wildlife Fund) and commercial operators.

EBM does not necessarily require a hierarchical decision-making structure, and it even has specific opportunities in the planning process for stakeholder involvement (see Chapters 6 and 7). However, many management decisions in ecotourism settings may involve conflict between the preferences of local people, international NGOs, and both local and foreign commercial operators. To the extent that the managing entity is weak, conflict among these competitors may be high and could limit management options to the lowest common denominator (i.e., the things that are easy to do, but may be ineffectual).

EBM urges more proactive decision making ("management by design" instead of "management by default"), and thus may require adaptation of a partnership model. Under this view, NGOs and commercial interests may deserve to be more than just stakeholders in the process, with elevated decision-making positions. Concurrently, however, they will also have increased responsibilities to look beyond their own interests, and consider the consequences of developing (or not developing) various types of infrastructure, education programs, or regulations.

North American protected areas are not without some of these same issues, but they may have more established governmental institutions to address them. The National Park idea, for example, was in large part an explicit attempt to minimize the influence of commercial or other private interests over the use of particularly special natural areas (Sellars 1997). Many ecotourism settings have similar resource values and the potential to provide important viewing experiences, but lack the institutional structure to decide what those values are and how to take appropriate steps to protect them.

The ecotourism idea suggests that there is a long-term interest among all parties to provide high-quality viewing experiences while protecting the natural resource. An area that is allowed to develop in response to economic and commercial considerations alone is likely to follow the path from emergence and discovery to saturation and decline (Prosser 1994). An area that is set aside only for highly involved tourists may not provide enough economic benefits for locals, who may then resent the park's presence and ignore its goals and regulations. EBM offers ways of finding compromises by urging the development of a diversity of experiential settings, but with each explicitly defined. If there is a common ground, defining desired types of experiences designed to be provided in perpetuity may help these partners to find it.

Conclusion

Internationally, ecotourism has been viewed as a "panacea": a way to conserve natural resources, protect fragile ecosystems, provide economic benefits to local communities, promote economic development in impoverished countries, and develop environmental and cultural sensitivity on the part of the travel industry and the traveling public (Ceballos-Lascurain and Johnsingh 1995; Honey 1999). Ecotourism's role in the conservation of wildlife and other natural resources has been widely acknowledged (Bowen and Draper 1995). It provides direct financial benefits for conservation by raising funds for environmental protection, research, and education; applying park entrance fees and receipts from tour company, airline, and hotel taxes to conservation efforts; and soliciting voluntary contributions to governmental and conservation organizations (Honey 1999). Ecotourism also provides direct financial benefits for indigenous communities, and in so doing, may reverse local negative attitudes toward wildlife. It creates political pressures both locally and internationally to conserve wildlife and other natural resources for current and future generations (Fillion et al. 1992).

While some argue that wildlife resources should be conserved solely for biological reasons, others suggest that wildlife should be saved because intact ecosystems are interesting and create benefits for people. The wildlife-ecotourism concept fits with this latter premise, suggesting that wildlife can be protected by providing enjoyable, low-impact and educational viewing experiences for tourists, while also supplying tangible economic benefits to local populations. This chapter demonstrates that the Experience-based Management concept can be applied to wildlife ecotourism to manage for these essential characteristics: ensuring minimal impacts on wildlife populations, providing high-quality, educational wildlife-viewing experiences through the use of trained guides, and supplying economic benefits for local populations. By defining desired ecotourism opportunities, developing indicators and standards for these opportunities, and linking actions to standards, EBM provides a framework for understanding and managing for sustainable ecotourism.

Summary Points

• International conservation of wildlife must be concerned with balancing wildlife protection with the rights of local communities to engage in traditional activities.

• The ecotourism concept suggests that wildlife can be protected by providing enjoyable, low-impact and educational viewing experiences for tourists, while also supplying tangible economic benefits to local populations.

• This chapter demonstrates that the Experience-based Management concept can be applied to wildlife ecotourism to manage for these essential characteristics: ensuring minimal impacts on wildlife populations, providing high-quality, educational wildlife-viewing experiences through the use of trained guides, and supplying economic benefits for local populations.

• By defining desired ecotourism opportunities, developing indicators and standards for these opportunities, and linking actions to standards, EBM provides a framework for understanding and managing for sustainable ecotourism.

Literature Cited

Bodmer, R. E. (1994). "Managing wildlife with local communities in the Peruvian Amazon: The case of the Reserva Comunal Tamshiyacu-Tahuayo." In D. Western and E. M. Wright (Eds.), *Natural Connections: Perspectives in Community-Based Conservation.* Washington, DC: Island Press.

Bogdonov, D., and D. Henry. (1995). "Ecotourism: A different approach to development using the principles of community economic development." In J. Bissonette and P. Krausman (Eds.), *Integrating People and Wildlife for a Sustainable Future.* Bethesda, MD: The Wildlife Society.

Bowen, J. T., and D. Draper. (1995). "Ecotourism and wilderness preservation: A symbiotic commensalism." In J. Bissonette and P. Krausman (Eds.), *Integrating People and Wildlife for a Sustainable Future.* Bethesda, MD: The Wildlife Society.

Boo, E. (1990). *Ecotourism: The Potentials and Pitfalls.* Baltimore, MD: World Wildlife Fund.

Bruner, E. M. (1991). "Transformation of self in tourism." *Annals of Tourism Research,* 18, 228-50.

Butler, R. (1992). "Alternative tourism: The thin edge of the wedge." Pp. 31-46 in *Tourism Alternatives.* Philadelphia, PA: University of Philadelphia Press.

Cater, E. (1994). "Introduction."In E. Cater and G. Lowman (Eds.), *Eco-tourism: A sustainable option?* New York: John Wiley and Sons, Ltd.

Ceballos-Lascurain, H. (1988). "The future of ecotourism." *Mexico Journal.* January 17.

Ceballos-Lascurain, H., and A. J. Johnsingh (1995). "Ecotourism: What works and what does not." In J. Bissonette and P. Krausman (Eds.), *Integrating People and Wildlife for a Sustainable Future.* Bethesda, MD: The Wildlife Society.

Damon, T. A., and C. Vaughn (1995). "Ecotourism and wildlife conservation in Costa Rica: Potential for a sustainable partnership." In J. Bissonette and P. Krausman (Eds.), *Integrating People and Wildlife for a Sustainable Future.* Bethesda, MD: The Wildlife Society.

Fillion, F. L., J. P. Foley, and A, J. Jacquemot (1992). The Economics of Global Ecotourism. Paper presented at the Fourth World Congress on National Parks and Protected Areas. Caracas, Venezuela. February 10-21.

Gauthier, D. A. (1993). "Sustainable development, tourism and wildlife." In J. G. Nelson, R. Butler, and G. Wall (Eds.), *Tourism and Sustainable Development: Monitoring, Planning, Managing.* Waterloo, Ontario: University of Waterloo, Heritage Resources Center Joint Publication Number 1.

Groom, M. J., R. D. Podolsky, and C. A. Munn (1991). "Tourism as a sustained use of wildlife: A case study of Madre de Dios, Southeastern Peru." In J. G. Robinson and K. H. Redford (Eds.), *Neotropical Wildlife Use and Conservation.* Chicago, IL: University of Chicago Press.

Hawkins, D. E. (1995). "Ecotourism: Opportunities for developing countries." In W. F. Theobald (Ed.), *Global Tourism: The Next Decade.* Oxford, England: Butterworth-Heinemann Ltd.

HaySmith, L., and J. D. Hunt (1995). "Nature tourism: Impacts and management." In R. Knight and K. Gutzwiller (Eds.), *Wildlife and Recreationists: Coexistence Through Management and Research.* Washington, DC: Island Press.

Honey, M. (1999). *Ecotourism and Sustainable Development: Who Owns Paradise?* Washington, DC: Island Press.

Hultsman, J. (1995). "Just tourism: An ethical framework." *Annals of Tourism Research, 22,* 553-67.

Hunter, C. (1997). "Sustainable tourism as an adaptive paradigm." *Annals of Tourism Research, 27,* 850-67.

The International Ecotourism Society. (1991). *The Ecotourism Newsletter.* Alexandria, VA: The Ecotourism Society. No. 1.

The International Ecotourism Society (1998). *Statistical Fact Sheet.* North Bennington, VT.

Lindberg, K. (1991). *Policies for Maximizing Nature Tourism's Ecological and Economic Benefits.* Washington, DC: World Resources Institute.

Pearce, P. (1982). *The Social Psychology of Tourist Behavior*. Oxford, England: Pergamon Press.

Prosser, R. (1994). "Societal change and the growth of alternative tourism." In E. Cater and G. Lowman (Eds.), *Eco-tourism: A sustainable option?* New York: John Wiley and Sons, Ltd.

Reinhold, L. (1993). "Identifying the elusive ecotourist." *Going Green: A Supplement to Tour and Travel News,* October 25, 36-37.

Saberwal, V. K. (1999). *Reconciling the Needs of Man and Wildlife in India*. American Society of International Law Wildlife Interest Group Occasional Paper No. 1. Berkeley, CA.

Sellars, R. W. (1997). *Preserving Nature in the National Parks*. New Haven, CT: Yale University Press.

Shackley, M. (1994). "The land of Lo, Nepal/Tibet." *Tourism Management,* 15, 17-26.

Shackley, M. (1996). *Wildlife Tourism*. Boston, MA: International Thompson Business Press.

Sirakaya, E., and R. W. McLellan (1998). "Modeling tour operators voluntary compliance with ecotrouism principles: A behavioral approach." *Journal of Travel Research,* 36, 42-55.

Tourism Canada (1989). *Pleasure Travel Markets to North America: United Kingdom, France, West Germany and Japan: Highlights Report.* Prepared for Tourism Canada and United States Travel and Tourism Administration by Market Facts of Canada Limited.

Valentine, P. S. (1990). "Ecotourism and nature conservation: A definition with some recent developments in Micronesia." *Tourism Management,* April, 107-15.

Van Den Berghe, P. L. (1994). *The Quest for the Other: Ethnic Tourism in San Cristobal, Mexico*. Seattle: University of Washington Press.

Wallace, G. N. (1996). *Toward a Principled Evaluation of Ecotourism Ventures*. Yale School of Forestry and Environmental Studies Bulletin Series, Number 99. New Haven, CT: Yale University.

Whittaker, D., and R. L. Knight (1998). "Understanding wildlife responses to humans: A need for greater clarity in research and management." *Wildlife Society Bulletin,* 26, 312-17.

Wight, P. A. (1994). "Environmentally responsible marketing of tourism." In E. Cater and G. Lowman (Eds.). *Ecotourism: A Sustainable Option?*New York: John Wiley and Sons.

Wight, P. A. (1996). "North American ecotourists: Market profile and trip characteristics." *Journal of Travel Research,* 24, 2-10.

World Tourism Organization. (1998). Ecotourism, now one-fifth of market. WTO Newsletter. January/February.

Zurick, D. (1995). *Errant Journeys: Adventure Travel in a Modern Age*. Austin: University of Texas Press.

Build an Experience and They Will Come: Managing the Biology of Wildlife Viewing for Benefits to People and Wildlife

R. Bruce Gill

Introduction

WILDLIFE-VIEWING MANAGEMENT (both biological and recreational management) is a slowly emerging professional discipline. Although there is an enormous body of professional literature describing the effects of biological management to benefit hunting and species conservation, there is virtually no comparable literature describing the effects of managing wildlife biology to produce wildlife-viewing opportunities and benefits. This essay will outline the *possibilities* for managing biological attributes of wildlife to produce viewing benefits.

Pierce and Manfredo (1997) reviewed the status of state and provincial wildlife-viewing programs throughout North America. They painted a picture of an anemic profession and an ambivalent practice. The profession is anemic because it lacks a distinct and focused philosophical paradigm. The practice is ambivalent because it confuses means and ends and because many professionals consider wildlife viewing and wildlife protection to be incompatible (Knight and Gutzwiller 1995). Consequently, wildlife-viewing programs are underappreciated by professionals, underfunded by agencies, and underattended by both managers and viewers.

What if the focus of wildlife viewing changed so the experience became the end, facilities and information became the means, and wildlife protection became a constraint? Better yet, what if wildlife viewing could be combined with wildlife preservation so people could enjoy fulfilling wildlife-viewing experiences and contribute to wildlife conservation at the same time? What if attributes of wildlife populations, habitats, and behavior could be manipulated intentionally to enhance viewing opportunities and experiences without adverse effects to wildlife? If we could and would do these things, I believe we would infuse professional wildlife-viewing management with renewed purpose, focus, and enthusiasm, and attract new participants.

Managing Wildlife Biology for Viewing Benefits: Concepts

Wildlife-viewing management seeks to provide people with sustainable opportunities to observe wildlife and to enhance their viewing experiences. Viewing expectations vary considerably among wildlife viewers. Some are satisfied simply with predictable wildlife-viewing opportunities. It is important to them that the likelihood of seeing wildlife on a given occasion is high. For these viewers, management should focus on manipulating animal distribution.

Other viewers prefer to see abundant wildlife. They are attracted to areas where wildlife is numerous. Management for these viewers should focus on manipulating absolute or relative animal abundance.

Viewers such as birders who keep lists of the various wildlife species they observe seek locations where they have a good chance to see a diversity of wildlife species. Alternatively, they seek a variety of sites each differing in wildlife species composition. These viewers require management activities that promote wildlife diversity.

Still others want not only to see wildlife, but also to seek opportunities to interact closely and intimately with individual animals or groups. Professional wildlife photographers, for example, seek opportunities where they can approach wildlife closely to photograph natural behaviors. In these circumstances, managers should manipulate wildlife so they tolerate people in close proximity without becoming disturbed or distressed. Potentially, wildlife-viewing managers can alter wildlife distribution, abundance, diversity, and proximity by manipulating wildlife food and water, habitats, populations, and behaviors.

Manipulating Food and Water

Food and water are among the most powerful of animal attractants. As a result, they can be used readily to manipulate distribution, abundance, diversity, and proximity of a wide variety of wildlife species. Backyard birders, for example, routinely use both food supplements and water to attract and concentrate birds. Experienced birders can attract a diversity of bird species by altering the variety of foods and types of feeders. Although food supplements occasionally have been used by wildlife biologists to prevent crop or property damage or to augment fledgling populations of endangered wildlife, supplemental feeding is generally discouraged by wildlife professionals (Boyd 1978; Beetle 1979; Gill and Carpenter 1985; Strickland and Crowe 1985). Nonetheless, supplemental feeding and watering are potentially useful tools for wildlife-viewing managers because they attract and concentrate wildlife like no other

Wildlife-viewing managers can alter the distribution, abundance, diversity, and proximity of wildlife by mmanipulating food and water, habitats, populations, and behaviors. (Photo by B. Gill)

management practice, and when wildlife associate food and water with positive interactions with people, they can become conditioned to tolerate viewers in very close proximity.

Supplemental Food. Supplemental feeding has a long history both as a research tool to study the effects of enriching and redistributing habitat nutrients on wildlife populations, and to reduce game damage. Supplemental feeding can alter wildlife distribution, abundance, diversity, and proximity in the following ways. Under experimental conditions, food supplements generally reduce the size of wildlife home ranges and attract both resident and transient wildlife.

The Wyoming Game and Fish Department routinely feeds elk at several feed grounds across the state, primarily to prevent damage to agricultural crops such as grass and alfalfa hay. Feeding lures elk away from agricultural areas and concentrates large numbers at designated feeding sites (Strickland and Crowe 1985).Feed grounds not only attract elk away from agricultural areas, they also attract considerable numbers of elk watchers each year. The National Elk Refuge near Jackson, Wyoming, for example, annually attracts nearly twenty-five thousand winter visitors who pay an average of about three dollars per person to ride sleighs to view elk (Boyce 1989).

Supplemental feeding in these circumstances: (1) concentrates and redistributes elk, reducing crop damage; (2) maintains elk at much higher densities than natural habitat can support because much of the natural habitat has been converted to cropland and human residences; and (3) provides an attractive wildlife-viewing opportunity because elk are concentrated with predictable distributions (Smith 2001).

Feeding deer in Colorado in the severe winter of 1983-84. Food is a powerful wildlife attractant. Supplemental feeding can be used to alter wildlife distribution, abundance, diversity, and behavior to create or enhance wildlife-viewing opportunities. (Photo by B. Gill)

Supplemental feeding also has been used to redistribute and concentrate birds. Salmon carcasses and carrion have been provided to wintering bald eagle populations to redistribute the birds and increase their survival and reproduction rates (Knight and Anderson 1990; McCollough et al. 1994). The effects of feeding on survival and reproduction were ambiguous, but feeding did attract large numbers of eagles to feed sites. Feeding also attracted other scavenging species such as American crows, common ravens, coyotes, and bobcats, increasing faunal diversity at feeding sites. All wildlife species visiting feed sites were easily viewed from blinds 20-30 meters away apparently without disturbing them.

If supplementally fed wildlife associate feeding with positive outcomes, they become remarkably habituated to feeders and other feed site visitors. Elk at the National Elk Refuge feed grounds permit sleighs full of visitors to approach within 10-15 meters apparently without altering normal behavior.

One couple from northern Minnesota began feeding bears inadvertently when bears were attracted to their backyard bird feeders. Over time, they intentionally fed otherwise wild bears until both bears and humans interacted naturally at feed sites. Eventually bears became so habituated that they interacted willingly and intimately with the human feeders, all without injury to bears or people and with only minor property damage which was eliminated subsequently as experience with bears revealed ways to prevent damage (Beckland 1999). These accounts illustrate supplemental feeding's potential to manipulate wildlife distribution and behavior for close, benign, and extraordinary viewing experiences.

Some argue that supplemental feeding attracts wildlife and creates conditions for nuisance encounters and property damage. In other circumstances where large and potentially dangerous wildlife such as bears associate supplemental food with people, feeding can create potential health and safety hazards. Each year hundreds of people are injured or killed because wildlife were fed improperly. Nonetheless, when supplemental feeding is managed properly, it need not result in negative outcomes for wildlife or people (Caputo 2002).

Following the establishment of Yellowstone National Park in 1872 and until the early 1970s, viewing bears at refuse dumps was an extremely popular tourist pastime. Bear viewing became so popular, in fact, that bleachers were erected to accommodate the tourists, and park rangers gave lectures while bears foraged among the nearby refuse. The use of human garbage as a supplemental bear food and attractant began inadvertently. While Yellowstone developed as a major tourist destination in the late 1800s and early 1900s, hotels developed within and around the park. It was common practice for hotel staff to dump garbage adjacent to hotels, and garbage attracted bears. Initially bear foraging was not discouraged because bears assisted with refuse disposal. In time, garbage at hotel sites became unsavory and bears became nuisances. Subsequently, garbage was transported to designated dumpsites, which attracted bears away from hotel areas and also attracted tourists who wanted to watch bears. Eventually bear watching at Yellowstone dumps became institutionalized as a tourist attraction and refuse feeding became an established practice with few negative consequences to bears or humans until humans began to approach bears and hand feed them (Schullery 1992).

Feeding Caveats and Considerations. Supplemental feeding operations must be designed and implemented carefully to avoid negative consequences to both wildlife and people. Indiscriminate wildlife feeding can be problematic. When wildlife are fed supplements that do not resemble the structure and nutrition of their natural diets, digestive and metabolic malfunctions can result. Supplemental feeding of chipmunks and marmots illustrates this point. Chipmunks and marmots depend upon stored body fat to survive hibernation, and both the quantity and quality of stored fat are important to their survival. Saturated fats found in animal tissues will not remain fluid at reduced body temperatures during hibernation, as do the unsaturated fats found in plant tissues. If hibernating animals are fed and metabolically store foods rich in saturated animal fats, these fats can impede effective hibernation and increase mortality rates in the hibernacula (Geiser and Kenagy 1993).

Mule deer likewise may languish when fed inappropriate foods. If mule deer are fed diets high in lignified or woody fiber, they are unable to reduce the particle size of fibrous foods sufficiently to promote effective passage through the digestive system. Reduced rates of food passage lead to impaction and ulceration of the digestive tract, resulting in death from starvation despite a filled digestive tract (Keiss and Smith 1966; Carpenter and Wallmo 1982).

Feeding can encourage the spread of contagious diseases among wild animals. A variety of avian disease outbreaks including salmonellosis, conjunctivitis, trichomoniasis, coccidiosis, aspergillosis, avian pox, and avian mange have been associated with bird-feeders (Brittingham and Temple 1988). Mammalian diseases encouraged by supplemental feeding include brucellosis, rabies, plague, and chronic wasting disease (McCorquodale and DiGiacomo 1985; Jenkins et al. 1988; Spraker et al. 1997). These problems can be mitigated to some extent by periodically cleaning and disinfecting feeders, by feeding only as much feed as animals consume daily, by dispersing or moving feeders to minimize animal concentrations at any single site, and by avoiding placement of feed directly on the ground.

Supplemental feeding can increase vulnerability of wildlife to predation by concentrating them in areas where hiding cover is

Black bear begging at the window of a car in Yellowstone National Park. Supplemental feeding can create wildlife-human conflicts; however, when supplemental feeding programs are designed and implemented carefully and thoughtfully, they can yield substantial wildlife-viewing benefits. (Photo by V. Barnes)

minimal. On the other hand, attraction of predators may be regarded by some as an added benefit to supplemental feeding.

Supplemental feeding of grazing and browsing wildlife can result in severe habitat degradation. White-tailed deer, for example, continue to browse even when supplemental foods are offered *ad libitum*. Supplemental feeding concentrates animals, thereby amplifying foraging effects on native browse, especially during severe winters (Doenier et al. 1997). The net effect of supplementing deer feed is to seriously degrade the ecological condition at feed sites for prolonged periods, even after feeding has ceased. Collateral browsing effects associated with supplemental feeding also can diminish floral and faunal diversity. In some natural areas, concentrations of large herbivores such as deer, elk, or moose eliminate preferred native forage species from a site or landscape altogether, reducing plant species diversity. Loss of plant diversity reduces animal diversity, impoverishing overall ecological diversity (Diamond 1992).

Problems with feeding unwholesome supplements can be surmounted by feeding diets that mimic the digestive and metabolic properties of natural diets. Zookeepers and wildlife rehabilitators and their respective journals (e.g., *Journal of Zoo Animal Medicine, Zoo Biology, Wildlife Rehabilitation Today*) are good sources of information regarding wholesome food supplements for wildlife. In addition, scientists who maintain captive wildlife for research experiments have published data on research diets that maintain wildlife in good health for prolonged periods (e.g., Ullrey et al. 1971; Baker and Hobbs 1985; Schwartz et al. 1985; Wild et al. 1994).

Allowing people to hand-feed supplements can result in injuries from bites and scratches. Large, aggressive animals such as bears are capable of inflicting serious, even fatal, injuries to people who attempt to feed them by hand. Prior to 1972, when the National Park Service began to aggressively manage conflicts between bears and people, an average of forty-five people per year were injured by black bears in Yellowstone National Park and grizzly bears injured an average of two people per year. Most of these injuries resulted when people attempted to hand-feed bears (Gunther and Hoekstra 1998).

Wildlife professionals generally discourage supplemental feeding of wildlife despite its potential to enhance wildlife-viewing opportunities. Health, nuisance, and economic problems associated with feeding and "unnaturalness" are commonly cited reasons to discourage supplemental feeding. Like most management tools, supplemental feeding is neither panacea nor pariah. Its use will depend upon the circumstances. If a wildlife-viewing area is isolated from other wildlife habitats, wildlife are too dispersed, natural forages are too sparse, or wildlife diversity is lacking, then supplemental feeding can be used effectively to provide predictable, enjoyable viewing opportunities, provided the benefits exceed ecological, social, and economic costs.

Supplemental Water. In arid environments, water can be a significant limiting factor for wildlife populations. Managing availability of water in arid environs can attract and concentrate some wildlife species and provide viewing opportunities. Mule deer and pronghorn were attracted to wastewater ponds during summer and fall at the Idaho National Engineering and Environmental Laboratory in southeastern Idaho. Pond use by mule deer peaked in September, while pronghorn use peaked in November. Both species visited ponds during daylight hours, but pronghorn were more likely to be seen in daytime than were mule deer, except in October when mule deer use was both diurnal and nocturnal (Cieminski and Flake 1997).

Avian species in arid environs are attracted to riparian sites both for water and cover. In central Arizona, bird densities at riparian sites were more than two-thirds higher than at adjacent desert habitats (Szaro and Jalke 1985). Both of these studies suggest that water development can improve wildlife-watching opportunities in desert habitats. Development of natural riparian sites is preferable to water tanks because riparian sites support greater abundance of vegetation and diversity of both plant species composition and canopy structure which in turn attract a diversity of bird species.

Manipulating Habitats

Habitats provide wildlife with food, water, and shelter, enabling them to conserve energy, avoid predators, and reproduce successfully. Management practices that amplify these habitat attributes attract and concentrate wildlife. Range and wildlife managers routinely employ the tools of prescribed fire, prescribed logging, brush removal, grazing, and revegetation to manipulate the distribution and abundance of livestock and wildlife (Krausmann 1996).

Prescribed Fire. Prescribed fire is a particularly effective way to alter wildlife distribution, abundance, and diversity (Krausman 1996). Fire alters wildlife distribution and abundance primarily by improving food quantity, quality, and availability. Fire typically removes standing plant tissue and releases nutrients such as nitrogen, carbon, and minerals bound in structural tissues of plants (Hobbs and Schimel 1984; Hobbs et al. 1991). Released nutrients stimulate nutritious plant regrowth, while removal of old growth

Mountain plover on Pawnee National Grassland. Prescribed fire can alter habitat structure and attract certain bird species. Mountain plovers are attracted to habitats featuring simple, open-canopy structures. Burning reduces vegetation density and height in grasslandf habitats, making them considerably more attractive to nesting plovers. (Photo by B. Gill)

increases visibility or accessibility of new plant tissues. Grazing wildlife (deer, elk, bison, bighorn sheep, and pronghorn antelope) are particularly attracted to burn sites for food. Prescribed fire can enhance wildlife-viewing opportunities by attracting and concentrating individuals.

In a typical example, prescribed fires were used to attract bison from unburned to burned sites at the Wichita Mountains Wildlife Refuge in Oklahoma. Prior to burning, a site comprising 28.8 percent of the annual bison range on the refuge yielded 8.8 percent of bison observations. One year after burning, 45.5 percent of bison observations occurred on the burn site. Bison were attracted from adjacent unburned habitats, not only altering their distribution, but also increasing their concentrations (Shaw and Carter 1990). Burning must be repeated periodically to maintain forage benefits to grazers.

Fire alters species diversity primarily by altering habitat structural characteristics and plant species composition. Bird species are attracted to habitats as much for canopy and understory structure as for food (Rottenberry and Wiens 1980). Burning tends to open canopies and simplify understory structure. Prescribed burns, therefore, can be used to attract bird species that prefer habitats that are less complex structurally. Burning is rarely an avian-management panacea because bird species differ in response to burn-modified habitats. Viewing managers who want to highlight targeted bird species must know the unique habitat requirements for that species to select a burn treatment providing the highest viewing dividends. For example, mountain plovers respond positively to grassland burns because they prefer simple, open canopy structures. Compared to unburned sites, burning can double mountain plover numbers (Svingen and Giesen 1999). Burning reduced grassland use by northern harriers in Texas, but increased use by American kestrels. Harriers are ambush hunters that hunt by flying low over vegetation then dropping suddenly on unsuspecting prey (Schipper et al. 1975). Vegetation cover favors this hunting strategy. Kestrels, on the other hand, prefer to hunt in open habitats, and their hunting success decreases with increasing vegetation height and complexity (Smallwood 1987; Toland 1987). If kestrels were the preferred species for viewing on a given site, burning would be an effective practice (Chavez-Ramirez and Prieto 1994).

Tree and brush removal. Mechanical removal of trees and shrubs (logging, cutting, chaining, etc.) can produce effects similar to prescribed fire because both habitat canopy and forage are altered when canopies are opened and understories are cleared. As with fire, wildlife species differ in their responses to tree and shrub removal. Clear-cut logging of boreal forests often favors deer, elk, and small mammals because logging encourages secondary succession of important forages (Potvin et al. 1999; St-Louis et al. 2000). On the other hand, logging adversely affects moose, pine marten, and spruce grouse because they prefer closed forests, and the degree of the impact increases with the size of the clear-cut. In general, a mosaic of clear-cut openings interspersed among uncut forest stands increases wildlife species diversity (Potvin et al. 1999).

Birds, too, respond to forest-cutting and shrub-removal treatments differently depending on the species. Biologists in western Arkansas compared bird responses on control areas with two cutting treatments; understory cutting only and understory plus overstory cutting. Indigo buntings, eastern wood-pewees, and white-breasted nuthatches were more abundant on plots where both understory and overstory vegetation was removed. Ovenbirds, worm-eating warblers, and Acadian flycatchers were more abundant on control plots. Tufted titmice were more abundant on plots where the understory was removed, but the canopy overstory was left intact (Rodewald and Smith 1998). These examples illustrate the importance of having clear objectives before implementing habitat-manipulation treatments. Wildlife species differ in their habitat requirements. Manipulating habitats to benefit one species inevitably penalizes others.

Grazing. Grazing can mimic some of the effects of prescribed fire, especially in grassland and shrubland habitats. Grazing removal of decadent or old-growth plant tissues stimulates regrowth of more nutritious tissues, and also may open stand canopies to increased sunlight, further stimulating plant growth. On the Isle of Rhum in Scotland, for example, grazing by cattle in winter attracted red deer to sites grazed by cattle. Cattle grazing removed standing dead foliage, exposing undergrowth to sunlight, which stimulated growth of new green forage. Red deer were attracted by the more nutritious regrowth on grazed sites. Not only did red deer concentrate on grazed sites, but ratios of calves per hind also were higher than on ungrazed sites, suggesting that grazed sites also supported higher red deer densities (Gordon 1988).

At Wind Caves National Monument, bison and pronghorn were attracted to prairie-dog colonies because prairie dogs removed standing dead biomass, stimulating plant regrowth. New forage tissues had higher concentrations of shoot nitrogen and were more digestible than similar species from ungrazed sites (Coppock et al. 1983a). Increased forage quality due to prairie-dog feeding attracted bison and pronghorn to prairie-dog colonies in preference to sites where prairie dogs were absent (Coppock et al. 1983b). Pronghorn and bison seemed particularly attracted to colony edges where plant composition and forage nutrition were greatest (Krueger 1986). Grazing by bison and pronghorn, in turn, contributed to stability of prairie-dog colonies by helping maintain close-cropped vegetation complexes preferred by prairie dogs (Coppock et al. 1983b; Cid et al. 1991). Foraging by an entire suite of grazing animals contributed ecological stability to prairie-dog communities, which in turn may have promoted a greater variety of wildlife species and increased wildlife numbers overall.

Revegetation. The potential of revegetation to attract wildlife is illustrated by unintended effects of planting attractive crops within the home ranges of wildlife. Elk, pronghorn, and geese are strongly attracted to winter wheat (Flegler et al. 1987; Torbit et al. 1993; Brelsford et al. 1998). Black bears and raccoons are attracted to ripening corn (Flemming 1983; Jonker et al. 1998). Mule deer seek out growing alfalfa (Austin et al. 1998).

Wildlife managers only recently, however, have begun to use large-scale revegetation as a tool to rehabilitate or reconstruct wildlife habitats. Biologists at the Rocky Mountain Arsenal National Wildlife Refuge have begun extensive vegetation reconstruction on sites that have been cleansed of pesticides and other toxic wastes. Disturbed sites are being reseeded to native grassland and riparian species with the ultimate objective of reintroducing wildlife species such as bison, pronghorn antelope, and swift fox that have been missing from the site for several decades.

Two substantially beneficial programs for wildlife that are seldom if ever used by wildlife-viewing managers are the Conservation Reserve Program (CRP) and mined-land-reclamation programs. Both programs have provisions to rehabilitate disturbed lands for wildlife. The CRP provides federal subsidies to private landowners to remove agricultural land from crop production. Mined-land-reclamation programs often have provisions to reclaim mine spoils so that the reconstituted spoil areas retain their wildlife values. Wildlife managers can influence both of these programs to revegetate with plant species that favor preferred wildlife species, and by participating on revegetation planning teams, can include wildlife-viewing objectives with other wildlife-management objectives (Dunn et al. 1993; Helm 1994; Carmichael 1997; Olson et al. 2000).

Manipulating Populations

Wildlife populations are characterized by their overall numbers, trajectory (increasing, decreasing, or stable), and sex and age structure or composition (proportions of males and females and the number of individuals within each age class in the population). Population size, trajectory, and sex and age structure depend upon the relationships between birth rates, death rates, and rates at which individuals emigrate and immigrate populations.

These characteristics of wildlife populations are important to wildlife viewing because they influence the viewing experience. Animal numbers affect the likelihood that they will be observed on any given visit. Elk are more likely to be observed in Rocky Mountain National Park than mule deer primarily because they are considerably more numerous. Physical and behavioral characteristics associated with sex, such as coloration, size, antlers, infant care etc., all influence viewing attraction and enjoyment. Opportunities to view natural birth and death sagas additionally contribute to attractiveness of viewing sites. To a degree, wildlife-viewing managers can actively manipulate population characteristics to enhance or sculpt wildlife populations to accomplish viewing objectives for specific wildlife populations or viewing areas.

Establishing or Augmenting Populations. The history of elk in Rocky Mountain National Park illustrates a dramatic wildlife-introduction success story with immense wildlife-viewing benefits. Elk within and around Rocky Mountain National Park had been hunted to extinction by 1900. Between 1913 and 1914, forty-nine elk were captured from Yellowstone National Park and translocated to Rocky Mountain National Park. Following release, elk numbers expanded rapidly to around six hundred by 1942, and continued to increase until today perhaps as many as three thousand elk reside in and around the park (Hess 1993). Visitor numbers grew concurrent with increasing elk numbers partly because elk were a major tourist attraction. On any given day in September and October, the park is crowded with visitors who line the roadsides to see and hear bull elk competing for harems.

Wildlife introductions such as those of elk into Rocky Mountain National Park and, more recently, wolves into Yellowstone National Park usually are conducted at large scales to conserve populations and metapopulations at landscape scales. Reintroductions at smaller scales may be as useful or more so to wildlife-viewing management. Recently, seventy acres of campus open space were reserved for burrowing owls at Mission College in Santa Clara, California. The site was historic owl habitat but development had all but eliminated nesting populations. Artificial nesting burrows were constructed to attract breeding pairs of owls, and vegetation was mowed frequently to simulate defoliated prairie dog colonies. Currently, up to twelve nesting pairs of owls reside on the reserve. New owl breeding pairs were attracted and owls have habituated to tolerate human activity around and near artificial nesting burrows. In one instance, a pair of burrowing owls established a successful nest immediately adjacent to a tennis court, which presumably reduced the vulnerability of owls to predators while unexpectedly providing people with extraordinary viewing opportunities (Holmes 1998).

Altering Food or Prey Abundance. Several studies have demonstrated that increasing food abundance can increase animal numbers both by attracting transient and immigrant animals and by increasing reproductive and survival rates of residents (Boutin 1990). Prior to the closure of dumps in Yellowstone National Park, from 50 to 75 percent of Yellowstone grizzlies obtained part of their diet from

Burrowing owlets in prairie dog burrow on Pawnee National Grassland. Wildlife population introductions can establish new populations or augment existing ones at either large or small scales. Burrowing owl populations have been established by introducing them into existing prairie dog colonies or by creating artificial burrows and mowing vegetation to simulate prairie dog colonies. (Photo by B. Gill)

garbage. Following the closure of Yellowstone dumps in the early 1970s, average litter size declined by 17 percent (Stringham 1983). Nearly 80 percent of this decline was associated with loss of supplemental dump food. The reverse effect was observed in kit fox populations at the Naval Petroleum Reserves in California after they were provided with food supplements. Feeding increased both reproductive and survival rates, which presumably increased recruitment rates and fox numbers overall (Warrick et al. 1999). These results suggest that management that increases food supplies to entire populations may result in two- to three-fold increases in animal numbers (Boutin 1990).

Urban open spaces are particularly attractive sites to enhance wildlife populations for viewing. Urban residents derive considerable viewing enjoyment from wildlife in urban open spaces (Gilbert 1982; Schauman et al. 1987; Simcox and Zube 1989). Moreover, urban open space that supports varied and abundant wildlife may elevate property values of surrounding residences (King et al.1991). Wildlife can be attracted to urban open space by designing habitats that encourage establishment of new populations (Laurie 1979; DeGraaf 1987; Rodiek 1987). Predator populations, for example, can be established and increased in response to habitat landscaping that increases prey species. Red foxes are attracted to sites that support small mammals and ground-nesting birds, while raccoons are attracted to sites that support ground-nesting birds, fishes, crustaceans, amphibians, and reptiles. Urban open spaces landscaped to provide wetland and riparian habitats and mid to tall grasslands are excellent habitats for prey of both red foxes and raccoons (Rosatte et al. 1991; Adkins and Stott 1998).

Manipulating Population Birth and Death Rates. Birth and death rates are the fuel and brakes of animal population dynamics. If birth and survival rates are increasing, populations increase and vice versa. Populations are stabilized when births equal deaths. Conservation biologists and game biologists have developed an array of practices and techniques to manipulate wildlife birth and death rates to increase, stabilize, or decrease population growth rates.

Artificial nesting structures have been used to establish and augment a variety of bird populations in habitats where they previously were absent, rare, or declining (Twedt and Henne-Kern 2001). Artificial nest structures that preclude predators effectively increased hatching success and survival of ducks and geese (Doty 1979). Artificial nest cavities increased nesting success and augmented local populations of the federally endangered red-cockaded woodpecker (Copeyon 1990). Various species of owls and raptors readily use artificial nesting structures (Smith and Belthoff 2001). Structures not only aid in establishing local populations, they also contribute to increased nesting success, juvenile survival, and serve as metapopulation dispersal sources (Hamerstrom et al. 1973; Postupalsky 1978; Snyder 1978; Bortolotti 1994; Gehlbach 1994; Johnson 1994; Moller 1994; Petty et al. 1994). Construction and careful placement of blinds at artificial nests sites could attract wildlife to selected sites and provide wildlife-viewing opportunities for species otherwise rarely seen.

Artificial den sites have been used infrequently to establish and augment mammal populations. However, urban red foxes select road culverts, drainpipes, tree-root cavities, and various other natural and manufactured cavities for natal dens. If young are reared successfully at a natal den site, the den may be used in successive years. Red foxes tolerate considerable human activity near den sites if they do not regard the activity as disruptive or threatening. These observations suggest that construction of artificial dens might be a useful management tool to manipulate red fox distribution and increase fox viewing opportunities.

Habitat manipulation can be an effective way to reduce predation on young animals. In Idaho, ratios of pronghorn fawns per one hundred does was directly related to vegetation cover, particularly during the birthing period, suggesting that fawn survival increased as vegetation increased. Management practices, such as manipulating time and intensity of livestock grazing, increase both the height and density of birthing and fawning cover and decrease vulnerability of neonatal pronghorns to predators (Autenrieth 1982).

Waterfowl refuges commonly are grazed by domestic livestock because refuge managers believe grazing removes old, rank vegetation and stimulates regrowth valued by ducks and other waterfowl as nesting habitat. However studies of grazing effects on duck-nesting ecology at the Monte Vista National Wildlife Refuge revealed that grazing had long-lasting and negative impacts on duck-nesting density and nesting success. Grazing cut nest densities by nearly half compared to ungrazed sites and nesting success at grazed sites declined by nearly 15 percent. Ducks prefer dense and structually complex habitats for nesting cover. Any management activity that reduces vegetation density and the complexity of canopy structure reduces nest numbers and hatching success and may increase the frequency of nest predation (Sugden and Beyersbergen 1987; Gilbert et al. 1996).

Logging forested habitats likewise can decrease loss of young animals to predators. Tree-crushing machines were used on the

Kenai National Moose Range in the mid-1970s to increase productivity of moose populations. On sites where trees were not crushed, black bears killed from 40 to 42 percent of radio-collared moose calves. In contrast, no radio-collared calves were killed by bears in forests where tree-crushing had occurred. Mature Kenai forests supported meager understory vegetation, providing moose calves with little effective cover. Tree crushing removed the canopy overstory, allowing increased light to penetrate and stimulate vigorous understory sprouting and plant growth. Increased understory vegetation provided more and better cover for neonatal moose calves, reducing predation from black bears (Schwartz and Franzmann 1983). In general, management practices that promote vegetation density and structural complexity also promote increased production and survival of newborns.

Often human-induced mortality is the most significant cause of wildlife deaths. Simply closing areas to hunting and trapping can significantly decrease overall mortality rates, resulting in stable or increasing populations. The North Carolina Wildlife Resources Commission, for example, established several black-bear sanctuaries in 1971 to protect bears from hunting and to serve as dispersal sources for adjacent hunted areas. Recent studies have compared survival rates of bears using sanctuaries with those of nearby hunted populations. Most of the observed bear mortality was attributed to hunter kills (legal and poaching). Mortality rates of adult males and subadults using sanctuaries were not different from those occupying hunted areas, probably because sanctuaries were too small to encompass entire home ranges of adult males. Subadults traveled widely on and off refuges because they had not yet established home ranges. Several female home ranges, in contrast, were completely encompassed by refuge boundaries. Consequently, females which spent all or most of their time within sanctuary boundaries experienced significantly lower mortality rates than females living in areas open to hunting (Powell et al. 1996; Beringer et al. 1998). An added advantage to protection from hunting is that animals in unhunted situations generally habituate much more readily to people.

Wildlife-viewing managers often confront problems of too many animals rather than too few (Diamond 1992). In these circumstances, managers may choose to manipulate population mortality rates. Numbers of both white-tailed deer and elk have increased dramatically over the past six decades. As deer and elk numbers increased, problems associated with intensive foraging have increased correspondingly. Intense forage removals by grazing animals can negatively impact other desired plant and animal

Red foxes have successfully colonized urban habitats throughout North America and Europe, providing countless hours if urban wildlife viewing. Red fox populations can be encouraged by developing urban open space to attract and foster abundant fox prey species such as rodents and ground-nesting birds. (Photo by B. Gill)

species. Intensive browsing of willows by elk in Rocky Mountain National Park, for instance, is suspected to have contributed to the demise of beavers (Hess 1993; Singer et al. 2000).

When natural areas are expansive, introductions of large predators may reduce and stabilize deer and elk numbers without additional human intervention. Additionally, the introductions may provide spinoff ecological and wildlife-viewing benefits. For example, wolves were reintroduced into Yellowstone National Park beginning in 1994 partly to obviate the necessity for periodic elk culling by Park Service personnel. Although it is premature to draw conclusions about long-term ecological consequences, several short-term outcomes are encouraging. First, elk have become the preferred and primary wolf prey species (Mech et al. 2001).

Second, as wolf numbers and predation intensity have increased, elk predator vigilance also has increased. This has had at least two subsequent consequences. It has created what Laundré et al. (2001) called a "landscape of fear" wherein the time elk spend watching for predators increased and the time they spend foraging decreased. Over the long haul this may amplify elk mortality rates via decreased physical condition. Additionally, the "landscape of fear" has altered elk distribution patterns as they seek out habitats where they are less vulnerable to predation. Three benefits are emerging. Elk use of heavily browsed stands of aspen and riparian shrubs is decreasing, allowing the browsed plants an opportunity to recover (Ripple et al. 2001); and elk use of open grasslands is increasing, making them more visible to tourists.

Third, as elk shift their distribution to more open and visible habitats, wolves should begin to hunt these areas more frequently. Consequently, they also will become more visible to roadside tourists. Already, tourists have responded to the wolf-viewing potential. Following the wolf reintroductions, Yellowstone National Park has experienced a surge in visits from tourists hoping to see wolves once again ranging freely after their prolonged absence.

Introductions of large predators are impractical for smaller natural areas. Culling (e.g., shooting of individuals) has been used as the tool of choice to control numbers of deer, elk, and other large grazing animals in natural areas where biological control by large predators is not feasible. However, culling is controversial even when potential ecological benefits can be demonstrated, in part because reductions historically have aimed to alleviate wildlife problems by lowering the density of entire populations of wildlife (Chase 1986; Porter et al. 1994; Porter 1997). Recent studies of deer suggest that selective culling of individuals might effectively reduce local deer numbers l and resolve insular ecological problems without the need to lower overall population density (McNulty et al. 1997).

Culling can provide coincident wildlife-viewing opportunities. Barnes (1967) used culled elk carcasses to attract and study black bears in Yellowstone National Park. Carcasses attracted black bears, grizzly bears, coyotes, ravens, magpies, and other scavengers and provided opportunities to study inter- and intra-specific feeding behaviors of multiple scavenging species. These results suggest that culling can be used not only to promote local diversity of plants

Coyote scavenging elk carcass. Culling of overabundant wildlife is a useful tool not only to manage animal numbers, but also to attract carnivorous or scavenging wildlife to carcasses, providing wildlife-viewing opportunities as a bonus. (Photo by B. Gill)

and animals, but also to provide viewing opportunities of scavenging guilds attracted by culled carcasses. If culling operations are focused intense, and of short duration; animal wariness may increase only minimally (Bend et al. 1999).

Wildlife contraception is emerging as an alternative to culling to manipulate wildlife numbers (Turner et al. 2001; Waddell et al. 2001). In controlled experiments, contraceptives have successfully, but temporarily, sterilized individuals of a variety of wildlife species, ranging from squirrels to elephants (McIvor and Schmidt 1996). It remains to be demonstrated, however, that contraceptives can effectively and economically control entire populations of wildlife over several years (Sinclair 1997).

Although not yet operational, parallel research is progressing to develop baits that can deliver contraceptives and vaccines orally to animals that spread communicable diseases (Bradley et al. 1997; Rossatte et al. 1997). These developments forecast the possibility of baits that deliver contraceptives and vaccines to simultaneously control overabundant wildlife, manage epizootics, and attract wildlife to viewing sites.

Adjusting Sex and Age Composition. In some situations, it may be desirable to manipulate the sex and age composition of wildlife populations. In antlered species, for example, a manager might want to encourage numerous large-antlered mature males for their viewing and photographic values.

Biologists at the Rocky Mountain Arsenal National Wildlife Refuge selectively culled does, young bucks, and bucks with poor antler conformation. Culling lowered and then stabilized overall deer numbers to prevent overpopulation and resulted in more bucks than does in the population, with a higher percentage of mature, large-antlered bucks than any other site in Colorado and perhaps in the nation.

Currently, it is technologically possible to alter sex ratios of animals *in vitro* and *in utero*. Animal-production researchers have successfully manipulated the sex of livestock litters by: (1) identifying sex of sperm or embryos and selecting for male or female sperm for *in vitro* for fertilization; or (2) identifying the sex of embryos and removing embryos of undesired sexes (Brand 1992; Reed 1993). These techniques are being seriously considered as management tools for the preservation and restoration of endangered species (Belden 1988). Whether they will prove to be economical or ethically acceptable to wildlife-viewing management remains to be seen.

Manipulating Wildlife Behavior

Behavioral manipulation of wildlife is of interest to wildlife-viewing managers for at least three primary reasons; (1) to allow viewers to approach wildlife more closely; (2) to allow viewers to view normal behaviors of undisturbed wildlife; and (3) to minimize viewing stresses and disturbances to wildlife. Three approaches to behavior manipulation—imprinting, foster-rearing, and habituation—have been used in various circumstances to allow researchers to make close observations of wildlife behavior with minimal observer disturbance. Seldom, however, have these approaches been used as wildlife-viewing management tools to contribute novel viewing opportunities and experiences.

Imprinting. Imprinting is a sudden and irreversible socialization process by which young animals learn to recognize parents and kin (Lorenz 1937). Imprinting young wildlife on humans rather than natural parents can be accomplished in at least two ways. Bird imprinting is accomplished by extracting fertilized eggs from nests and artificially incubating them. Immediately upon hatching, young birds receive all food and care from surrogate human "parents." Young birds imprint the human surrogates as natural parents and relate to them as kin. Mammal imprinting is similar except that newborns are removed from dams between twenty-four and forty-eight hours after birth to allow for one or more nursing bouts to assure adequate colostrum intake, but before newborns are irreversibly imprinted upon natural parents. Surrogate human parents subsequently bottle-feed and care for the newborns until they imprint on humans.

Imprinted animals are invaluable to wildlife research and potentially useful to wildlife education and nature instruction. Wildlife educators, for example, could release imprinted wildlife into selected habitats and allow animals to demonstrate intricacies of diet and habitat selection, scent marking, grooming, and social behaviors, etc. while instructors lectured on their adaptive significance. Imprinted animals also are useful to demonstrate unique anatomical characteristics of each species and explain their evolutionary life-history significance.

As early as 1948, hand-reared, human-imprinted white-tailed deer were used to describe and evaluate important deer habitat and forages in the Missouri Ozarks. Imprinting and fosterrearing tamed the deer to the extent that researchers could closely observe (from between 6 inches and 3 feet) the dynamics of forage selection. Details of diet selection were observed that otherwise would have gone unnoticed at greater distances (Dunkeson 1955). Subsequently, the "tame animal" foraging technique has expanded to a variety of other mammals, including mule deer (Bartmann et al. 1982), elk (Baker and Hobbs 1982), pronghorn (Schwartz et al. 1976), moose (Regelin et al. 1987), bighorn sheep and mountain goats (Dailey et al. 1984), caribou (White and Trudell 1980), and musk oxen (Frisby et al. 1984).

Imprinted birds have been used much less frequently as research tools. In a notable exception, researchers in Pennsylvania artificially hatched and hand reared wild turkeys to evaluate brooding habitats (Healy and Goetz 1974). Poults were imprinted on human handlers and trained to respond to simulated adult turkey calls so they would return to handlers after release. When released into natural habitats, hand-reared turkeys appeared to behave the same as wild turkeys in all respects except that they tolerated people observing feeding and other behaviors at close range unaltered by the observers' presence. Among other insights, researchers observed details of behavioral reactions to avian predators that would not otherwise have been possible.

Foster rearing. Foster rearing has been used most often to re-establish or augment populations of declining or endangered wildlife or to nurture orphaned or injured wildlife until they can be released back into the wild. In contrast to the methods used in imprinting, foster parents usually take care to prevent imprinting. An important goal of foster rearing is to release animals back to the wild that are not more susceptible to human predation or inclined to remain near people, increasing human-wildlife conflicts.

However, foster rearing can be combined with imprinting in certain circumstances and create viewing opportunities. If imprinted and foster-reared animals are trained to live in the wild independent of human assistance, they can be released just as other foster-reared animals are. Limited anecdotal evidence suggests that when imprinted, foster-reared animals are released into the wild, they attract and are attracted by other individuals of the same species, especially during breeding. Foster-reared animals are intensely habituated to humans, and once they associate with other wild kin, wild associates also tend to habituate. If foster-reared animals are radio-collared before being released into the wild, they and associated kin can be relocated more or less at will to observe seasonal behaviors, habitat use and other events of viewing interest (DeBruyn 1999).

Wallmo (Wallmo and Neff 1970) may have been one of the first to recognize the potential of foster-reared wildlife to nutritional research when he observed that the feeding behavior of foster-reared pronghorn antelope was similar to that of wild pronghorn with which they associated in a large enclosure. Both wild and foster-reared individuals selected similar diets, used similar habitat patches, and otherwise appeared to behave comparably. Because of these similarities, Wallmo concluded that foster-reared individuals could be used to quantify diet selection of wild pronghorns in much greater detail because foster-reared individuals could be followed and closely observed whereas wild pronghorns could not.

Outdoor photographers and naturalists, however, were among the first to report the viewing benefits of foster-reared animals (Crisler 1958). Chris and Lois Crisler, who worked on consignment as wildlife photographers for Walt Disney films, realized early that foster-reared animals had entertainment potential. On one particular assignment they traveled to the Brooks Range in Alaska to film migrating caribou and associated wildlife. They wanted to photograph wolves hunting caribou, but did not want to rely entirely on serendipity to provide photo opportunities. They obtained wolf pups, captured from natal dens, and foster reared them. They taught the wolves to hunt and kill wild prey, including caribou. Intitially the foster-reared wolves were locked in holding pens each night to assure they did not escape into the wild. Eventually, the holding pen door was left open full time, allowing the wolves come and go at will. As the foster-reared wolves learned to hunt and kill wild prey, they contacted wild wolves and eventually were assimilated into wild wolf packs. Despite ranging free with wild wolves, the foster-reared wolves did not lose their attachment to humans. They returned home periodically, accompanied by wild wolves. Uncharacteristically, the wild wolves remained with the foster-reared wolves even in close proximity to the humans and allowed the Crislers to observe and photograph them much closer than would have been possible absent the influence of the foster-reared wolves (Crisler 1958). These and other anecdotal examples hint at the potential for foster-rearing wildlife and releasing them into the wild to create viewing and educational opportunities and to hasten the habituation of wild individuals of the same species.

Habituating wild animals. Habituation of wildlife has been defined as a "waning of a response to a repeated stimulus that is not associated with either a positive or negative reward"(Knight and Temple 1995). Wild animals generally avoid close association with humans because they regard people as predators. Nonetheless, when wildlife experience prolonged contact with people without negative consequences, they can habituate (Lord et al. 2001). Deliberate habituation of wildlife to tolerate close human presence, though new to wildlife-viewing management, has an extended history as a wildlife research tool.

Bighorn sheep ewe and lamb resting in alpine meadow. With patience and persistence, wildlife can be habituated to tolerate people in close proximity without being disturbed or distressed, providing intimate opportunities to view natural wildlife behaviors. (Photo by B. Gill)

Geist (1971) habituated free-ranging wild bighorn sheep in Banff National Park, Alberta, Canada, by visiting them daily until they approached to accept salt rewards offered by hand. Each visit would begin with the observer first approaching the sheep until the sheep recognized the observer as a source of positive rewards. As bighorns associated observers with salt rewards, the sheep initiated encounters in search of salt. So long as the observer fed salt, bighorns would allow close human association. Gradually salt rewards were reduced so that bighorns would behave normally, but still permit observers to intermingle without distracting or disturbing normal sheep behaviors. Once human presence was accepted among free-ranging bighorn social groups, researchers were afforded remarkably detailed observations of behaviors of individual and social groups of bighorns.

Subsequent researchers used similar habituation techniques to study the behavior of free-ranging mountain gorillas and chimpanzees (Fossey 1986; Goodall 1986). Virtually all aspects of the social ecology of these primates were observed, recorded, and reported. In addition, photographers were included in study teams and visually recorded a variety of wild-primate behaviors for television programming, combining wildlife research with a kind of virtual wildlife-viewing experience that attracted sizeable viewing audiences.

Other researchers were able to habituate free-ranging wild bears and wolves so that mothers with dependent young tolerated close observation by researchers. Bear researchers were able to see and record grooming, nursing, and instructional behaviors of free-roaming bears without harm or disruption to either bears or observers (Rogers and Wilker 1990; DeBruyn 1999). Mech (1988) was even able to habituate wild wolves to permit him to accompany them while he rode an off-road vehicle. The mobility of the off-road vehicle permitted him to accompany wolves as they hunted and eventually killed musk oxen.

Even though researchers who studied habituated animals occasionally allowed associates and other observers to accompany them during their wildlife studies, Alaskan bear biologists apparently were the first to extend wildlife habituation to promote public wildlife viewing. Free-ranging brown bears inhabiting the McNeil River State Game Sanctuary were slowly and deliberately habituated to tolerate the close presence of wildlife viewers apparently without disturbing normal bear behaviors and with minimal safety risks to viewers (Aumiller and Matt 1994) (See also Chapter 16.)

At least three general circumstances facilitate and expedite wildlife habituation: (1) initial food provisioning; (2) human reinforcing behaviors that are consistent, repetitious, and neutral or positive; and (3) recognition and comprehension of and appropriate responses to wildlife behaviors. Food is the attractant that draws wildlife to people or retains wildlife in a locale in the presence of people. Food can be naturally provided, as it is for bears at the McNeil River State Game Sanctuary, or it can be provided by the habituator. Once wildlife are initially habituated, food rewards should be withdrawn gradually to encourage normal wildlife behaviors in the presence of human observers.

Trust or predictability is essential to habituation because wildlife avoid frightening or chaotic situations. Consistent positive or neutral human behaviors convey predictability, and predictable behaviors that wildlife associate with positive or neutral rewards promote habituation. Loud noises, unpredictable rapid body movements, or darting behavior may frighten wildlife and cause flight or even defensive attacks. In any case, these behaviors retard or inhibit habituation. Wildlife, not humans, should determine what, when, where, how, and how often viewing events will occur.

Wildlife exhibit characteristic behaviors to disturbing situations or novel stimuli. Swift fox, for example, increase yawing rates when mildly disturbed. Behaviors that communicate alarm, stress, annoyance, and dominance must be identified and correctly interpreted so habituators and observers can reinforce habituation with appropriate responses (Geist 1971; Fossey 1986; Goodall 1986; Rogers and Wilker 1990; Aumiller and Matt 1994; Mech 1997; DeBruyn 1999).

Adaptive Resource Management

The foregoing discussion of how attributes of individual animals, wildlife populations, and habitats might be manipulated to provide viewing benefits admittedly has been optimistic. Clearly, outcomes can be negative as well as positive for both wildlife and people. Supplemental feeding of bears, for example, can result in human injuries, increase attractive nuisances, and cause property damage (Gunther 1994). Yet the examples from Minnesota and Alaska suggest that, with careful planning and management, outcomes can be predictable and overwhelmingly positive.

Short-term benefits might ultimately be offset by long-term problems. Manipulations that concentrate wildlife might augment viewing opportunities in the short term but destroy habitats or spread debilitating diseases in the long term. Manipulating wildlife habitats, populations, and behaviors is such a new concept to wildlife-viewing management that many of the consequences, short or long term, cannot be anticipated accurately. Uncertainty in science and application requires an adaptive approach that treats applications as experiments that are not expanded until rigorous evaluations demonstrate efficacy and obvious benefits. This "learning by doing" approach has been called adaptive resource management (Lancia et al. 1996). Active wildlife-viewing management is an emerging professional discipline without a robust scientific base to demonstrate efficacy, beneficence, safety, and public acceptability. Faced with these circumstances, prudence urges an adaptive, evaluative paradigm for active wildlife-viewing management before management practices are expanded into general practice.

Conclusion

Wildlife-viewing management is a slowly emerging wildlife management discipline. It is slow to emerge partly because it is under-funded, under-staffed, and under-appreciated by wildlife agencies. It is also slow to emerge because it is held to an artificial and unrealistic standard of naturalness that discourages overt manipulations of wildlife and condemns viewing activities that disturb wildlife. The naturalness standard is artificial because it stems from the subjective values of wildlife professionals. Wildlife managers themselves have manipulated and disturbed wildlife for years by killing, capturing, collaring, prodding, and poking them, and invading their lives, justifying the disturbances because they were "in the interests of science." The standard is unrealistic because human influences are so pervasive that virtually everything people do perturbs and disturbs wildlife.

The challenge for wildlife-viewing management is to manipulate wildlife in ways that provide fulfilling viewing experiences while managing and mitigating disturbances.

This essay attempts three broad accomplishments. It attempts to open the wildlife profession to a new discipline of an active wildlife-viewing management that manipulates wildlife and habitats to provide benefits to people and wildlife. It attempts to open the public to wildlife-viewing opportunities hitherto unavailable on a large scale. It attempts to open the eyes of wildlife professionals to wildlife-viewing possibilities that foster human enrichment and wildlife conservation.

Summary Points

• Wildlife-viewing management is anemic because it is passive rather than active, permissive rather than manipulative. Active wildlife-viewing management would seek to manipulate wildlife distribution, abundance, diversity, and behavior to provide wildlife-viewing opportunities and benefits.

• Standard tools of game management and conservation biology can be adapted and applied successfully to wildlife viewing management programs. These include supplementing food and water, prescribed fire, selective removal of trees and brush, grazing, revegetation, wildlife species introductions or translocations, artificial nest and den construction, culling, contraception, and habituation.

• Imprinting, foster rearing, or habituating wildlife to tolerate close human presence is perhaps one of the most exciting and promising tools to provide wildlife-viewing opportunities without stressing or disturbing wildlife.

• Wildlife-viewing management will be as successful as it is imaginative. People are attracted to and seek to enjoy wildlife for diverse reasons. The challenge to wildlife-viewing managers is to free their imaginations from their biases to develop visionary viewing opportunities and benefits.

• Several skills and practices that wildlife-viewing managers will require to generate wildlife-viewing opportunities can be borrowed from sister disciplines like game and nongame management and conservation biology. Nonetheless, wildlife viewing seeks to provide people with exceptional benefits. Sooner or later wildlife-viewing managers must develop unique research foundations and management practices to evolve as a distinct discipline and engage the singular challenges of wildlife viewing. Better sooner than later.

- Active wildlife-viewing management is an emerging professional discipline without a robust scientific base to demonstrate efficacy, beneficence, safety, and public acceptability. Adaptive resource management is a recent paradigm that, in the face of considerable uncertainty, recommends management experiments instead of management prescriptions. Like scientific experiments, management experiments should be carefully designed, replicated, and evaluated to promote "learning by doing." Adaptive resource management should be a fundamental paradigm for active wildlife-viewing management.

Literature Cited

Adkins, C. A., and P. Stott (1998). "Home ranges, movements and habitat associations of red foxes (*Vulpes vulpes*) in suburban Toronto, Ontario, Canada." *Journal of Zoology, London*, 244, 335-46.

Aumiller, L. D., and C. A. Matt (1994). "Management of McNeil River State Game Sanctuary for viewing brown bears." *International Conference for Bear Research and Management*, 9, 51-61.

Austin, D. D., P. J. Urness, and D. Darin (1998). Alfalfa hay crop loss due to mule deer depredation. *Journal of Range Management*, 51, 29-31.

Autenrieth, R. E. (1982). "Pronghorn fawn habitat use and vulnerability to predation." *Proceedings of the Pronghorn Antelope Workshop*, 10, 121-31.

Baker, D. L., and N. T. Hobbs (1982). "Composition and quality of elk summer diets in Colorado." *Journal of Wildlife Management*, 46, 694-703.

Baker, D. L., and N. T. Hobbs (1985). "Emergency feeding of mule deer during winter: tests of a supplemental ration." *Journal of Wildlife Management*, 49, 934-42.

Barnes, V. G. Jr. (1967). Activities of black bears in Yellowstone National Park. M.S. Thesis. Fort Collins, CO: Colorado State University.

Bartmann, R. M., A. W. Alldredge, and P. H. Neil (1982). "Evaluation of food choices by tame mule deer." *Journal of Wildlife Management*, 46, 807-12.

Becklund, J. (1999). *Summers with Bears*. New York: Hyerion.

Beetle, A. A. (1979). "Jackson Hole elk herd: a summary after 25 years of study." In M. S. Boyce and L. D. Hayden-Wing (Eds.), *North American Elk: Ecology, Behavior and Management*. Laramie, WY: University of Wyoming.

Belden, R. C. (1988). "Florida panther reintroduction feasibility study." *Mountain Lion Workshop*, 3, 52.

Bender, L. C., D. E. Beyer, Jr., and J. B Haufler (1999). "Effects of short-duration, high-intensity hunting on elk wariness in Michigan." *Wildlife Society Bulletin*, 27, 441-45.

Beringer, J., S. G. Seibert, S. Reagan, A. J. Brody, M. R. Pelton, and L. D. Vangilder (1998). "The influence of a small sanctuary on survival rates of black bears in North Carolina." *Journal of Wildlife Management*, 52, 727-34.

Bortolotti, G. R. (1994). "Effect of nest-box size on nest-site preference and reproduction in American kestrels." *Journal of Raptor Research*, 28, 127-33.

Boutin, S. (1990). "Food supplementation experiments with terrestrial vertebrates: patterns, problems, and the future." *Canadian Journal of Zoology*, 68, 203-20.

Boyce, M. S. (1989). *The Jackson Elk Herd: Intensive Wildlife Management in North America*. New York: Cambridge University Press.

Boyd, R. J. (1978). "American elk." In J. L. Schmidt and D. L. Gilbert (Eds.), *Big Game of North America: Ecology and Management*. Harrisburg, PA: Stackpole Books.

Bradley, M. P., L. A. Hinds, and P. H. Bird (1997). "A bait-delivered immunocontraceptive vaccine for the European red fox (*Vuples vulpes*) by the year 2002?" *Reproduction, Fertility, and Development*, 9, 111-16.

Brand, A. (1992). "Animals in biotechnology—state of the art." In J. W. James (Project Manager), *Biotechnical Innovations in Animal Productivity*. Oxford, England: Butterworth-Heinemann.

Brelsford, M. J., J. K. Peek, and G. A. Murray (1998). "Effects of grazing by wapiti on winter wheat in northern Idaho." *Wildlife Society Bulletin*, 26, 203-8.

Brittingham, M. C. and S. A. Temple (1988). "Avian disease and winter bird feeding." *The Passenger Pigeon*, 50, 195-208.

Caputo, R. (2002). "Motherbearman." *National Geographic* 201 (3), 88-101.

Carmichael, D. B., Jr. (1997). "The Conservation Reserve Program and wildlife habitat in the southeastern United States." *Wildlife Society Bulletin*, 25, 773-75.

Carpenter, L. H., and O. C. Wallmo 1982. "Habitat evaluation and management." In O. C. Walmo (Ed.), *Mule and Black-tailed Deer of North America*. Lincoln, NB: University of Nebraska Press.

Chase, A. (1987). *Playing God in Yellowstone*. Boston, MA: The Atlantic Monthly Press.

Chavez-Ramirez, F., and F. G. Prieto (1994). "Effects of prescribed fires on habitat use by wintering raptors on a Texas barrier island grassland." *Journal of Raptor Research*, 28, 262-65.

Cid, M. S., J. K. Detling, A. D. Whicker, and M. A. Brizuela (1991). "Vegetational responses of a mixed-grass prairie site following exclusion of prairie dogs and bison." *Journal of Range Management*, 44, 100-105.

Cieminski, K. L., and L. D. Flake (1997). "Mule deer and pronghorn use of wastewater ponds in a cold desert." *Great Basin Naturalist*, 57, 327-37.

Copeyon, C. E. (1990). "A technique for constructing cavities for the red-cockaded woodpecker." *Wildlife Society Bulletin*, 18, 303-11.

Coppock, D. L., J. E. Ellis, J. K. Detling, and M. I. Dyer (1983a). "Plant-herbivore interactions in a North American mixed-grass prairie. 1. Effects of black-tailed prairie dogs on intraseasonal above-ground biomass and nutrient dynamics." *Oecologia*, 56, 1-9.

Coppock, D. L., J. E. Ellis, J. K. Detling, and M. I. Dyer (1983b). "Plant-herbivore interactions in a North American mixed-grass prairie. 2. Responses of bison to modification of vegetation by prairie dogs." *Oecologia*, 56, 10-15.

Crisler, L. (1958). *Arctic Wild*. New York: Harper and Brothers, Publishers.

Dailey, T. V., N. T. Hobbs, and T. N. Woodard (1984). "Experimental comparisons of diet selection by mountain goats and mountain sheep in Colorado." *Journal of Wildlife Management*, 48, 799-806.

DeBruyn, T. D. (1999). *Walking with Bears*. New York: The Lyons Press.

DeGraaf, R. M. (1987). "Urban wildlife habitat research—application to landscape design." In L. W. Adams and D. L. Leedy (Eds.), *Integrating Man and Nature in the Metropolitan Environment: Proceedings of a National Symposium on Urban Wildlife*. Columbia, MD: National Institute for Urban Wildlife.

Diamond, J. (1992). "Must we shoot deer to save nature?" *Natural History*, 8/92, 2-8.

Doenier, P. B., G. D. DelGiudice, and M. R. Riggs (1997). "Effects of winter supplemental feeding on browse consumption by white-tailed deer." *Wildlife Society Bulletin*, 25, 235-43.

Doty, H. A. (1979). "Duck nest structure evaluations in prairie wetlands." *Journal of Wildlife Management*, 43, 976-79.

Dunkeson, R. L. (1955). "Deer range appraisal for the Missouri Ozarks." *Journal of Wildlife Management*, 19, 358-64.

Dunn, C. P., F. Stearns, G. R. Guntenspergen, and D. M. Shapre (1993). "Ecological benefits of the Conservation Reserve Program." *Conservation Biology*, 7, 132-39.

Flegler, E. J., Jr., H. H. Prince, and W. C. Johnson (1987). "Effects of grazing by Canada geese on winter wheat yield." *Wildlife Society Bulletin*, 15, 402-5.

Flemming, A. A. (1983). "Preferential feeding by raccoons on maize." *Georgia Journal of Science*, 41, 109-14.

Fossey, D. (1986). *Gorillas in the Mist.* Boston, MA: Houghton Mifflin.

Frisby, K., R. G. White, and B. Sammons 1984. "Food conversion efficiency and growth rates of hand-reared muskox calves." *Biological Papers of the University of Alaska.* Special Report 4, 196-202.

Gehlback, F. R. (1994). "Nest-box versus natural-cavity nests of the eastern screech-owl: an exploratory study." *Journal of Raptor Research*, 28, 154-57.

Geiser, F., and G. J. Kenagy. (1993). "Dietary fats and torpor patterns in hibernating ground squirrels." *Canadian Journal of Zoology*, 71, 1182-86.

Geist, V. (1971). *Mountain Sheep: A Study in Behavior and Evolution.* Chicago, IL: University of Chicago Press.

Gilbert, D. W., D. R. Anderson, J. K. Ringelman, and M. R. Szymczak (1996). "Response of nesting ducks to habitat and management on the Monte Vista National Wildlife Refuge, Colorado." *Wildlife Monographs*, 131, 1-44.

Gilbert, F. F. (1982). "Public attitudes toward urban wildlife, a pilot study in Guelph, Ontario." *Wildlife Society Bulletin*, 10, 245-53.

Gill, R .B. and L. H. Carpenter (1985). "Winter feeding: a good idea?" *Proceedings of the Western Association of Fish and Wildlife Agencies*, 65, 57-66.

Goodall, J. (1986). *The Chimpanzees of Gombe. Patterns of Behavior.* Cambridge, MA: Harvard University Press.

Gordon, I. J. (1988). "Facilitation of red deer grazing by cattle and its impact on red deer performance." *Journal of Applied Ecology*, 25, 1-10.

Gunther, K. A. (1994). "Bear management in Yellowstone, 1960-93." *International Conference on Bear Research and Management*, 9, 549-60.

Gunther, K. A., and H. E. Hoekstra (1998). "Bear-inflicted human injuries in Yellowstone National Park, 1970-1984." *Ursus*, 10, 377-84.

Hamerstrom, F., F. N. Hamerstrom, and J. Hart (1973). "Nest boxes: an effective management tool for kestrels." *Journal of Wildlife Management*, 37, 400-403.

Healy, W. M. and E. J. Goetz (1974). "Imprinting and video-recording wild turkeys—new techniques." *Transactions of the Northeast Section of The Wildlife Society*, 31, 173-82.

Helm, D. J. (1994). "Establishment of moose browse on four growth media on a proposed mine site in southcentral Alaska." *Restoration Ecology*, 2, 164-79.

Hess, K., Jr. (1993). *Rocky Times in Rocky Mountain National Park. An Unnatural History.* Niwot, CO: University Press of Colorado.

Hobbs, N. T. and D. S. Schimel (1984). "Fire effects on nitrogen mineralization and fixation in mountain shrub and grassland communities." *Journal of Range Management,* 37, 402-5.

Hobbs, N. T., D. S. Schimel, C. E. Owensby, and D. S. Ojima (1991). "Fire and grazing in the tallgrass prairie: contingent effects of nitrogen budgets." *Ecology,* 72, 1374-82.

Holmes, B. (1998). "City planning for owls." *National Wildlife,* 36 (6), 46-53.

Jenkins, S. A., B. D. Perry, and W. G. Winkler (1988). "Ecology and epidemiology of raccoon rabies." *Review of Infectious Diseases,* 10, 620-25.

Johnson, P. N. (1994). "Selection and use of nest sites by barn owls in Norfolk, England." *Journal of Raptor Research,* 28, 149-53.

Jonker, S. A., J. A. Parkhurst, R. Field, and T. K. Fuller (1998). "Black bear depredations on agricultural commodities in Massachussetts." *Wildlife Society Bulletin,* 26, 318-24.

Keiss, R., and B. Smith (1966). "Can we feed deer?" *Colorado Outdoors,* 15 (2), 1-8.

King, D. A., J. L. White, and W. W. Shaw (1991). "Influence of urban wildlife habitats on the value of residential properties." In L. W. Adams and D. L. Leedy (Eds.), *Wildlife Conservation in Metropolitan Environments.* Columbia, MD: National Institute for Urban Wildlife.

Knight, R. L. and D. P. Anderson (1990). "Effects of supplemental feeding on an avian scavenging guild." *Wildlife Society Bulletin,* 18, 388-94.

Knight, R. L., and K. J. Gutzwiller (1995). *Wildlife and Recreationists: Coexistence through Management and Research.* Washington, DC: Island Press.

Knight, R. L. and S. A. Temple (1995). "Origin of wildlife responses to recreationists." In R .L. Knight and K. J. Gutzwiller (Eds.), *Wildlife and Recreationists: Coexistence through Management and Research.* Washington, DC: Island Press.

Krausman, P. R. (1996). *Rangeland Wildlife.* Denver, CO: The Society for Range Management.

Krueger, K. (1986). "Feeding relationships between bison, pronghorn, and prairie dogs: an experimental analysis." *Ecology,* 67, 760-70.

Lancia, R.A., C. E. Braun, M. W. Collopy, R. D. Dueser, J. G. Kie, C. J. Martinka, J. D. Nichols, T. D. Nudds, W. R. Porath, and N. G. Tilghman (1996). "ARM! For the future: adaptive resource management for the wildlife profession." *Wildlife Society Bulletin,* 24, 436-42.

Laundré, J. W., L. Hernandez, and K. B. Atkendorf (2001). "Wolves, elk, and bison: reestablishing the "landscape of fear" in Yellowstone National Park, U.S.A." *Canadian Journal of Zoology* 79, 1401-9.

Laurie, I. (1979). *Nature in Cities: The Environment in Design and Development of Urban Green Space.* New York: John Wiley and Sons.

Lord, A., J. R. Waas, J. Innes, and M. J. Whittingham (2001). "Effects of human approaches to nests of northern New Zealand dotterels." *Biological Conservation* 98, 233-40.

Lorenz, K. Z. (1937). "The companion in the birds' world." *Auk*, 54, 245-73.

McCollough, M. A., C. S. Todd, and R. B. Owen, Jr. (1994). "Supplemental feeding program for wintering bald eagles in Maine." *Wildlife Society Bulletin*, 22, 147-54.

McCorquodale, S. M., and R. F. DiGiacomo (1985). "The role of wild North American ungulates in the epidemiology of bovine brucellosis: a review." *Journal of Wildlife Diseases*, 21, 351-57.

McIvor, D. E. and R. H. Schmidt (1996). *Annotated Bibliography for Wildlife Contraception: Methods, Approaches, and Policy.* Logan, UT: Utah State University.

McNulty, S.A., W. F. Porter, N. E. Mathews, and J. A. Hill (1997). "Localized management for reducing white-tailed deer populations." *Wildlife Society Bulletin*, 25, 265-71.

Mech, L. D. (1988). *The Arctic Wolf: Ten Years with the Pack.* Stillwater, MN: Voyageur Press, Inc.

Mech, L. D., D. W. Smith, K. M. Murphy, and D. R. McNulty (2001). "Winter severity and wolf predation on a formerly wolf-free elk herd." *Journal of Wildlife Management* 65, 998-1003.

Moeller, A.P. (1994). "Facts and artefacts in nest-box studies: implications for studies of birds of prey." *Journal of Raptor Research*, 28, 143-48.

Olson, R. A., J. K. Gores, D. T. Booth, and G. E. Schuman (2000). "Suitability of shrub establishment on Wyoming mined lands reclaimed for wildlife habitat." *Western North American Naturalist*, 60, 77-92.

Petty, S. J., G. Shaw, and D. I. K. Anderson (1994). "Value of nest boxes for population studies and conservation of owls in coniferous forests in Britain." *Journal of Raptor Research*, 28, 1134-42.

Pierce, C. L. and M. J. Manfredo (1997). "A profile of North American wildlife agencies' viewing programs." *Human Dimensions of Wildlife*, 2, 27-41.

Porter, W. F. (1997). "Ignorance, arrogance, and the process of managing overabundant deer." *Wildlife Society Bulletin*, 25, 408-12.

Porter, W. F., M. A. Coffey, and J. Hadidian (1994). "In search of a litmus test: wildlife management in the U.S. national parks." *Wildlife Society Bulletin*, 22, 301-6.

Postupalsky, S. (1978). "Artificial nesting platforms for ospreys and bald eagles." In S. A. Temple (Ed.), *Endangered Birds. Management Techniques for Preserving Threatened Species*. Madison, WI: University of Wisconsin Press.

Potvin, F., R. Courtois, and L. Belanger (1999). "Short-term responses of wildlife to clear-cutting in Quebec boreal forests: multi-scale effects and management implications." *Canadian Journal of Forest Research*, 29, 1120-27.

Powell, R. A., J. W. Zimmerman, D. E. Seaman, and J. F. Gilliam (1996). "Demographic analyses of a hunted black bear population with access to a refuge." *Conservation Biology*, 10, 224-34.

Reed, K. C. (1993). "Pre-determination of progeny sex in livestock: embryo sexing by rapid biopsy and duplex PCR." In K. J. Beh (Ed.), *Animal Health and Production for the 21st Century*. Melbourne, Australia: CSIRO Publications.

Regelin, W. L., M. E. Hubbert, C. C. Schwartz, and D. J. Reed (1987). "Field test of a moose carrying capacity model." *Alces*, 23, 243-84.

Ripple, W. J., E. J. Larsen, R. A. Renkin, and D. W. Smith (2001). "Trophic cascades among wolves, elk, and aspen on Yellowstone National Park's northern range." *Biological Conservation* 102, 227-34.

Rodewald, P. G., and K. G. Smith (1998). "Short-term effects of understory and overstory management on breeding birds in Arkansas oak-hickory forests." *Journal of Wildlife Management* 62, 1411-17.

Rodiek, J. (1987). "A general approach to landscape design for wildlife habitat." In. L. W. Adams and D. L. Leedy (Eds.), *Integrating Man and Nature in the Metropolitan Environment: Proceedings of a National Syposium on Urban Wildlife*. Columbia, MD: National Institute for Urban Wildlife.

Rogers, L. L., and G. W. Wilker (1990). "How to obtain behavioral and ecological data from free-ranging, researcher-habituated black bears." *International Conference on Bear Research and Management*, 8, 321-27.

Rosatte, R. C., M. J. Power, and C. D. MacInnes (1991). "Ecology of urban skunks, raccoons, and foxes in metropolitan Toronto." In L. W. Adams and D. L. Leedy (Eds.), *Wildlife Conservation in Metropolitan Environments*. Columbia, MD: National Institute for Urban Wildlife.

Rosatte, R. C., C. D. MacInnes, R. T. Williams, and O. Williams (1997). "A proactive prevention strategy for raccoon rabies in Ontario, Canada." *Wildlife Society Bulletin*, 25, 110-16.

Rottenberry, J. T., and J. A. Wiens (1980). "Habitat structure, patchiness, and avian communities in North American steppe vegetation: a multivariate analysis." *Ecology*, 61, 1228-50.

Schauman, S., S. Penland, and M. Freeman (1987). "Public knowledge of and preferences for wildlife habitats in urban open spaces." In L. W. Adams and D. L. Leedy. (Eds.), *Integrating Man and Nature in the Metropolitan Environment: Proceedings of a National Symposium on Urban Wildlife*. Columbia, MD: National Institute for Urban Wildlife.

Schipper, W. J. A., L. S. Burma, and P. Bossenbroek (1975). "Comparative study of hunting behavior of wintering hen harriers (*Circus cyaneus*) and marsh harriers (*Circus aeruginosus*)." *Ardea*, 63, 1-29.

Schullery, P. (1992). *The Bears of Yellowstone*. Worland, WY: High Plains Publishing Co.

Schwartz, C. C., J. G. Nagy, and S. M. Kerr (1976). "Rearing and training pronghorns for ecological studies." *Joournal of Wildlife Management*, 40, 464-68.

Schwartz, C. C. and A. W. Franzmann (1983). "Effects of tree crushing on black bear predation on moose calves." *International Conference on Bear Research and Management*, 5, 40-44.

Schwartz, C. C., W. L. Regelin, and A. W. Franzmann (1985). "Suitability of a formulated ration for moose.: *Journal of Wildlife Management*, 49, 137-41.

Shaw, J. H., and T. S. Carter (1990). "Bison movements in relation to fire and seasonality." *Wildlife Society Bulletin*, 18, 426-30.

Simcox, D. E., and E. H. Zube (1989). "Public value orientations towards urban riparian landscapes." *Society and Natural Resources*, 2, 229-39.

Sinclair, A. R. E. (1997). "Fertility control of mammal pests and the conservation of endangered marsupials." *Reproduction, Fertility, and Development*, 9, 1-16.

Singer, F. J., L. C. Zeigenfuss, and D. T. Barnett (2000). "Head to head: elk, beaver, and the persistence of willow in national parks." *Wildlife Society Bulletin*, 28, 451-53.

Smallwood, J. A. (1987). "Sexual segregation by habitat in American kestrels wintering in southcentral Florida: vegetative structure and responses to differential prey availability." *Condor*, 89, 842-49.

Smith, B. L. (2001). "Winter feeding of elk in western North America." *Journal of Wildlife Management* 65, 318-26.

Smith, B. W., and J. R. Belthoff (2001). "Effect of nest dimensions on the use of artificial burrow systems by burrowing owls." *Journal of Wildlife Management* 65, 318-26.

Snyder, N. F. R. (1978). "Increasing reproductive success by reducing nest site limitations: a review." In S. A. Temple (Ed.), *Endangered Birds. Management Techniques for Preserving Threatened Species.* Madison, WI: University of Wisconsin Press.

Spraker, T. R., M. W. Miller, E. S. Williams, D. M. Getzy, W. J. Adrian, G. G. Schoonveld, R. A. Spowart, K. I. O'Rourke, J. M. Miller, and P. A. Merz (1997). "Spongiform encephalopathy in free-ranging mule deer (*Odocoileus hemionus*), white-tailed deer (*Odocoileus virginianus*), and Rocky Mountain elk (*Cervus elaphus nelsoni*) in northcentral Colorado." *Journal of Wildlife Diseases*, 33, 1-6.

Strickland, M. D., and D. M. Crowe (1985). "Winter feeding: a good idea?" *Proceedings of the Western Association of Fish and Wildlife Agencies*, 65, 67-71.

St-Louis, A., J-P Ouellet, M. Crête, J. Maltais, and J. Huot (2000). "Effects of partial cutting in winter on white-tailed deer." *Canadian Journal of Forest Research* 30, 655-61.

Stringham, S. F. (1983). "Effects of climate, dump closure, and other factors on Yellowstone grizzly bear litter size." *International Conference on Bear Research and Management*, 6, 33-39.

Sugden, L. G., and G. W. Beyersbergen (1987). "Effect of nesting cover density on American crow predation of simulated duck nests." *Journal of Wildlife Management*, 51, 481-85.

Svingen, C., and K. Giesen (1999). "Mountain plover (*Charadrius montanus*) response to prescribed burns on the Commanche National Grassland." *Journal of the Colorado Field Ornithologists*, 33, 208-12.

Szaro, R. C., and M. D. Jakle (1985). "Avian use of a desert riparian island and its adjacent scrub habitat." *The Condor*, 87, 511-19.

Toland, B. R. (1987). "The effect of vegetative cover on foraging strategies, hunting success, and nesting distribution of American kestrels in central Missouri." *Journal of Raptor Research*, 21, 14-20.

Torbit, S. C., R. B. Gill, A. W. Alldredge, and J. C. Liewer (1993). "Impacts of pronghorn grazing on winter wheat in Colorado." *Journal of Wildlife Management*, 57, 173-81.

Turner, J. W., I. K. M. Liu, D. R. Flanagan, A. T. Rutberg, and J. K. Kirkpatrick (2001). "Immunocontraception in feral horses: one inoculation provides one year of infertility." *Journal of Wildlife Management* 65, 235-41.

Twedt, D. J., and J. L. Henne-Kern (2001). "Artificial cavities enhance breeding bird densities in managed cottonwood forests." *Wildlife Society Bulletin* 29, 680-87.

Ullrey, D. E., H. E. Johnson, W. G. Youatt, L. D. Fay, B. L. Schoepke, and W. T. Magee (1971). "A basal diet for deer nutrition research." *Journal of Wildlife Management*, 35, 57-62.

Waddell, R. B., D. A. Osborn, R. J. Warren, J. C. Griffen, and D. J. Kesler (2001). "Prostaglandin $F_{2\alpha}$-mediated fertility control in captive white-tailed deer." *Wildlife Society Bulletin* 29, 1067-74.

Wallmo, O. C., and D. J. Neff (1970). "Direct observation of tamed deer to measure their consumption of natural forage." In H. A. Paulsen, Jr. and E. H. Reid (Cochairmen), *Range and Wildlife Habitat Evaluation: A Research Symposium*. U.S. Department of Agriculture, Forest Service Miscellaneous Publication No. 1147. Washington, DC: U.S. Government Printing Office.

Warrick, G. D., J. H. Scrivner, and T. P. O'Farrell (1999). "Demographic responses of kit foxes to supplemental feeding." *The Southwestern Naturalist*, 44, 367-74.

White, R. G., and J. Trudell (1980). "Habitat preference and forage consumption by reindeer and caribou near Atkasook, Alaska." *Arctic and Alpine Research*, 12, 511-29.

Wild, M. A., M. W. Miller, D. L. Baker, N. T. Hobbs, R. B. Gill, and B. J. Maynard (1994). "Comparing growth rates of dam- and hand-raised bighorn sheep, pronghorn, and elk neonates." *Journal of Wildlife Management*, 58, 340-47.

Economic Considerations in Wildlife-viewing Planning

Peter J. Fix, John B. Loomis, and Michael J. Manfredo

Overview

MANAGING FOR RECREATION on public lands often involves making trade-offs. At a broad level, deciding that recreation will take place may imply that other activities, such as logging or mining, might not be able to continue at their current level, or perhaps not at all. Within recreation, there are also trade-offs to consider. For example, allowing hunting in an area may preclude an effective program of wildlife viewing (Vaske et al. 1995). Even within a particular activity, there are still trade-offs that must be taken into account. Within wildlife viewing itself, participants may seek different types of experiences. For example, one viewer may seek a high-solitude experience with little development, while another may seek an opportunity with a high level of development (e.g., interpretive information and hardened trails). At some sites, it would be difficult to provide both types of experiences. (Some sites may allow the opportunity for both due to physical features of the site.)

Economics provides a framework for evaluating these trade-offs by providing guidance for making comparisons between them. These economic factors include the economic impacts to the community, the benefits to society resulting from different uses of public land, and the benefits received by participants from different focuses of recreation management. However, economic analysis is infrequently conducted in wildlife-viewing management.

This chapter will illustrate how economics can help professionals in wildlife-related fields to analyze the trade-offs of providing various wildlife-viewing opportunities. Specifically, concepts and analytical tools will be presented that can be directly applied to Experience-based Management (EBM). This will be accomplished by introducing the concept of economic impacts to a community, illustrating how to measure them, and giving several examples of how to use this concept. The concept of benefits to the recreationist will then be introduced, followed by an illustration of how to measure them, and examples of how to use this concept. The examples will illustrate comparisons of wildlife viewing to private industries and uses, wildlife viewing to other recreation activities, and different types of viewing experiences. The examples will be based on questions that are likely to arise in EBM. These two sections will be followed by a discussion of how a change in the price to visit a wildlife-viewing site may change visitation and revenue to the managing agency.

Economic Impacts to the Community

There are many different uses of public land. Some of these uses are extractive, while others may focus on limited types of recreation (e.g., wilderness areas), or providing habitat protection (e.g., wildlife refuges). Each of these uses has an economic impact to the region where the activity takes place. The economic impact may result from an industry in the area or from tourists visiting the area for recreation. The impact may be relatively small (e.g., a wilderness area located in a remote area) or relatively large (e.g., a large logging or mining company in a small community), but all activities will have an economic impact.

Often, different uses cannot be provided at the same time. The economic impact associated with each activity can become a concern when deciding which use should occur. Communities that perceive they are at risk of losing jobs and income will be concerned over potential management actions. Likewise, a community that stands to gain jobs and income may lobby for particular management actions. Economic analysis provides a systematic way to compare the economic impacts from different activities.

Basis for Economic Impacts to a Community

The basis of the economic impact to a community revolves around *spending*, either by visitors (e.g., wildlife viewers) to an area or employees of an industry (e.g., logging). This section will focus on economic impacts from wildlife viewing. The analysis is, however, the same for estimating the economic impacts of private industry or different recreation activities.

When wildlife viewers visit an area, some will spend money on various goods and services such as gas, food, lodging, guided tours, and souvenirs. This flow of money from the wildlife viewer to the businesses is referred to as *direct effects*. The direct effects create income for the businesses providing services to the wildlife viewers (WV businesses). This income is used by the WV business owners to pay the costs of doing business such as rent, supplies, utilities, insurance, and employees, as well as providing a source of return on their investment (i.e., profit). The initial spending thus becomes income for two additional groups: businesses that provide the supplies and services to WV businesses and employees of the WV businesses. This process repeats as the supporting businesses use their income for paying the costs of doing business and their employees purchase household items. This process of spending repeats several times and is defined as the indirect effects.

The direct effects plus the indirect effects divided by the direct effects is known as the *multiplier*. Studies of outdoor recreation have found multipliers ranging from 1.6 to 2.7 at different recreation areas (Loomis and Walsh 1997). A multiplier of 2.7 means that for every $1 of direct effects, the total effects in the community are $2.70. The reason multipliers differ is because, as the initial spending circulates, part of the money leaves the local economy, due to businesses purchasing supplies out of the region, employees spending their incomes outside the region, and taxes. Money leaving the economy (generally the size of a county or larger) is referred to as a *leakage*. Different local economies have different leakages depending on the degree of local development (e.g., the less developed the economy, the more money that will leak out of the economy).

As money circulates through the economy, jobs are needed to supply the goods and services that are purchased. There are *direct jobs* needed to support the initial transaction between the wildlife viewer and the business. These workers demand goods and services and, in turn, support additional jobs. It is possible for several layers of jobs in an area to be supported by the spending of wildlife viewers. There is also a multiplier for the jobs that result from the expenditures in the area, the *employment multiplier*, defined as the total employment divided by the direct employment.

Regional Economic-impact Analysis

An analysis that calculates local economic impacts is referred to as a *regional economic-impact analysis*. The first step in conducting such an analysis is to define the study area, usually the economic area surrounding the viewing site. Generally, the smallest study area is the county level. For larger sites such as a national park, the study area may comprise all counties surrounding the park. A state wildlife agency might be interested in the impact to the entire state and, thus, the state would become the study area. The larger the study area, the fewer the leakages, the larger the multiplier, and the greater the economic impact. However, the larger the study area, the less applicable the results are to specific sites.

After the study area has been defined, wildlife viewers visiting the area must be surveyed with respect to their expenditures *within the study area*. It is important to measure not only the amount of the expenditures but *what* goods and services were purchased. This is critical because different categories of goods and services have different multipliers associated with them. Some services may be labor intensive resulting in most income staying in the area; others may have a large leakage to vendors outside the study area. As will be explained in the next step, different multipliers apply to these different categories of goods and services.

After the spending data have been gathered, the total impacts are estimated by applying the appropriate multipliers for the geographic region and the type of good or service that was purchased. Table 12-1 shows hypothesized business categories with spending data and multipliers for a county surrounding a viewing site. Although not specific to any particular county, the multipliers are based on previous research conducted in western U.S. counties (Fix 1996). It is important to recognize that Table 12-1 shows an *income multiplier*, and thus the multiplier is applied to direct income. There is also an *output multiplier* that would be applied to spending.

The initial spending of $1,340,000 results in $1,445,900 of income to business owners and employees in the area. Table 12-1 highlights the importance of measuring what goods and services were purchased. For example, purchases of fuel do not contribute as much to the local economy as those for hotels and lodging. The overall economic impact to the area is derived from the multipliers of the industries in the area. Generally, one multiplier cannot be applied to total spending in the area; individual multipliers *must* be applied to the different spending categories in the area. Since multipliers vary by spending category, one can determine what *type* of new spending will have the greatest effect on the economy. In the hypothesized economy in Table 12-2, an additional dollar spent by those who stay overnight or purchase a meal at a restaurant will have the greatest economic impact to the region.

Multipliers are obtained through a process called an *input-output analysis*, which examines the relationship between industries to

Table 12-1. Hypothesized Spending Data and Income Multipliers for Wildlife Viewing

Category	Spending	Leakage	Direct income	Income multiplier	Total income
Hotels & lodging	$400,000	$150,000	$250,000	2.19	$547,500
Eating & drinking	$450,000	$170,000	$280,000	2.18	$610,400
Retail sales	$300,000	$180,000	$120,000	1.8	$216,000
Fuel	$190,000	$130,000	$60,000	1.2	$72,000
Totals:	$1,340,000	$630,000	$710,000		$1,445,900

Table 12-2. Comparison of Wildlife Viewing to Off-Highway Vehicle Use

Activity	Visitors	Potential for economic impacts
Off-highway-vehicle use	Mostly from local area	There is very little economic impact from this activity. Since most visitors reside in the local area, they would be purchasing goods and services in the area if they were not driving off-highway vehicles.
Wildlife viewing	Most reside outside the local area	There is a high potential for the visitors to the wildlife-viewing area to inject new spending into the area. The potential viewers reside outside of the area so their spending represents new income. They would not be likely to stop in the town if there was no viewing area.

determine how changes in business activity in one industry affect other industries. Conducting an input-output analysis can be time consuming and expensive. However, multipliers can be obtained from government agencies, such as the U.S. Department of Commerce, and private companies, such as the Minnesota Implan Group. Some states have also developed their own input-output analysis.

The two commonly reported statistics from a regional economic impact analysis are *jobs supported*—the number of jobs in an area supported by the spending, both direct and indirect—and *income generated*—the additional personal income received by people in the area. (See Box 1)

Applications of Economic Impacts to Communities

There are several applications for estimates of the economic impacts from wildlife viewing to a community: comparing the impacts of wildlife viewing to extractive uses of the habitat, comparing the impacts of wildlife viewing to other recreation activities, and comparing the impacts of managing for different types of wildlife-viewing experiences.

Example 1: Comparing the economic impact of wildlife viewing to off-highway-vehicle driving. A rural town's planning commission and several conservation organizations propose that a nearby area of a national forest adjacent to the town be designated as a viewing site with interpretive information and trails. The town is located on a major route between two national parks. However,

> ## Box 1. Cautions when using a Regional Economic-impact Analysis
>
> A few notes of caution should be mentioned with respect to estimating the economic impacts to a community. When conducting a regional economic impact analysis, it is important to distinguish between visitors residing in the study area and those visiting from outside of the study area. Economic theory suggests that participants residing within the study area are not bringing new money into the region, but rather transferring money from one segment of the economy to another. If the local residents were not participating in wildlife viewing, they usually would have spent their money in the region on some other good or service. This is not likely to be an issue when conducting a regional economic impact analysis for a sparsely populated rural county with a popular viewing site. However, it becomes an issue that needs to be considered when conducting a statewide analysis.
>
> A second note of caution is that impacts to a community are only one tool for analyzing tradeoffs on public land. As will be highlighted in a following section, impacts to the community are at the expense of the wildlife viewer, they are giving up part of their income to pay for these expenses. When using certain accounting stances, the impacts to a community are simply a transfer of spending from area to another. In these situations, only the *benefits to the participant* should be considered.
>
> The final note of caution is that the social costs of increasing viewers in the area must be considered. There may be impacts such as increased crime, the need for additional infrastructure, and a loss of social fabric associated with increased visitors to the area.

few tourists traveling this route stop in the town as there are few attractions. The planning commission expects the wildlife-viewing area to provide an economic stimulus to the town. Currently the area proposed for a viewing site has a few old logging roads that are used for driving off-highway vehicles and some mountain biking. These activities would be banned if the area were to be designated as a viewing site.

A few principles from regional economic-impact analysis can help in determining if the impacts to the town from wildlife viewers would offset any loses from displacing the current users. Interviews

A highly developed viewing area may result in the greatest economic impact to the community, but may not maximize the economic benefits to the viewer. (Photo by Pete Fix)

with the off-highway-vehicle users found that most reside in the county where the town and proposed viewing site are located. Therefore, they do not represent a new source of revenue to the area. The viewers that are expected to stop at the viewing site will be from outside of the local area and therefore do represent an injection of spending into the area. It is expected that the wildlife-viewing site will greatly increase the number of visitors stopping in the town to dine at restaurants and purchase goods in the local shops (Table 12-2).

Example 2: Comparing the economic impact of different wildlife-viewing experiences. A particular planning region has a diversity of wildlife available for viewing. Currently there is no development other than a roadside pull-out. Three wildlife-viewing management alternatives are being considered for this site. Each of these alternatives focuses on providing a different type of experience. The first would focus on providing a learning experience for generalist and occasionalist viewers. A highly developed viewing area would be necessary to achieve this goal, which would include a fully staffed visitor center. The visitor center would provide pamphlets on what animals are expected to be in the area, educational wildlife videos would be shown, and there would be a trained staff available to answer questions and provide guided tours.

The second alternative would focus on providing an opportunity to view wildlife while hiking. While self-guided educational opportunities would be available, the emphasis would be on viewing while hiking. To provide this experience, a relatively extensive hiking trail system would be developed.

The third alternative would focus on a providing a high-solitude experience. It would provide opportunities to photograph, sketch, or paint wildlife in an extremely natural setting. This experience would attract high-involvement and creative viewer typologies. To achieve this goal, there would be minimal facilities in the area (Table 12-3).

Different groups are lobbying for each of the alternatives. An active wildlife-viewing group is lobbying heavily for alternative 3. Although the generalist and occasionalist viewers are not as organized, the local businesses are lobbying for alternative 1. The wildlife-viewing group feels strongly about obtaining their desired experience. The local community's economy is experiencing slow growth and people would like to increase tourism spending in the area. Conducting a regional economic-impact analysis for the county surrounding the viewing site will determine the economic impacts of each alternative. Such an analysis may be as follows.

Different spending patterns are expected to be associated with the viewers attracted by the different alternatives. For alternative 1, not only would there be more visitors, but the per-visitor spending is expected to be higher (Table 12-4). The viewers attracted would be more likely to purchase goods and services, dine at the local restaurants, and stay overnight in hotels and motels. The hiking viewers are expected to have lower per-visitor spending than alternative 1 visitors. Many in this group would be self-sufficient, bring their own food, visit for the day only, and not stop in town. Very few of the high-involvement and creative viewer groups would purchase goods and services in the local community, and they would be more likely to camp if staying overnight. In general, many of them would avoid development when possible.

Not only is the per-visitor spending greatest with alternative 1, more spending occurs in categories such as hotels and restaurants. These categories have larger multipliers than spending on goods such as gasoline purchases. This further increases the disparity in economic impact between the alternatives. Alternative 1 would maximize the economic impact to the community. However, the social costs to the community (e.g., more visitors in area) must be examined to determine if this is the best economic development strategy for the community.

Benefits to the Recreationist

The economic impacts to the community accrue to the businesses and residents in the area surrounding the site, *at the expense of the participant.* The expenditures represent a cost to the participant and a benefit to the community. What the participants value, and derive

Table 12-3. Description of Three Wildlife-Viewing Alternatives

	Experience	Development	Expected visitation
Alternative 1	A generalist and occasionalist experience focused on learning. The viewing experience will be easy to obtain and focus on groups. This can be a "social experience."	The site will have a high level of development. This will include a staffed visitor center, paved parking, hardened trails, and guided tours.	This site will be designed to accommodate a high volume of visitors. It is expected that many commercially operated bus tours will stop in the area.
Alternative 2	This experience focuses on viewers who like to combine wildlife viewing with hiking, though educational opportunities still exist.	The development will focus on a relatively extensive interpretive trail system. There will be a gravel parking lot with minimal facilities (e.g., pit toilets).	There is expected to be moderate use in the area, attracting mainly hikers. Some high-involvement viewers will visit; however most will find this area too crowded.
Alternative 3	This experience will focus on the high-involvement and creative viewers. It will provide high solitude and the opportunity to photograph and sketch wildlife without being disturbed by other viewers.	To ensure the high-solitude experience is available, there will be very little development. There will be no maintained trails or interpretive information.	This site is expected to attract a small but dedicated group of viewers. The visitors will spend long periods of time on site.

Table 12-4. Economic Impacts of Different Wildlife-viewing Management Alternatives

	Per-visitor spending	*Visitors/year*	*Total spending*	*Total income w/ multiplier[1]*
Alternative 1	$20	600,000	$12,000,000	$15,600,000
Alternative 2	$15	350,000	$5,250,000	$6,300,000
Alternative 3	$10	160,000	$1,600,000	$1,840,000

1. Please note this is an income multiplier, applied to direct income and different spending categories as in Table 12-1, but is not shown in this table.

benefit from, is their desired experience (e.g., having the opportunity to photograph or sketch, being with family and friends, etc.). However, in order to obtain this experience they must incur some cost. The cost of obtaining a wildlife-viewing experience can vary greatly, from a tank of gas and a few hours time to airfare and hotels. However, the price paid to obtain the experience may not accurately reflect its value to the participants.

Economists are interested in measuring the total value people place on goods and services. The *total economic value* is the maximum amount the consumer is willing to pay to obtain the good (in this case, the wildlife-viewing experience). This concept can be illustrated graphically (Figure 12-1). Figure 12-1 shows that the consumer is willing to pay $25 for the first unit of this good, $20 for the second, $15 for the third, and so on. The reason the consumer

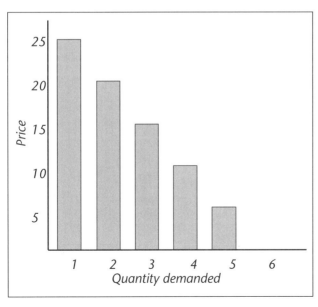

Figure 12-1. Maximum willingness to pay for goods and demand curve. Bars show willingness to pay for each unit of the good.

is willing to pay less and less for each additional unit is because less and less satisfaction is received from the consumption of each additional unit of the good due to satiation. This is referred to as the law of diminishing marginal utility. The maximum willingness to pay (WTP) for each unit of the good traces out the demand curve.

This total WTP can then be compared to the price the consumer paid, to determine the benefit they are receiving from consuming the good: the difference between the total value placed on the good by the consumer and the amount of money they had to give up to obtain the good. This difference is called *consumer surplus*, which is formally defined as follows: "The difference between a person's maximum willingness to pay and the price they paid" (Varian 1990). A very simplified example: you are willing to pay up to $250 for a particular good you have been debating purchasing (perhaps binoculars). You go to purchase these binoculars with $250 dollars in your pocket and find out they are on sale for $199. You leave the store with the binoculars *and* $51 in your pocket. This $51 is consumer surplus; it is a real benefit to you as it can be used to purchase other goods and services.

The concept of consumer surplus is critical in economics because it allows the measurement of the value of publicly provided goods. Publicly provided goods are referred to as *non-market goods* as opposed to a *market goods* (Box 2). The point of this distinction is that we rarely know the value that people place on wildlife viewing: the price is intentionally set below the participant's probable maximum WTP and there are no "market signals" to indicate the value of the experience. In other words, wildlife-viewing experiences

---

Box 2. Market vs. Non-market good

At this point the distinction between two different types of goods should be made: *market* and *non-market* goods. Market goods are traded in private markets and their price and the attributes of the products change often in response to demand. Therefore, the consumer is faced with different prices for different products. Non-market goods' price is not determined by the forces of supply and demand. Wildlife viewing falls under the category of non-market goods. Most often it is provided at no cost to participants, or when there is a fee it is usually set below the market clearing price (this ignores costs the consumer may have incurred to get to the site). Thus, there can be large amounts of consumer surplus associated with non-market goods.

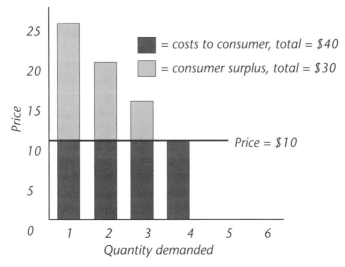

Figure 12-2. Price, cost, and consumer surplus.

do not accumulate in a warehouse because the price was set above consumers' maximum WTP and they did not sell. Consumer surplus allows a way of measuring the value people place on wildlife-viewing experiences.

Consumer surplus can be shown graphically on a demand curve. Figure 12-2 shows a market price of $10 for the demand curve traced out in Figure 12-1. At this price the consumer will purchase four units of the good, for a total cost of $40. On the first unit, $15 dollars in consumer surplus was gained ($25-$10); $10 was gained on the second unit, and $5 on the third unit, for total consumer surplus of $30. In Figure 12-1, the total WTP for four units was $70. In this example, the total WTP for four units is still $70. However, the consumer was only required to pay $40 of the $70 and received a $30 net benefit.

This concept allows a direct comparison of the value of wildlife viewing to consumptive uses and other recreation activities and market goods. Consumer surplus has been recommended as the preferred measure of the economic benefits of outdoor recreation by an interagency committee of the U.S. Government (U.S. Water Resources Council 1980; U.S. Department of the Interior 1986). This allows wildlife viewing to be put into commensurate terms with other recreation activities and non-recreation uses. The commensurate terms are dollars. This provides a way to compare the benefits of wildlife viewing to competing industrial or recreational uses.

While a higher price will result in increased *revenue* per visitor, *setting the price too high may reduce customers to the point that total revenue actually* decreases. *(Photo by Mike Manfredo)*

Measuring the Benefit to the Viewer

In order to utilize the concept of consumer surplus, the maximum WTP to view wildlife in a particular area must be measured. Although there are several methods to obtain WTP values, this chapter will focus on the two most popular approaches: the *contingent valuation method* and the *travel cost method.* These methods have been widely used in recreation and offer flexibility to measure the maximum WTP in many different situations. These two techniques are most applicable to EBM and can capture the uniqueness of certain situations such as high-solitude versus low-solitude wildlife- viewing in a specific area.

Contingent Valuation Method

The contingent valuation method (CVM) is perhaps the most commonly used technique to estimate the maximum WTP for outdoor recreation, including wildlife viewing. It has been used to estimate the benefits of viewing birds (Cooper and Loomis 1991; Swanson 1993) and for funding programs that provide wildlife-viewing opportunities (Jakus et al. 1997). The CVM uses a survey instrument to elicit a person's maximum WTP for the activity or issue in question (Mitchell and Carson 1989). This is done by creating a realistic scenario and then, through one of several formats, asking a question regarding the respondent's willingness to pay.

An example of a scenario may be: "Currently at [the local viewing area] there is not an entrance fee. The budget for the area is being cut and without additional funds, the facilities in the area will begin to deteriorate which may result in the area being closed to visitors." After the scenario is developed, several different types of questions can be used to elicit the maximum WTP: open-ended, closed-ended, and iterative-bidding. The open-ended question format would complete the scenario by asking "What is the most you would pay for a daily entrance fee to [the wildlife-viewing area]?" This type of question format has the advantage of directly eliciting the maximum WTP. However, this is not usually how a market operates and people may not be familiar with giving their maximum WTP. The closed-ended question format would complete the scenario by asking: "Would you pay x as a daily entrance fee to visit [the wildlife-viewing area]?" where x represents a range of prices, randomly assigned to the respondents. The respondents would be more familiar with this format, as it is very similar to how a market works. However, since it does not directly elicit the maximum WTP, various statistical techniques must be used to estimate it. The iterative-bidding format would ask respondents if they were willing to pay a specified dollar amount and then "bid" (up or down depending on the response) until the maximum is reached.

It is essential when conducting a CVM to present a realistic scenario. The scenario can be presented with words or pictures depending on the specifics of the resource being valued and the survey instrument used. The survey instrument can be administered by mail, telephone, or on-site, if the situation permits. For example, the issue may be the value of additional development at a wildlife-viewing area. Respondents could be shown pictures of the current conditions and the proposed development (visitor center, interpretive signs, etc.) and asked if they would pay x more to visit if the development were in place. Current visitors could be sampled on-site and random households could be sampled through the mail. The main advantage of the CVM is that it is extremely flexible; depending on the scenario developed it can be used to estimate the economic value of current situations, variations of current situations, or possible future management actions.

The end result of the CVM is the net value of wildlife viewing to the participants. The aggregate benefits to the viewers can be compared to the costs of the project to see if it passes a benefit-cost test; and the benefits to the viewers, net of the costs to provide, can be compared to the benefits of other uses of the area to see which provides the greatest benefits.

Travel Cost Method

The travel cost method (TCM) is another technique used to estimate the consumer surplus of recreation activities. This method has been applied to estimate the benefits of whale viewing along the California coast (Loomis et al. 1999). The TCM has the advantage of using actual behavior by the respondents to estimate the consumer surplus of the activity (as opposed to the hypothetical nature of the CVM). The TCM uses the visitors' costs to reach the site as a proxy for the price of the recreation activity. These travel costs are then plotted against the total number of trips taken per year to the site to estimate a demand curve (quantities demanded at different prices). The consumer surplus is calculated as the area under the demand curve and above the price. Figure 12-3 illustrates the TCM by converting the previous examples of demand curves into yearly trips to a wildlife-viewing area with different travel costs. As in the previous example of consumer surplus, if the travel costs to the site are $10, this individual will take four trips with total costs of $40 and consumer surplus of $30.

Applications of Economic Benefits

There are several applications for the above mentioned concepts. Consumer surplus can be used to evaluate and compare wildlife viewing to competing consumptive uses and other recreation activities. It can also be used to compare the values of different types of wildlife-viewing settings.

Example 3: Comparing the benefits of wildlife viewing to commercial industry. An excellent example of using estimates of consumer surplus to compare wildlife viewing to consumptive uses is a study conducted by Swanson (1993). This study examined the trade-off between the commercial harvest of salmon and using spawned-out chum salmon to provide food for eagles, supporting eagle populations in a popular viewing area. The study used the CVM to estimate the net WTP of viewing eagles in the Skagit River

---

Box 3. Variations of the TCM

There are two variations of the TCM, the individual and zonal, which are commonly used. While both methods are acceptable, they have slightly different data gathering needs and data analysis requirements. The reader interested in learning more details about the TCM or CVM can reference Loomis and Walsh (1997) for more details on the specifics of these two methods.

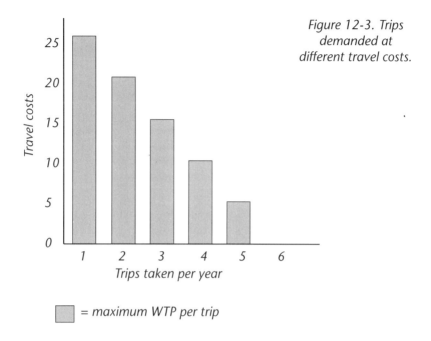

Figure 12-3. Trips demanded at different travel costs.

Trips taken per year

= maximum WTP per trip

Bald Eagle Natural Area (SRBENA) in Washington state, which was found to be $56 per trip. This value was then compared to the commercial value of salmon. Optimizing the benefits of viewing and the commercial harvest results in a salmon harvest of 413 per day. This results in daily spawning numbers of 750, supporting 197 eagles and 209 daily visitors to the SRBENA.

Example 4: Comparing the benefits of wildlife viewing to commercially provided recreation. As another hypothetical example, an alpine (downhill) ski area has proposed an expansion into a popular wildlife-viewing area. The land in question is federally owned. The viewing site attracts a wide variety of visitors, ranging from specialists to generalists. The owners of the ski resort argue for jobs created, the need to stay competitive, and an improved experience for the skiers. The proponents of the viewing area believe something of great importance to many people will be lost, but are having difficulty expressing this loss in terms comparable to the ski area. An economic comparison in this situation compares the net benefits from continuing to manage the land as a viewing area versus the net benefits if the ski area is allowed to expand.

Currently the visitors to the viewing area receive large benefits, as the site provides a high-quality experience for a wide range of viewing typologies. Suppose the current gross benefits to the viewers were estimated with a CVM and found to be $1,600,000 annually.

Table 12-5. Net Benefits of Wildlife Viewing versus a Ski Area Expansion

	Left as viewing area	Ski area developed
Gross benefits	$1,600,000	$10,500,000
Costs[1]	$600,000	$10,000,000
Social Net Benefits:	$1,000,000	$500,000

1. The costs include the travel costs to the participants and the annual operating costs.

The viewing area costs $600,000 annually to operate, resulting in $1,000,000 in net benefits. The gross value the skiers place on the ski area expansion is large, $10,500,000 annually. However, alpine skiing is a very costly activity to provide, costing $10,000,000 annually. These costs get passed on to the consumer in the form of lift-ticket prices and through the pricing of other services. Therefore, the *net* benefit associated with the ski area expansion is only $500,000 (Table 12-5).

In this situation, economic efficiency would dictate that the area should remain a viewing area. In this situation it is appropriate to look only at the net benefits to the recreationists. The land in question is federal land, which requires that a national accounting stance be used when evaluating benefits and costs. At the national level, the jobs and income from the ski area expansion are just a transfer from one economic sector to another. The benefits to the recreationists, however, represent a real gain to the economy.

Example 5: Comparing the benefits of different types of wildlife-viewing experiences. In Example 2, the economic impacts to the community of three different wildlife-viewing management alternatives were analyzed. However, this is only one aspect of analyzing these different alternatives. The benefits received by the participants in the different management alternatives should also be analyzed. The benefits received by the viewers need to be net of the costs of providing each alternative. It can then be determined which alternative provides the greatest net benefits to viewers.

Alternative 1 attracts the greatest number of visitors, 600,000 annually. However, these visitors are mostly occasionalists and generalists. Their main motivation is being with family and friends, having a change of pace, and escaping life's demands. Although visiting the viewing area adds enjoyment to their trip, it is not valued very highly. The average maximum net WTP, over and above their costs (e.g., once the visitors' travel costs have been subtracted from their gross WTP), to visit the viewing area for these visitors is

$15 per visit. This results in a total annual *net* value to the visitors of $9,000,000. Alternative 2 attracts 350,000 visitors annually. These visitors are more involved in viewing, but it is as part of another activity, in this case hiking. These visitors value the experience higher than visitors under alternative 1. However, there are many substitute hiking sites close by, lowering the maximum WTP to visit this site. The average maximum net WTP to visit this site is $25. The total annual *net* benefit to the viewers is $8,750,000.

Alternative 3 attracts the fewest number of visitors, 160,000 annually. However, these visitors are highly involved in wildlife viewing. Management alternative 3 provides a unique experience for these viewers. There is a relatively large maximum net WTP for these visitors, on average $50 more than their current costs. This results in annual *net* benefits to the viewers of $8,000,000.

The costs of providing each type of viewing experience also need to be taken into account. To determine which management alternative provides the greatest benefits to society, the benefits need to be *net* of costs. For these three management alternatives, 1 is the most costly to provide followed by 2 and 3. Examining the benefits net of the costs to provide each shows that alternative 3 provides the greatest benefits to society (Table 12-6).

The above examples show that regional economic-impact analysis and examining the benefits to the viewer in a benefit/cost analysis may sometimes result in different conclusions. In the above examples, management alternative 1 resulted in the greatest economic impact to the community, whereas alternative 3 resulted in the greatest net benefits to viewers. The resolution of these conflicting results is in the accounting stance used. At a statewide level, spending from *resident* visitors and the resulting income created and jobs supported are at the expense of another region within the state. Of course, a reduction in nonresident visitors should be counted as a loss to the state's economy. At a national level, jobs and income are a transfer from one region of the country

Table 12-6. Comparison of Benefits and Costs of Providing Different Viewing Experiences

	Alternative 1	Alternative 2	Alternative 3
Annual net benefits to viewers	$9,000,000	$8,750,000	$8,000,00
Annual agency costs to provide	$4,000,000	$3,000,000	$1,000,000
Annual Social Net Benefits:	$5,000,000	$5,750,000	$7,000,000

Private businesses often fill a wildlife-viewing niche that the public sector cannot. This createss jobs and income in the area. (Photo by Mike Manfredo)

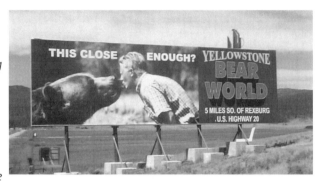

to another and, thus, are not a benefit to the economy. The exception to this rule is if there is persistent high unemployment in the area surrounding the project site.

If this viewing site is located on state land, the jobs supported by resident visitors simply become a transfer of economic activity from one segment of the economy to another. In this situation, it would have to be determined how many visitors are residents. If the viewing site is located on federal land, the jobs supported are a transfer from one economic area to another. The benefits accruing to the viewers are the appropriate unit of analysis in this situation. If the viewing site is located on land owned by the local community, the economic impacts to the community would be a real gain to them and would be the appropriate unit of analysis.

Using Economics to Determine the Optimal Price

Besides measuring the impacts to the community and the benefits to the participants, economics can also be used to determine the optimal price for wildlife-viewing programs. A manager's concern may be to establish a source of funding for a program through a fee. However, a higher price does not necessarily imply greater revenue. Too high a price may cause a reduction in participation that the price increase will not offset. This would lower the total revenue.

The tool that economists use to determine the effect of a price increase on revenue is the *price elasticity of demand*, which is formally defined as "The percentage change in quantity demanded resulting from a 1 percent change in price along a given demand curve" (Hyman 1988 p. 148). Elasticity is expressed as a number. For example, an elasticity of .5 means that a 1 percent change in price will result in a .5 percent change in demand.

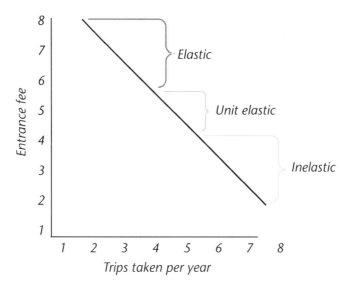

Figure 12-4. Demand curve and elasticity.

There are three general categories of price elasticity of demand: elastic, unit, and inelastic (referred to with respect to demand). Elastic demand occurs when a 1 percent change in price results in a change in quantity demanded that is greater than 1 percent. The corresponding value of elasticity is greater than 1. Unit elastic demand occurs when a 1 percent change in price results in a 1 percent change in quantity demanded. The corresponding absolute value of elasticity is 1. Inelastic demand occurs when a 1 percent change in price results in a less than 1 percent change in quantity demanded. The corresponding absolute value of elasticity is less than 1.

The elasticities are important because they predict the relationships between price and revenue. With elastic demand, an *increase* in price will *decrease* total revenue, as the decrease in demand is larger than the increase in price. However, a *decrease* in price will *increase* revenue. A price change with unit elastic demand will be offset by the change in demand and revenue will stay the same. Inelastic demand will result in increased revenue when prices are increased and decreased revenue with price decreases.

Several factors will influence the price elasticity of demand. The first is the availability of substitute sites. If there are many suitable wildlife-viewing sites near each other, the demand for any particular site will be very sensitive to price (elastic demand). If no substitute wildlife-viewing sites are nearby, demand will be more inelastic.

Quality also affects the price elasticity of demand. A high-quality viewing experience will be more price inelastic. The percentage of income spent on the good will also influence the price elasticity of demand. If wildlife viewing represents a small percentage of the viewer's income, it is not likely to be very sensitive to price.

The largest influence on the price elasticity of demand, however, is the initial price before the change occurs. Demand will be less sensitive to price at low dollar amounts than at high dollar amounts. In other words, a 20 percent increase in price will have different effects on demand if the initial price is $5 or $40. A linear demand curve has a segment with each type of elasticity. At the segment of the demand curve associated with higher prices, the demand is elastic. The segment of the demand curve associated with lower prices has inelastic demand. The middle segment of the demand curve has unit elasticity (Figure 12-4). The same price increase will have different effects on demand depending from where on the demand curve the price change is initiated. All of these different factors affecting demand should be accounted for when determining a pricing policy (Box 4). Table 12-7 shows the effects of a price

---

Box 4. Calculating Price Elasticity of Demand

The price elasticity of demand is calculated as the absolute value of the percent change in quantity demanded divided by the percent change in price. Thus the relationship between quantity demanded and price must be known. This relationship can be estimated with the CVM or TCM. With both of these techniques, regression is generally used to estimate the relationship between quantity demanded and price, taking into account the other factors affecting demand such as income and the price of substitutes. The regression coefficient for the price variable shows the relationship between quantity demanded and price. However, if the equation estimated is linear, the estimated relationship is in *units* not *percent*. Therefore the price coefficient must be substituted into the elasticity formula as follows:

Price elasticity = price coefficient * (mean price / mean quantity)

The mean price and mean quantities are obtained from the output of the regression equation estimated. If the estimated equation is semi-log or double log, the price coefficient can be interpreted as the elasticity.

Table 12-7. Elasticity of Demand and the Effects on Visitation and Revenue

Change in price	Elasticity		Visitors +/-	Revenue
$2 to $4	Inelastic	.25	8,000 to 6,000 = -2,000	$16,000 to $24,000 = +$8,000
$6 to $8	Elastic	1.5	4,000 to 2,000 = -2,000	$24,000 to $16,000 = -$8,000

increase on visitation and revenue corresponding to Figure 12-4. It is clear that an increase in price may not always increase revenue.

Conclusions

Economics can assist in applying Experience-based Management to wildlife-viewing decisions. Economic tools can measure the economic impacts to the community resulting from managing for different wildlife-viewing experiences, as well as the benefits received by the viewers seeking these different experiences. Besides being used to compare different types of wildlife-viewing experiences, these two pieces of information can be used to compare wildlife viewing to extractive uses of the resource and different recreation activities that are possible in the planning area.

Summary Points

• Many wildlife-viewing-management decisions have some tie to economics. Economic principles can provide useful tools to the professionals faced with these decisions.
• Three economic concepts are relevant to management decisions regarding wildlife viewing: impacts to the local community, benefits to the participants, and pricing.
• Economic impacts to the local community are the income generated and jobs supported by wildlife viewers visiting an area. Depending on the accounting stance (with a fully employed economy), they may be a transfer of economic activity from one area to another, with no new income created.
• Benefits to the recreationist are the gain to the participant over and above the cost of the experience. The benefits to the recreationists are a real gain to the economy and can be directly compared to competing, non-recreation uses of the resources or other recreation activities.
• Estimating demand curves can assist in determining the relationship between price, visitation, and revenue.

Literature Cited

Cooper, J., and J. Loomis (1991). "Economic value of wildlife resources in the San Joaquin Valley: Hunting and viewing values." In A. Dinar and D. Zilberman (Eds.), *The Economics and Management of Water and Drainage in Agriculture.* Boston, MA: Kluwer Academic Publishers.

Fix, P. J. (1996). The economic benefits of mountain biking: Applying the TCM and CVM at Moab, Utah. Unpublished Master's Thesis, Colorado State University, Fort Collins, CO.

Hyman, D. N. (1988). *Modern Microeconomics: Analysis and Applications,* 2nd ed. Boston, MA: Richard D. Irwin, Inc.

Jakus, P. M., J. M. Fly, and B. Stephens (1997). "Estimating Tennessee residents' willingness to pay for teaming with wildlife." *Human Dimensions of Wildlife* 2:3, 16-26.

Loomis, J. B., and R. G. Walsh (1997). *Recreation Economic Decisions: Comparing Benefits and Costs.* State College PA: Venture Publishing Inc.

Loomis, J., S. Yorizane, and D. Larson (1999). *A Travel Cost Method Analysis of Whale Watching along the California Coast: Evaluation of Mutli-destination Trip Bias.* Department of Agricultural and Resource Economics, Colorado State University, Ft. Collins, CO.

Mitchell, R. C., and R. T. Carson (1989). *Using Surveys to Value Public Goods: The Contingent Valuation Method.* Washington, DC: Resources for the Future.

Swanson, C, S. (1993). Economics of non-game management: Bald eagles on the Skagit River Bald Eagle Natural Area, Washington. Unpublished Ph.D. Dissertation. Department of Agricultural Economics and Rural Sociology, Ohio State University.

U.S. Department of the Interior (1986). "Natural resource damage assessments; Final rule." *Federal Register* 51:148, 27674-753.

U.S. Water Resources Council (1980, April 14). "Principles, standards, and procedures for planning water and related land resources." *Federal Register 45*:73, 25301-48.

Varian, H. R. (1990). *Intermediate Microeconomics: A Modern Approach,* 2nd ed. New York: W.W. Norton Co.

Vaske, J. J., K. Wittmann, S. Laidlaw, and M. P. Donnelly (1995). *Human-wildlife Interactions on Mt. Evans.* Project report for the Colorado Division of Wildlife. Human Dimensions in Natural Resources Unit Report Number 18. Fort Collins, CO: Colorado State University.

Information and Education for Managing Wildlife Viewing

Alan Bright and Cynthia L. Pierce

The Importance of Communication in Wildlife-viewing Programs

RECENTLY, GROWING INTEREST in wildlife viewing has become an important component of an agency's "product mix." Yet when asked about the nature of existing wildlife-viewing programs, agencies often respond by stating that they have published a viewing guide (Pierce and Manfredo 1997). This suggests a belief by both agencies and the public that the role of viewing programs is to develop brochures, displays, and other informational materials about when and where to encounter wildlife for viewing. This belief is reinforced when agencies measure the success of a viewing program by the numbers of publications distributed to the public, displays created, and viewing sites constructed, without addressing the effectiveness of these efforts.

For example, state wildlife-viewing guides have become a recognizable product and symbol associated with agency viewing programs. However, two questions are often overlooked when they are produced. First, how does the viewing guide fit within the overall plan for the agency's viewing program? Second, what are the objectives for producing the guide and how can they be addressed using good principles of communication and information design? Most viewing guides follow a typical format, describing wildlife that may be viewed in the state and where. While this information meets specific needs, it is also advantageous to identify whether other information should be included (e.g., general guidelines on viewing), whether there are ways to increase interest (e.g. including other photos and pictures of viewers in the field), and to provide an evaluation of the guide's effectiveness.

Communication can inform, educate, and persuade audiences. The most common function of communication for wildlife agencies is to inform the public of management practices and to educate about biological processes (Stout and Knuth 1993). Wildlife-recreation information often describes available opportunities, regulations, ethics, and techniques for participating (Simcox and Hodgson 1993) and may even work to change public beliefs and attitudes toward management methods or goals. Wildlife-viewing programs are frequently expected to emphasize information delivery through such methods as viewing platforms or trail guides to help people identify animals. This expectation may be one reason viewing programs are seen as having an educational emphasis. However, the outputs for all wildlife-recreation activities are the experiences provided. Meeting the diversity of experience expectations and providing appropriate communications to complement them can be best served by the Experience-based Management (EBM) approach to planning and service delivery. Within this process, information and education serve as tools to facilitate and enhance viewing experiences.

Developing an Information and Education Program for Wildlife Viewing

Experience-based Management provides a process for determining what experiences recreationists desire and identifying the resources that can be provided to meet this demand. Segmenting wildlife viewers by the types of experiences desired allows communication planners to design appropriately targeted information. An effective communication plan requires that wildlife agencies address five phases, each of which addresses a question in the development of the plan:

Phase 1. Determine Objectives. Why is information being provided?

Phase 2. Identify Target Audience(s). To whom will it be provided?

Phase 3. Develop the Message. What kinds of information will be provided?

Phase 4. Select Media and Implementation. How and when will it be provided?

Phase 5. Evaluation. How will achievement of communication objectives be measured?

Under an EBM approach, answers to these questions are based on the segmentation of wildlife-viewing experiences. This approach should include the adoption of a service philosophy that emphasizes the importance of determining and providing information viewers actually desire. While the agency may have specific communication goals, the public's receptivity will be improved if the information addresses their needs and preferences for message delivery. EBM facilitates effective communication by helping to identify experiences, as well as the information and communication methods desired by viewers. Table 13-1 provides an example of an outline for a communication plan, designed to target different viewing experiences, based on the wildlife-viewing typology described in Chapter 5. This chapter describes the development, implementation, and evaluation of an information and education

Table 13.1. Example of a Communication Plan Outline for a Wildlife Viewing Opportunity Typology, Developed for a Site Along the Front Range, Colorado.[1]

Objectives
Highly Specialized Hiking/Wildlife Viewing
Increase specific knowledge of species and viewing techniques.

High Specialized Wildlife Photography
Increase knowledge of particular species, where and when to best view these species.

Nature Touring
Increase general learning. Other benefits provided are difficult to control (e.g., enhancement of family togetherness).

General Interest Wildlife Viewing/Hiking
Increase general learning. Provide information how wildlife viewing experiences can fulfill broader satisfactions (e.g. stimulation, social opportunities, relief from stress)

General Interest Wildlife Viewing/Biking
Increase general knowledge. Provide information how wildlife viewing experiences can be combined with biking to fulfill a broad array of satisfactions (e.g., exercise, needs for social opportunities).

Audience
Highly Specialized Hiking/Wildlife Viewing
Learning and teaching motivate this group. Socialization is important. Usually associated with a referent group, such as naturalists, birdwatchers. Moderate to high experience. High specialization and commitment.

High Specialized Wildlife Photography
Photographs, sketches, and paints. Motivated by opportunities for creativity and the opportunities to develop skills, achieve, use equipment, explore and experience autonomy. Moderate to high experience. Highly involved and committed.

Nature Touring
Observes and enjoys nature but does so at a distance. Does not intrude on environment. Participation is moderate to occasional. Generalist type of experience. High demand for information and education.

General Interest Wildlife Viewing/Hiking
Observes and enjoys nature with moderate entrance into the natural environment. Benefits realized from wildlife viewing can be realized through a number of substitute activities. Occasional to moderate participation. Generalist experience desired. Demand for information and education.

table continues

General Interest Wildlife Viewing/Biking
Combine bike touring with wildlife viewing. Observes and enjoys nature but low entrance into the environment. Accepts moderate degree of naturalness. Occasional to high participation in biking. Moderate to low visits to a site. Range from generalist to specialist. Not committed to wildlife viewing

Messages
Highly Specialized Hiking/Wildlife Viewing
Messages should provide detailed info on specific species, where to view (especially opportunities for off-trail hiking), natural history, and management activities. This group is likely to be responsive to cognitive appeals.

High Specialized Wildlife Photography
Messages should provide info on specific species, where and when to view (especially opportunities for solitary and off-trail experiences), natural history and management activities. Likely to be responsive to cognitive appeals.

Nature Touring
Messages should be quite general, requiring low cognitive effort, conveyed through symbols, pictures, dioramas, or in brief phrases and simple language

General Interest Wildlife Viewing/Hiking
Messages should be general, require low cognitive effort, entertaining, conveyed through symbols, pictures, in brief phrases and simple language. Messages mostly about wildlife in general with some high profile or symbolic species covered.

General Interest Wildlife Viewing/Biking
Messages should be general, require low cognitive effort, conveyed through symbols, pictures, and simple language. Messages about wildlife in general with some high profile or symbolic species covered. Rules and regulations help define appropriate contact with wildlife.

Channels/ Media
Highly Specialized Hiking/Wildlife Viewing
Continuously updated information such as newsletters, phone line. Trails with minimal interpretive signing.

High Specialized Wildlife Photography
Continuously updated information (e.g., newsletters, phone line. Trails with minimal interpretive signing. Information at low use, designated trails.

Nature Touring
Messages provided at nature centers, during bus or train tours, roadside information.

General Interest Wildlife Viewing/Hiking
Interpretation at nature trails, visitor/nature centers, simple materials. Accepts structured experiences (less need for naturalness). Social contact advantageous.

General Interest Wildlife Viewing/Biking
Interpretation at nature trails, visitor/nature centers, simple materials. Accepts structured experiences (low need for naturalness). Social contact advantageous.

Evaluation
Highly Specialized Hiking/Wildlife Viewing
Very conducive to both quantitative and qualitative evaluations since this groups can provide information on likes/dislikes, knowledge, and need for information.

High Specialized Wildlife Photography
Extremely conducive to both quantitative and qualitative evaluations since this groups can usually provide information on likes/dislikes, knowledge, and need for information.

Nature Touring
Done onsite or before/after experiences (e.g., at visitor centers). Long term effects can be determined using mail or phone surveys.

General Interest Wildlife Viewing/Hiking
Evaluation should employ simple methods and focus on broad benefits. This group is unlikely to demonstrate lasting knowledge change or strong commitment to wildlife viewing.

General Interest Wildlife Viewing/Biking
Because individuals are mobile, evaluation can be difficult. May wish to record personal information and contact by phone/mail later. Onsite evaluation may be acceptable.

---

[1] Adapted from Manfredo et al. (2000).

plan using the five phases noted above. This process helps in constructing a communication plan for wildlife-viewing programs.
The following sections describe a process for creating an information and education plan for a viewing program using an EBM approach. Each phase in the process is described, the importance of each phase is explained, and examples for addressing the phases are illustrated.

In practice: To help see how this process can evolve, we can use a simple example of a wildlife-viewing program planner who is creating an information plan. For our example, assume the viewing planner wants to provide information that focuses on creativity experiences, specifically opportunities for photography and painting/sketching. How will she proceed to create communications targeted specifically to these types of experiences?

Phase 1. Determine the Objectives of the Information Campaign

This phase addresses the question: *Why are you providing information?* Objectives describe the reasons for developing the campaign strategy and establish the standards by which success of the campaign can be measured. Objectives may also indicate the connection between the information program and the service philosophy of the wildlife agency. In information campaigns, we are typically interested in targeting changes in beliefs or knowledge, attitudes, or behaviors. Objectives should be realistic and measurable, they should precede decisions about messages or media, and may specify a date for completion.

There are a variety of potential objectives for an information and education program related to wildlife viewing. Objectives might include:

(1) To enhance public knowledge of the connection between suitable wildlife habitat and the health of the wildlife species;

(2) To educate the public on the role of the wildlife agency in managing wildlife habitat;

(3) To inform the public about available wildlife-viewing opportunities;

(4) To facilitate identification of wildlife viewed by the visitor;

(5) To provide information that helps facilitate participation and meet viewer needs for desired experiences; and

(6) To obtain support for current and/ or proposed wildlife-management practices.

There are a variety of messages, media, and reasons for providing information to wildlife viewers and recreationists, so it is extremely important to have clear goals and objectives before developing communication materials. (Photos by Cynthia Pierce)

Although determining the objectives of the information campaign is described as the starting point of the process, this phase is often done concurrently with subsequent phases: decisions about the objectives of an information campaign depend on who the target audience will be, what the desired response is to the information campaign, and what media are available to the agency for disseminating the information.

> **In practice**: *Phase 1. Objectives.* The wildlife-viewing planner knows she wants to address creative experiences in this particular information campaign. She also knows she would like to have this project completed this year and her performance objective is to illustrate progress and success on this project. She develops the following objective: *To increase, by x percent, the viewing public's awareness of "creative" viewing opportunities available to them, within one year.* This specifies a targeted change in knowledge, for a specific audience, by a specified time.

Phase 2. Identify Target Audience(s) in the Campaign

This phase addresses the question: *To whom will the information be provided?* This involves identifying specific target audiences and determining whether those audiences are relevant to the objectives of the information campaign. This information is used to develop the strategy for providing information about wildlife viewing (Manfredo and Larson 1993).

Criteria for viable target segments

The public may vary from several wildlife-viewing segments to only one. Regardless of the number of segments, it is important to determine what segments are relevant as target audiences for a particular information and education program. For an audience to be relevant, it should exhibit five characteristics (Grunig 1989; Kotler and Andreasen 1987).

First, target audience segments must be *pertinent* to the goals and objectives of the agency and the information campaign. Providing information about specific wildlife-viewing experiences should be provided to audience segments who are either already interested or have been identified as potentially interested. Second, the agency should be able to measure the characteristics by which the segment is based (*measurability*). A segment would be measurable, for example, if it was based on age, experiences desired, or previous documented visits to a viewing site. Third, the segment must be

Specialized viewers, such as artists, will have very different experience expectations and information needs from generalists.. (Photo by Cynthia Pierce)

reachable with communications (*accessibility*). For an on-site campaign, will the target audience be likely to visit the site and, as a result, be exposed to the information? Fourth, the segment should be significant enough to merit spending resources on the information program (*substantiality*). Traditionally, this suggests the segment be profitable. However, a better viewpoint would be to consider the benefits derived from providing the information to the segment, regardless of size. Fifth, the most crucial criterion is that segments possess *differential responsiveness*. If different segments respond the same to a communication strategy (that is, they are looking for the same wildlife-viewing experience), they should not be treated as separate segments. If a segment does not meet the above five criteria, it should be dropped as a target audience or combined with other segments for the information program.

Describing target audiences

Under an EBM framework, experiences themselves are segmented. Within these experience types, target audiences may be described in terms of *target descriptors* or characteristics of the segment. Target descriptors can relate to geographic, sociodemographic, psychographic, behavioral, and/or temporal characteristics (see Table 13-2 for examples). Descriptors can be identified by studies of viewers within experience types.

Strategies for reaching the target audience(s)

There are several kinds of strategies for reaching target audience(s) (Figure 13-1) (Andreasen 1995). In an *undifferentiated strategy*, the target audience is treated as one group with similar needs and interests. For example, a single interpretive program about the

Table 13-2. Target Descriptors and Breakdown Variables

Target Descriptors	Typical Breakdown Variables	Examples
I. Geographic	Residence, region	Colorado, Front Range
II. Sociodemo-graphic	Population size, age, sex, number of children, marital status, income, occupation, education, national origin/ethnicity, religion	Male, 40-50 years old
III. Psychographic	Life style, needs, personality, values	Extrovert, Wildlife Protection Orientation
IV. Behavioral	User status, usage rate, experiences sought, benefits sought, loyalty status, skill level, specialization, usage readiness	Takes 2-3 trips per year for wildlife viewing. Life listing birder. Seeks solitude experiences.
V. Temporal	Time of day, time of year, day of week, length of time, seasonal considerations	Fall. Weekends.

agency's role in managing wildlife habitat could be made available at all wildlife-viewing sites, regardless of the nature of the experience that the site provides. In this case, there is no differentiation of the audience by specific experiences desired, effectively treating all wildlife viewers as a single target audience.

Under a *differentiated strategy*, distinct target audiences with varying needs and interests receive information that differs in content and/or media channel. While the goal of this strategy is to reach all relevant target audiences, the methods used and information provided each segment would differ. For example, information about where to view specific wildlife species along with the natural history of those species might be provided for a highly specialized hiking/wildlife-viewing segment in a newsletter while more general information about viewing locations might be provided for all other members of the public in a Sunday newspaper.

In a *concentrated strategy*, differences among target audiences and their needs and interests are recognized with information targeted to only some segments while other segments would be ignored for a particular campaign. For example, an information campaign about wildlife-viewing experiences might be developed and focused solely

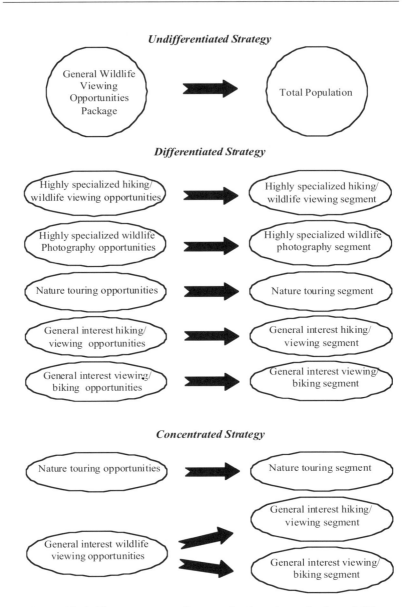

Figure 13-1. Three target audience selection strategies for wildlife-viewing education.

on ecotourists while the remainder of the public would receive no information.

Each of these strategies can be used in developing an information and education campaign for wildlife viewing. The decision about an appropriate strategy is based primarily on the nature of opportunities available, wildlife-viewing audiences, and available funds. For example, an agency may choose a generalized approach because it does not have enough resources to design several specific campaigns.

Defining the target audience response to the campaign

Once the target audience(s) have been identified, desired responses to the information should be determined, consistent with the objectives developed in phase 1. The desired responses to an information program can be described using the AIDA model, which addresses four levels of response to an information campaign (Fine 1991): attention (including awareness and knowledge), interest, desire, and action. The model is based on the idea that, in order to have their behavior influenced, recipients of information must first go through the phases of attention, interest, and desire. However, not all information campaigns have behavior change as their ultimate goal. Some campaigns may simply have, as a primary objective, increasing knowledge.

The first response level is *attention*. Attention focuses primarily on increasing the target audience's awareness and knowledge of the wildlife agency, issues related to wildlife management, and wildlife-viewing opportunities.

The second response level is *interest*. This stage attempts to motivate the audience to learn more. This may be done by focusing on the experiences and resources available and the benefits to the target audience in taking advantage of specific wildlife-viewing opportunities.

The third response level is *desire*. The target audience evaluates the nature of the wildlife-viewing opportunities and whether they want to take advantage of them. This stage moves the audience from simple interest to more active consideration. At this stage, the target audience becomes committed to the idea of taking part in wildlife-viewing opportunities or to thinking about the information provided. This involves continued emphasis on the experiences and benefits of the opportunities as well as addressing perceived constraints to participation.

The final response level is *action*. At this stage, the audience is encouraged to take some action, including actually participating in wildlife-viewing activities. Information should continue to focus on the benefits of participation as well as removal of perceived constraints to participation.

In practice: *Phase 2. Audience.* The wildlife-viewing planner has decided to target highly specialized viewers interested in photography and art (Table 13.1). To help target the information campaign, she gathers information about this particular segment, including characteristics (e.g., age, income) and interests. She finds that these highly involved users apparently enjoy more solitary experiences when viewing wildlife (preferring not to be disturbed when capturing their subjects on film or canvas), tend to be in their forties, and are primarily interested in information that will help them understand the habits of the wildlife they hope to see. A small segment is novices who appreciate information on the easiest opportunities to photograph popular species. The information planner decides on two audience segments, novices and experts (a differentiated approach), with most efforts directed toward experts, and focusing on increasing awareness of opportunities (attention).

Phase 3. Develop the Message

This phase addresses the question: *What kinds of information will you provide?* The extent to which a target audience receives, pays attention to, understands, remembers, and/or responds to an information and education campaign depends on the content and delivery of the message.

Message content and type

The desired audience response, the characteristics of the wildlife-viewing experience, and the interests of the target audience shape the content of the message. For example, the highly specialized hiking/wildlife-viewing audience might be interested in detailed information about specific wildlife species and where the most solitary viewing experiences are found. The general-interest wildlife-viewing/hiking audience might be interested in messages about high-profile wildlife species designed more to entertain than inform. Table 13-1 provides an example of message content that might be appropriate for different wildlife-viewing target audiences.

Message type refers to the general way in which the message is presented. For example, it may simply be informative, presenting straight facts about a specific wildlife species or wildlife-viewing opportunity. On the other hand, the message may take the form of

Messages need to be developed with both the agency's and viewers' objectives and interests in mind.(Photo by Cynthia Pierce)

an argument, with reasons for (and possibly against) certain wildlife-viewing behaviors or management activities. Message types may use symbols (such as Smokey Bear) or psychological appeals in order to influence the response of the target audience (see Box 1 for a description of different types of messages).

Characteristics of Messages

Message execution is the general feel of the message and focuses on message style, tone, words, order, and format. A variety of execution styles may be used to disseminate information about wildlife-viewing opportunities. Suppose that the wildlife-management agency has developed viewing areas to see a variety of wildlife species. To encourage the public to take advantage of these opportunities, the message style may concentrate on creating a particular mood, such as illustrating how much a family can enjoy these experiences (see Box 2 for examples of message execution styles).

The *tone* of the message may be serious (e.g., the dangers to joggers in an area adjacent to cougar habitat) or light and humorous (e.g., the playful brawling of bear cubs). Related to tone are the *words* (written or spoken). Words that are memorable and capture the viewer's attention should be used, such as when headlines and slogans are used to provoke interest.

The *order* of ideas involves two considerations (Kotler and Andreasen 1987). The first consideration is *one- vs. two-sided arguments*. Will a message be more effective if it provides one side of an argument or if both sides are presented? Two-sided messages appear to be more effective when the target audience is (a) opposed to the communicator's position, (b) highly educated, and (c) likely

Box 1. Types of Messages

Information. Presentation of straight facts about a program

Example: "The best times to view bighorn sheep are in the morning and evening."

Argument. Use of logical arguments to support reasons for participation in a program or activity

Example: "Everyone wants a chance to see the wildlife, so please don't frighten the animals by approaching them too closely."

Motivation with psychological appeals. Incorporation of emotional appeals to enhance the attractiveness of a program or activity

Example: "Hearing the regal call of the bull elk creates a thrilling experience for the park visitor. Because early fall is a popular time for participating in these opportunities, you'll want to plan ahead for your visit."

Repeat-assertion. Use of hard-sell approaches that involve restatement of information

Example: "Don't forget, feeding of wildlife is unsafe, unhealthy, and unlawful."

Command. Presentation as a direct order with the intention to influence the behavior of a target audience

Example: "Do not feed the wildlife!"

Symbolic association. Use of a symbol to subtly promote information about a program or activity

Example: The watchable-wildlife binoculars symbol or a camera icon is sometimes used to indicate good places for viewing or photographing wildlife.

Imitation. Association of a program or activity with an individual, group, or situation that is desirable to the target audience

Example: Celebrity and organizational endorsements.

Source: O'Sullivan (1991) and Simon (1971).

Box 2. Examples of Message Execution Styles

Fantasy. As an individual sits at home surrounded by cranky children, she imagines herself sitting, with a few good friends, in a rustically developed viewing structure, camera in hand, watching a herd of bighorn sheep.

Mood. A young family strolling through a wildlife-viewing area is experiencing quality time together, enjoying nature and viewing the wildlife around them.

Musical. Film of the flight of eagles accompanied by classical music.

Lifestyle. A wife says to her husband that instead of spending the weekend hard at work at the office or at home, he might enjoy going camping and hiking at the wildlife refuge. He agrees, and the next frame shows him going to work on Monday morning refreshed and ready for another week at the office.

Technical expertise. Several wildlife professionals are shown discussing the best places and times to view elk during the fall.

Testimonial evidence. A group of three or four people is shown talking about the positive experiences they have had viewing wildlife at a viewing area and the importance of supporting an upcoming ballot initiative to increase funding to develop more viewing areas.

Source: O'Sullivan (1991)

---

to be exposed to other information counter to the communicator's position. The second consideration is the *order of presentation*. Should the communicator present the strongest arguments first or last? Presenting the strongest arguments first establishes attention and interest in the position, especially when it is doubtful the target audience will attend to the whole message, as is often the case in written communications such as newspapers or brochures.

Formatting can affect an information campaign. In print media, message size and position are often combined with the use of illustrations, color, and shapes in order to increase attention to the information. On-site interpretive displays may utilize a variety of print, pictorial, and audio components to enhance the effectiveness of information. For radio spots, the choice of words and voice qualities such as speech rate and rhythm are important. The complexity of these formatting considerations emphasizes the need to consider using consultants with expertise in the area of developing messages and media.

The effectiveness of the information program

The most effective way to obtain the desired response from the target audience is to ensure that they pay attention to and think about the message. Following are several factors that influence whether an individual will do so (Bright and Manfredo 1993).

Comprehension of the message. Make sure that the language used in the message is consistent with the knowledge and education level of the target audience. Technical jargon should be avoided unless it is an important part of the message and is adequately explained. Message comprehension also suffers if the audience is distracted while tending to the message. For example, appeals to avoid feeding wildlife may not be heeded if delivered using road signs where individuals are too distracted to pay attention.

The use of verbal and nonverbal components. Where possible, combine visual or audio (nonverbal) elements of an information campaign with the written or spoken word (verbal). This increases the ability of the target audience to comprehend and retain information. For example, a list of rules and regulations regarding appropriate viewing behavior may be more effective if visual components (e.g., pictures showing individuals complying) are combined with written messages.

Repeating the message. Repeating a message affects comprehension and persuasiveness. This is done by presenting the same message several times, repeating it in different formats, or using several sources to present the message. However too much exposure to a message can cause it to loses its effectiveness. Changing the form or vehicle of the message without altering the content can help.

The relevance of information. It is important to ensure that information is relevant to the target audience. For example, highly specialized wildlife viewers might want messages that provide them with knowledge and information about the natural history of unique wildlife species or where to seek high-solitude experiences.

The prior knowledge and experience of the target audience. The complexity of a message should be geared toward the knowledge and experience level of the audience for which it is intended. For example, knowledgeable and experienced people may be familiar with basic information, and hence less motivated to read it. On the other hand, individuals with low levels of knowledge and experience may not be able to comprehend complex messages.

The quality and presentation of the message. The quality and presentation of arguments influence an audience's response to the information. It stands to reason that strong convincing arguments should be used when providing information about a wildlife-management issue to the public. Providing information in a condescending (e.g., patronizingly simple) manner or assuming that the public views the situation in a certain way (e.g, agencies often erroneously assume that wildlife viewers are not in favor of hunters) may have the opposite effect to that intended. For example, if an agency is involved in closing access to a traditional outdoor-recreation area in order to preserve wildlife habitat, well-thought-out reasons for such closure should be included in the information program.

Source credibility. Two aspects of source credibility are *expertise* and *trustworthiness*. If the target audience perceives the agency as having expertise or knowledge related to the issue and also as trustworthy in providing information, they are more likely to consider the information and accept it. Although expertise can be supported by evidence (e.g., data, degrees, years of experience), trust is often difficult to build and may fluctuate and vary from issue to issue.

Attitude toward the information program. Similar to source credibility, if the audience likes the method or format by which information is presented, they will be more likely to (a) think carefully about information presented and/or (b) respond positively to the information.

In practice: *Phase 3. Messages.* The viewing planner develops messages for creative experts that describe places with opportunities for solitude, while messages for novices describe accessible areas with concentrations of popular wildlife. The messages are targeted for the level of knowledge and interest of each group. Messages for novices have a lighter tone and include more visual elements. Messages for experts provide more information on species' life histories and have a more serious tone.

Phase 4. Select the Media and Implementation

This phase addresses the question: *How and when will the information be provided?* Although we describe selection of the media strategy as a phase following development of the message, it is likely that these two phases will be addressed simultaneously; while determining message content, the media planner should also consider what media would be best to present those messages.

Selection of media

There are many categories of media for an information campaign (e.g., television, radio, print), each with advantages and disadvantages based on cost and ability to reach the target audience. The main considerations in determining media type are (1) the media habits of the target audience, (2) the suitability of the messages to the media type (e.g., long messages would be inappropriate for billboards), (3) the timing and complexity of the message, and (4) the cost. Table 13-3 describes the advantages and disadvantages of several types of media.

To decide which media should be used, factors such as circulation, costs for different advertisement sizes, color options (if applicable), advertisement positions (e.g., within a newspaper or magazine), and quantities of insertions (e.g., for brochures in a newspaper) should be considered. In addition, qualitative characteristics of the media like credibility, prestige, geographic reach (e.g., different editions of the same magazine or newspaper), reproduction quality, and psychological impact are important (Kotler and Andreasen 1987).

Two of the most important considerations are reach and frequency. *Reach* is the number of individuals exposed to a particular message. For an interpretive display at a roadside wildlife-viewing location, reach would be described as the number of people who pull over and are exposed to the information in the display. Paid research services, including Arbitron, A. C. Neilsen, Leading National Advertisers, Inc., and Mediamark Research Inc. supply information on vehicle-exposure data for TV and radio in geographic areas,

text continues on page 297

Besides traditional media, such as print, potential viewers can receive information from new technologies, including video, cd-rom, television and cable stations tailored to wildlife programs, and the increasing possibilities of the Internet. (Photos by Cynthia Pierce)

Table 13-3. Media Types, Examples, Advantages and Disadvantages.

Examples	*Advantages*	*Disadvantages*

Television (broadcast, cable, videocassette)

Examples	*Advantages*	*Disadvantages*
Public service announcements (PSAs)	High population reach	Can be expensive
Paid advertisements	In-depth examination of issues	Low audience selectivity
News and feature stories on issues with wildlife management professionals	Potential for timely and flexible coverage	
Themes in entertainment programming	Combines sight, sound, and motion in the message	
Educational programs	High audience attention	

Radio

Examples	*Advantages*	*Disadvantages*
PSAs	High frequency	Lower attention than TV
Paid advertisements	Low cost (though rate structures are non-standardized)	
News and feature stories	High audience selectivity	
Broadcast/information channels near recreations site	Potential for timely and flexible coverage	
Audio tape tours	Can create stimulating messages	

Newspapers

Examples	*Advantages*	*Disadvantages*
PSAs	Broad reach for a variety of target audiences	Short life
Paid advertisements	Excellent in communicating complex issues	
News and feature stories	Potential for timely and flexible coverage	
	Good local audience coverage	
	High credibility in newspapers of record	

table continues

Magazines

Examples	Advantages	Disadvantages

PSAs
Paid advertisements
News and feature
  stories
Newsletters
	Highly selective
	  geographically
	  and demographically
	High credibility and
	  prestige
	Long life
	Good pass-along
	  readership
	Excellent in
	  communicating
	  complex information
		Can be expensive
		Long lead time

Direct mail

Examples	Advantages	Disadvantages

Brochures
Pamphlets
	Opportunity for
	  personalization
	Allows for selectivity,
	  flexibility, and control
		Can be expensive
		"Junk mail" image

Billboards

Examples	Advantages	Disadvantages

Paid advertisements	High geographic
	  selectivity
	Relatively inexpensive
	  and effective
	High repeat exposure
		Low audience
		  selectivity
		Simple messages only

Displays

Examples	Advantages	Disadvantages

Interpretive displays in
  visitor centers
Viewing platform
  displays
Interpretive trails
Museum displays
	Reaches specific
	  audience of interest
	May be maintained
	  and updated on a
	  continuous basis
	Can provide interactive
	  and detailed
	  information
		Static media is usually
		  of less interest
		Requires skill in
		  display design or
		  hiring of consultant
		Tends to become
		  forgotten and
		  updated
		  infrequently

Internet site

Examples	Advantages	Disadvantages
Agency/organization web site Virtual visits (opportunities to experience the activity online)	High reach and frequency May be maintained and updated on a continuous basis May be used in conjunction with other campaign media Relatively inexpensive and effective n-depth and creative examination of issues	Low selectivity that can be increased by referencing in more selective media vehicles Requires experienced personnel to maintain and monitor

Interpersonal

Examples	Advantages	Disadvantages
Field contacts Onsite programs Tours	Extremely interactive Can be quite inform-ational and persuasive for short-term change Audience tends to be very interested in messages	Personnel requirements Effectiveness varies by personnel skill, etc. Limited access (it cannot be provided continuously)

Based on Alcalay and Taplin (1989), Bagozzi (1986), Brown and Einsiedel (1990), Kotler (1982), and O'Sullivan (1991).

specific magazines, and newspapers. But remember that different media vehicles vary in how *selective* they are in reaching specific audiences. For example, a magazine devoted to bird watching would likely reach a smaller, more selective audience than would a general-interest magazine found on a newsstand.

Frequency is the number of times the target audience will be exposed to the message. For example, a roadside billboard along a high-use commuting route would provide high frequency for individuals commuting along that route. On the other hand, an interpretive display at an interstate rest area would likely have low frequency since it is unlikely that many travelers would return to that rest area later. Frequency can be controlled somewhat for certain media, for example, by repeatedly running a radio message. It is important to remember, however, that reach and frequency only indicate how many individuals are receiving the communication and how often. It does *not* imply that information will actually be attended to, processed, remembered, or acted upon.

Media timing and implementation

The next step in developing the media strategy is to decide on the timing and implementation of the campaign. Implementing the information campaign involves disseminating the information in the vehicles selected to the intended target audience(s). The primary challenge is to ensure that:

- There is a clear delegation of responsibilities,
- Tasks are spelled out along with a timeline for their accomplishment, and
- There is high attention to detail.

As part of the planning process, it is advisable that the media planner develop a matrix that organizes the media vehicle used with the timing of the campaign. Several timing patterns may be considered. One option is seasonal timing. Most wildlife-based recreation experiences, such as viewing migratory birds or hunting elk, take place during a particular time of year. Information campaigns about specific wildlife-related recreation experiences should be timed to help target audiences plan for trips.

Similarly, short-run timing distributes efforts during a brief period. For example, information about elk-viewing opportunities during the fall might be included as a supplement to the Sunday newspaper. Alternatively, effort may be spread out using daily public service announcements aired on the radio throughout the week as well as on weekends; a specific interpretive display at a wildlife-viewing center that is in place for the entire viewing season is another example. A burst pattern includes several exposures in succession with no exposure between bursts. For example, the elk-viewing supplement in the Sunday newspaper might be followed up by small-scale advertisements in the newspaper or on radio or television periodically over the following three weeks to remind the public of the opportunities. This pattern can create high attention and interest yet retains reminder advantages. Table 13-4

Table 13-4. Example matrix planner connecting media vehicle with timing of information campaign

	Week 1							Week 2						
Media Vehicle	S	M	T	W	Th	F	S	S	M	T	W	Th	F	S
Newspaper ad	X					X	X	X					X	X
Internet site	X	X	X	X	X	X	X	X	X	X	X	X	X	X
Radio spot (PSA)					X	X						X	X	
Direct mail brochure		X							X					
T.V. ad			X		X					X		X		

provides an example matrix planner that connects the media vehicle used with the timing of an information campaign.

In practice: *Phase 4. Media.* Different media are chosen for the two target audiences based upon the messages and the best way of reaching these groups. Creative novices tend to seek general information from visitor centers or viewing sites. They also want items with souvenir value and maps. Folded maps describing likely viewing sites for their interests are developed, along with displays at relevant viewing sites. Creative experts want more specific information about sights with low visitation. Media chosen for this group include a phone line describing current accessibility and animal sightings at certain areas, as well as written pieces inserted in the agency's magazine, with press releases distributed to other relevant outside publications. The planner may also schedule personal presentations for local wildlife-photography clubs.

Phase 5. Evaluate the Information Campaign

This phase addresses the question: *How will you measure achievement of your communication objectives?* Unfortunately, evaluation of the communication program or message delivery is typically overlooked, but unless an evaluation is conducted, the organization will not be able to determine whether or not their efforts were successful, especially in terms of meeting objectives. This type of evaluation is called summative. Evaluation of different stages of the information campaign is called formative evaluation.

Formative evaluation

Summative evaluation examines the effects of an information and education program against its goals and objectives. However,

It is important to consider how, or if, viewers will actually use information and what will improve or detract from its effectiveness. Having clear objectives can help determine whether messages were effective. (Photos by Cynthia Pierce)

research may be conducted to improve message delivery in developing an information and education campaign, not just at the end when evaluating the outcomes of the campaign. This type of research is called *formative evaluation* (Atkin and Freimuth 1989). Formative evaluation can be further divided into pre-production research and pre-testing.

Pre-production formative research. Pre-production formative research gathers information about target audiences. It helps with (a) identification and description of target audiences, (b) determination of the desired audience responses, and (c) selection of the most effective media for reaching target audiences.

Pre-testing. The second phase of formative evaluation is pre-testing or copy-testing. This is the process of measuring audience reactions to preliminary versions of messages. There are two primary stages at which pre-testing may be used: concept development and message execution.

In developing the concept, pre-testing can provide direction and save time and money by narrowing down choices before the production of the message is complete. For example, in an information campaign designed to enhance support for a tax to protect wildlife habitat, pre-testing might show that an emphasis on the protection of a threatened wildlife species would be more effective than focusing on an increase in wildlife-viewing opportunities.

Pre-testing may also be used to test message execution. The purpose of this stage of pre-testing is to predict how effectively a message will elicit the desired responses from the target audience(s). More specifically, pre-testing can (a) assess the level of attention toward a message, (b) measure comprehension, (c) identify strong and weak points of the message, (d) determine the personal relevance of the message, and (e) gauge sensitive or controversial elements of the message.

Methods of conducting formative evaluation. Several research methods are used in formative evaluation. Focus-group interviews and self-administered questionnaires help in both pre-production and pre-testing. Individual in-depth interviews are often used to address sensitive issues or when a focus group is not practical. Often, these are done at a location frequented by individuals from a specific target audience such as a particular wildlife-viewing area. Self-administered questionnaires are useful for both the concept-development and message-execution stages of pre-testing, and are more structured than focus groups or in-depth interviews. They also allow testing of messages in the environment in which they will be delivered.

Summative or impact evaluation

The most common type of program evaluation is *summative or impact evaluation*. It is conducted once the media messages have been developed and the plan implemented. There are three reasons that summative evaluation of an information campaign is conducted (Brown and Einsiedel 1990). The first reason is *accountability*. For example, government organizations such as wildlife-management agencies are naturally interested in justifying public funds spent on information campaigns by demonstrating their effectiveness. Second is *replication and generalizability*; that is, the extent to which the information campaign can be repeated or revised, and whether it could be used in other situations. Third is an *assessment of organization objectives*. Summative evaluation can determine whether the agency is meeting its overall goals and objectives and/or the need to establish new objectives. Three evaluation models may be used in an impact evaluation: the *advertising approach*, the *impact-monitoring approach*, and the *experimental approach*. Table 13-5 summarizes the advantages and disadvantages of each impact evaluation approach.

Table 13-5. Advantages and Disadvantages of Impact Evaluation Approaches.

Advantages	Disadvantages
Advertising approach	
Sensitive to changes in attention, interest, and desire	Does not address change in behavior
	Cannot make causal inferences
Easy and inexpensive to administer	Low relevance to policy issues
Appropriate for new programs	
Appropriate for assessing audience perceptions of information campaign	
Impact-monitoring approach	
Sensitive to changes in behavior	Insensitive to changes in attention, interest, and desire
High relevance to policy issues	
Easy and inexpensive to administer	Cannot make causal inferences
Experimental approach	
High validity	Expensive
Can make causal inferences	Requires high level of expertise
High relevance to policy issues	Ethical issues related to withholding information from groups as part of research design

Based on Brown and Einsiedel (1990) and Flay and Cook (1989)

The advertising approach to impact evaluation. The advertising approach surveys a random sample of the target audience. The survey is designed to learn whether the target audience (a) recalls and recognizes the message, (b) likes the campaign materials, and (c) intends to act on their new knowledge and attitudes (Flay and Cook 1989). Identifying the relationship between exposure to the information and evidence of audience response assesses campaign effectiveness. This approach is appropriate for evaluating audience attention, interest, and desire.

The impact-monitoring approach to impact evaluation. The impact-monitoring approach tracks the responses that occur as a result of the information campaign. This can include monitoring the number of requests for information about a particular wildlife-viewing opportunity or actual visitation to a site at periodic intervals following the campaign. This provides the agency with an assessment of immediate and long-term effects of the campaign. The impact-monitoring approach is appropriate for assessing the effects of an information campaign on action or behavior, rather than attention, interest, or desire.

The experimental approach to impact evaluation. While the impact-monitoring approach can determine changes in behavior, it is sometimes difficult to determine whether changes are attributable to the campaign itself. The experimental approach systematically compares groups exposed to the information campaign to those not exposed to it. This helps identify whether exposure to the campaign is the cause of behavior. For example, a multi-community campaign might be conducted whereby newspaper media are used to provide information about a wildlife-viewing opportunity to one community, newspaper and interpersonal communication are used for a second community, and no information is provided to a third community. The evaluators would examine differences in attention, interest, desire, and action between the three communities.

Other considerations for summative evaluation. In addition to evaluating direct impacts, the process by which the campaign was administered should be assessed. *Process evaluation* provides information about which elements of the information campaign worked. To evaluate the process, the following questions should be addressed:

(1) Did materials reach their audience (reach)? How often did they reach their audience (frequency)? Were they understood? Were they used as intended?

(2) Did the media vehicles present the information when and where planned?

(3) Did the target audience respond in the ways intended?

(4) What else occurred in the target audience's environment, over and above the information campaign, that might have affected attention, interest, desire, and action?

> **In practice:** *Phase 5. Evaluation.* In developing the messages and media for the information campaign, the planner uses formative evaluation by having focus groups evaluate the folded maps and displays. After a personal presentation to a wildlife-art class at a local university, the planner asks students to call the new phone line. Shortly afterwards, questionnaires are sent to those who have called in, to get their reactions. Prior to implementation, baseline information is gathered by surveying samples of expert and novice creative viewers on their awareness of relevant opportunities. Sometime after program implementation, at the end of the program year, new surveys are conducted to determine whether a change in awareness has taken place and the program objectives have been accomplished!

Summary Points

Given the importance of effective communication regarding wildlife-viewing opportunities, agencies can improve their information and education outreach efforts by ensuring they are planned and purposeful, and meeting real needs. EBM helps provide a structure for developing communication plans by using a wildlife-viewing typology, based on the varying experiences different segments of the public desire. This typology can serve as the basis for the communication plan, by providing a framework for identifying target audiences, developing objectives and messages for these audiences, selection of media, and conducting evaluations. If there is a systematic approach, the campaign will be more likely to meet its objectives and result in improvements in future communications with stakeholders and interested publics. The information and education campaign should address the following phases.

Phase 1. Determine the objectives of the information campaign. The first step is to determine *why information is being provided.* Objectives provide a specific, measurable description of what the information and education campaign should achieve, and standards for evaluating success.

Phase 2. Identify target audience(s) in the campaign. The second step is to determine *whom information will be provided to.*

EBM can be used to identify target audiences based on (a) the experience outcomes that different segments are looking for; (b) the kinds of activities the segments desire; and (c) the social, managerial, and environmental settings that would best provide these experiences. Agencies should be sure that audiences could be realistically segmented. Identifying the desired audience response focuses on what you want your target audience to do with the information.

Phase 3. Develop the message. This step determines *what kind of information will be provided.* In constructing messages, it is important to consider the message content, execution, style, tone, and ordering of ideas. Factors affecting message effectiveness include comprehension, the combination of verbal and nonverbal components, message repetition, relevance of information, prior knowledge of the audience, the credibility of the source, and the audience's attitude toward the ad or other message.

Phase 4. Media and Implementation. This step determines *how and when the information will be provided.* Media selection is based upon the most cost-efficient and effective ways of reaching target audiences. Each media type has advantages and disadvantages in terms of reach, frequency, selectivity, cost efficiency, etc., that should be considered in light of the goals, objectives, and available budget of the agency and its information program. There should also be careful consideration of the timing of media and message placement.

Phase 5. Evaluate the information campaign. Finally, *the achievement of communication objectives should be measured.* Formative evaluation provides research to help in developing the information campaign, including testing of messages and media. Summative evaluation examines the effectiveness and efficiency of the information campaign after it has been completed. Evaluation helps to (a) exhibit accountability of the campaign, (b) allow for improvement and replication of the information campaign in the future, and (c) determine how the wildlife-viewing information fits in with overall agency goals.

Literature Cited

Alcalay, R., and S. Taplin (1989). "Community health campaigns: From theory to action." In R.E. Rice and C.K. Atkin (Eds.), *Public Communication Campaigns*, 2nd edition. Newbury Park, CA: Sage Publications.

Andreasen, A. R. (1995). *Marketing Social Change: Changing Behavior to Promote Health, Social Development, and the Environment.* San Francisco, CA: Jossey-Bass.

Atkin, C. K., and V. Freimuth (1989). "Formative evaluation research in campaign design." In R.E. Rice and C.K. Atkin (Eds.), *Public Communication Campaigns*, 2nd edition. Newbury Park, CA: Sage Publications.

Bagozzi, R. P. (1986). *Principles of Marketing Management*. Chicago, IL: Science Research Associates.

Bright, A. D., and M. J. Manfredo (1993). "An overview of recent advancements in persuasion theory and their relevance to natural resource management." In A. W. Ewert, D. J. Chavez, and A. W. Magill (Eds.), *Culture, Conflict, and Communication in the Wildland-Urban Interface*. Boulder, CO: Westview Press.

Brown, J. D., and E. F. Einsiedel (1990). "Public health campaigns: Mass media strategies." In E.B. Ray and L. Donohew (Eds.), *Communication and Health: Systems and Applications*. Hillsdale, NJ: Lawrence Erlbaum Assoc.

Fine S. H. (1991). *Social Marketing: Promoting the Causes of Public and Nonprofit Agencies*. Boston, MA: Allyn and Bacon.

Flay, B. R. and T.D. Cook (1989). "Three models for summative evaluation of prevention campaigns with a mass media component." In R.E. Rice and C.K. Atkin (Eds.), *Public Communication Campaigns*, 2nd edition. Newbury Park, CA: Sage Publications.

Grunig, J. E. (1989). "Publics, audiences and market segments: Segmentation principles for campaigns." In C. T. Salmon (Ed.), *Information Campaigns: Balancing Social Values and Change*. Newbury Park, CA: Sage Publications.

Kotler, P. (1982). *Marketing for Nonprofit Organizations*. Englewood Cliffs, NJ: Prentice-Hall.

Kotler, P., and A.R. Andreasen (1987). *Strategic Marketing for Nonprofit Organizations*, 3rd edition. Englewood Cliffs, NJ: Prentice-Hall.

Manfredo, M. J., and R. A. Larson (1993). "Managing for wildlife viewing recreation experiences: An application in Colorado." *Wildlife Society Bulletin*, 21(3), 226-36.

O'Sullivan, E. L. (1991). *Marketing for Parks, Recreation, and Leisure*. State College, PA: Venture Publishing.

Pierce, C. L., and M. J. Manfredo (1997). "A profile of North American wildlife agencies' viewing programs." *Human Dimensions of Wildlife*, 2(3), 27-41.

Simcox, D. E., and R. W. Hodgson (1993). "Strategies in intercultural communication for natural resource agencies." In A. W. Ewert, D. J. Chavez, and A. W. Magill (Eds.), *Culture, Conflict, and Communication in the Wildland-Urban Interface*. Boulder, CO: Westview Press.

Simon J. L. (1971). *The Management of Advertising.* Englewood Cliffs, NJ: Prentice-Hall.

Stout, R. J., and B. A. Knuth (1993). "Using a communication strategy to enhance community support for management." In J. B. McAninch (Ed.), *Urban Deer: A Manageable Resource?* Proceedings of a Symposium held at the 55th Midwest Fish and Wildlife Conference, St. Louis, Missouri, December 12-14, 1993.

Marketing Wildlife-viewing Experiences

Mark Damian Duda and Steven J. Bissell

Introduction

The purpose of this chapter is to provide an overview of the principles of marketing as they apply to wildlife-related recreation. The examples we will show are mostly concerned with wildlife viewing, but can be applied to any wildlife-related experience. Most of our experience is with public agencies and we use many examples from that context. However, marketing wildlife viewing by the private sector is also a major consideration and the outline of the marketing process we provide here is equally applicable to private or public organizations.

The term "marketing" is one of the most misused and most misunderstood terms within fish and wildlife recreation. Often equated with hard selling, cheap selling, trickery, or just simply promotion, many fish and wildlife managers and administrators shy away from learning what marketing really is and how utilizing a marketing approach can contribute to their goals of developing successful wildlife-viewing programs.

Marketing is not hard selling, as some detractors think. It is not even selling. In fact, marketing is the opposite of selling. As marketing expert Phil Kotler (1980) notes, selling focuses on the needs and desires of the seller (in this case, the agency or organization). Marketing, on the other hand, focuses on the needs and desires of the constituent or customer (in this case, the wildlife viewer). As business manager Peter Drucker observes, "Marketing is the whole business seen from the point of view of its final result, that is, from the customer's point of view" (quoted in Kotler 1980). Whether agencies or organizations are simply beginning to develop wildlife-viewing programs, attempting to increase funding for wildlife-viewing programs, developing a wildlife-viewing guide or brochure, interested in enhancing the satisfaction of wildlife viewers, trying to minimize conflicts between wildlife viewers and other outdoor recreationists, or attempting to alter harmful or injurious behavior such as wildlife feeding, habitat trampling, or the tendency for some wildlife viewers to try and get too close to animals, a marketing approach will make efforts toward these ends more successful.

Marketing Defined

Marketing describes a number of different activities. In this chapter, marketing is a step-by-step process that begins with people (the markets) and ends with programs, products, services, and strategies. Note that this is the opposite of how many programs, products, and services are developed; they start with a program, product, or service, and then look for constituents and customers to use them. Too many times, problems are identified and solutions immediately developed. But how does one know that the chosen solution is the best solution? And as importantly, what kind of thinking and process went into that solution? For every problem, there are hundreds of potential solutions and hundreds of variations on a program, product, or service. How is the best solution chosen? How is the best, most effective wildlife-viewing program, product, or service developed?

Marketing is a process that assists agencies and organizations in making the right decisions because it takes them through a series of smaller decisions and information-gathering procedures that assist in reaching larger decisions. By following a marketing approach, the "what to do" part (strategies, programs, products, and services) emerges from the process. Solutions do not have to be pulled out of thin air. A marketing approach takes out the guesswork. It leads to the most effective decisions and the development of the most appropriate wildlife-viewing management strategies, programs, products, and services.

Within the context of fish and wildlife recreation, marketing is the deliberate and orderly step-by-step process of first defining what it is—exactly—that one is trying to achieve; understanding and defining different groups of constituents (markets) through research; and then tailoring programs, products, and services to meet those needs through the manipulation of the marketing mix, also called the four Ps—product, price, place, and promotion.

The purpose of marketing in an agency or organization is to better meet both the goals and objectives of the organization and the needs of its wildlife viewers. A marketing approach to experience-based wildlife-viewing recreation will enhance the ability of the agency to protect and conserve wildlife and habitat while still providing wildlife viewers with quality wildlife-viewing programs, products, and services.

Five Principles of the Marketing Process

There are five important points about the marketing process as we have defined it above. First, and perhaps most importantly, although marketing is customer driven, it is always within the context of the mission statement and goals of the organization. Within the context of fish- and wildlife-viewing programs, it should be understood that marketing is always practiced within the constraints of resource protection and ethical and appropriate wildlife-viewing behavior. Marketing is not, and should not be viewed as "giving wildlife viewers everything they want," regardless of the ramifications on the resource. This is why the marketing process begins with the organization's mission statement as well as programmatic goals and objectives.

Second, this definition of marketing assumes that different groups of wildlife viewers require different programs, products, and services. For example, research clearly indicates that residential wildlife viewers are quite different from nonresidential wildlife viewers. One important difference is the age of participants. People who participate in wildlife viewing around their home are more likely to be older. While 32 percent of U. S. residents sixty-five years and older participated in residential wildlife viewing, only 6 percent of the same age group participated in nonresidential wildlife viewing (U. S. Department of the Interior 1997). Highly specialized bird watchers are a different market from more casual wildlife viewers (Kellert 1996). In fact, market segmentation and the concept that different wildlife viewers seek different wildlife-viewing experiences are fundamental to Experience-based Management (EBM), which defines recreation demand in terms of the desired psychological outcomes, activities, and settings associated with participation. Different wildlife viewers have different desired outcomes and these outcomes necessitate different management strategies. Third, this definition of marketing in no way implies that marketing always means "doing more of something" or "getting more people to do something." The marketing process is a systematic planning process that assists in achieving the goals and objectives defined by the agency or organization. For example, marketing can be used to either increase or decrease the number of wildlife viewers at a wildlife-viewing site. In most cases, overcrowding at wildlife-viewing sites is isolated geographically and temporally; that is, overcrowding usually occurs only in a few of the best viewing spots at certain times of the day, or certain days of the week. Marketing can be used to better manage demand through the manipulation of the marketing mix—product, price, place, and promotion. For example, demand could be managed by making viewing at the more popular wildlife-viewing sites more expensive at peak times or, if there is no charge, to begin requiring payment (price manipulation). Other methods could include making less-popular sites less expensive or free (another manipulation of price), developing alternative viewing areas (product and place manipulation), or emphasizing other alternative viewing spots or species (promotion manipulation). The marketing process can be used to enhance the satisfaction of wildlife viewers, in this case by spreading out demand, but also can be used as a conservation tool by spreading out and managing demand. The science of marketing is in no way antithetical to fish-and wildlife-conservation goals, but can and should be used as a means of protecting the resource from wildlife viewers. Again, the term does not always mean more of something. In fact, the objective of one type of marketing, "counter marketing" is to destroy demand, such as when wildlife viewers are feeding wildlife such as ground squirrels or white-tailed deer.

Fourth, this definition of marketing should be used as a structure for and process in decision making in a variety of contexts; from spreading out wildlife-viewing demand to promoting different types of wildlife viewing to identifying an equitable allocation of the resource. Marketing is simply a process, and the process and structure are extremely flexible.

Finally, marketing is sometimes viewed as synonymous with promotion, or even more incorrectly, advertising. However, promotion, especially advertising, is only one part of the entire marketing process, and comes at the end of the process. This is because many different decisions need to be made before one gets to decisions about promotion: What is it that one is trying to achieve? Who is one trying to influence? And specifically, what is the product (or program or service), price, and place that has been developed to meet this market's needs? In this context, promotion simply becomes informative and educates a specific market that the agency or organization has a program, product, or service that meets their needs.

The Concept of Marketing in Wildlife Recreation

Marketing is a new concept to many fish and wildlife managers, but familiar to the private sector dealing with wildlife-related recreation. Thus many questions about the use of marketing will occur within natural-resource organizations and agencies. For example, one question is, "Should natural-resource organizations and agencies market?" The answer is that most agencies and organizations should take a marketing approach when interacting with their constituents. Marketing is providing programs, services, and products to wildlife viewers based on their needs within the constraints of resource protection. An approach that focuses strictly on attempting to create a demand for a product or attempting to create interest in a program, product, or service or promoting a program through increased promotional efforts is only part of marketing. All of the promotional efforts in the world will not create a demand for a product that does not fit a market.

Creating or Meeting Demand

Another question commonly asked by agencies is, "Should agencies meet demand for wildlife recreation or create demand?" We think that agencies should both meet current demand and develop programs to meet latent demand. Survey research indicates that there is high latent demand for wildlife-viewing opportunities. For example, among Maryland residents, while 21 percent of the population had actually taken a wildlife-viewing trip, 62 percent said they would be interested in doing so. This represents a major latent demand for wildlife viewing (Responsive Management 1993).

Additionally, it appears that it is in the best interest of the fish- and wildlife-recreation profession to have many active users. Active users are active constituents. In the past, fish and wildlife agencies' most active constituents have been their most active users—anglers and hunters. In the future, these agencies' most active constituents will also include the wildlife viewer. As Riley (1985) notes, "Reverence and respect are hard to teach ... we cannot expect the voters of tomorrow to support conservation measures if they are not active users of our natural resources."

Moreover, research indicates that children who participate in wildlife-related activities such as wildlife viewing, fishing or hunting know more about wildlife, appreciate wildlife more, are less fearful of wild animals, and exhibit less anthropomorphic tendencies toward wildlife than children who do not participate (Kellert and Westervelt 1980; Westervelt and Llewellyn 1985). In short, wildlife-related recreation is an excellent wildlife-education tool. The fear that promoting fish- and wildlife-recreation use will place undue stress on the resource has little merit. Professional wildlife managers can mitigate potential conflicts through creative and intelligent management of wildlife and people through the use of the marketing process. The benefits of creating active users clearly outweigh any potential disadvantages.

Some critics of marketing in natural-resource agencies and organizations claim that marketing and promotion is the realm of private industry; that public agencies and nonprofit organizations have no legitimate role in engaging in these activities. However, this is not a prevalent viewpoint. Public agencies have a long history of marketing and promoting a variety of activities and products. The State of Florida actively markets and promotes tourism, the consumption of Florida seafood, and park visitation, to name a few. The State of Virginia actively markets and promotes Virginia as a tourist destination and the State of California actively markets its wine industry. There are hundreds of examples of governmental agencies and nonprofit organizations marketing their programs, products, and services. These agencies and organizations need to learn from the private sector and to assist the private sector in the marketing of wildlife-viewing recreation.

The Costs of Marketing

Another common perception is that "marketing costs too much; the money should be spent on more tangible programs ... Agencies have more important activities to spend money on." Financial planners are quick to point out when investing, "It's not what it costs, it's what it pays." Relatively small amounts of money invested in marketing can reap large benefits in numerous ways: increased financial support from wildlife viewers; decreased wildlife-viewing impacts; decreased habitat destruction at wildlife-viewing sites; and increased political support from wildlife viewers. Additionally, utilizing a marketing approach can make more efficient use of funds that are available. Money spent on ill-conceived programs, products, and services is money wasted. The development of a nature center that is not visited, a brochure that is not read, or a wildlife-viewing site that people do not stop at is money wasted.

Public Opinion, Marketing, and Policy

Engaging in marketing does not mean that natural-resource agencies and organizations should let public opinion dictate wildlife-viewing policy. For example, market research may reveal that public attitudes on issues are quite different from those of the agency. If public attitudes are detrimental to the resource, the agency now has baseline data upon which to base educational and public-relations efforts. The market research will allow the organization to base these efforts on a solid foundation of fact. On the other hand, if survey research reveals that public attitudes toward an issue are different from current regulations or policies but within the boundary of resource protection, policy and regulation changes could be initiated as positive public relations.

The value of effective marketing arises from the notion that a product, program, or service cannot be all things to all people. Yet government agencies were set up to serve all people. On its face, then, here is an inherent dilemma for public-agency marketing. This argument would hold a grain of truth in a perfect, theoretical world. However, government agencies have always had to choose among projects with various levels of benefits to different groups. It is a fact that with limited resources, agencies must pick and choose programs, products, and services. Agencies cannot do all things; they also cannot be all things to all people. Good planning coupled with a marketing approach compels agencies to choose highest-priority projects and to choose target groups. Marketing will make the limited dollars available for limited projects more effective by targeting specific markets or market niches rather than attempting to assume that all markets are important, or, worse, that all products and services are equally attractive and require equal marketing.

Examples from the Private Sector

The private sector has been more actively involved in marketing wildlife-viewing recreation than have public agencies or organizations. In some cases this has involved a major change in the business of these groups. For example, in Virginia, outfitters that had previously specialized in adventure kayaking, switched to dolphin watching in late summer to accommodate a growing market in that wildlife-viewing experience. Other private operators such as tour companies followed suit. As a result, dolphin watching in Virginia has grown with little impact on the resource and no major effort from public agencies.

A simple Web search on LookSmart.com (http://www.looksmart.com/) for ecotourism produced wildlife-viewing opportunities from Jordan to Namibia, from Montreal to the North West Province, and two major publications about wildlife viewing in the private sector. The use of the Web as a marketing tool is growing rapidly in both the private and public sectors. All of the principles of the marketing process we discuss in this chapter apply to marketing using the Web.

Other examples of private-sector wildlife viewing include the growing whale-watching opportunities on both coasts of the United States. In Africa, wildlife viewing has become a major economic influence, with some countries such as Kenya deriving a significant portion of national revenue from wildlife-viewing-related tourism.

Nonprofit and nongovernmental organizations have a long history of marketing products in order to promote and finance their organizations. This marketing experience is no less important than their experience in working with management agencies on resource-protection issues. Agencies should not be reluctant to use the expertise of these private-sector wildlife-recreation and -protection groups in developing marketing programs.

The Marketing Process

A successful marketing approach is really a rational planning system. It addresses the issue of an organization's or agency's current status through a situation assessment, then develops precise objectives, determines the best way to achieve the objectives, and, finally, evaluates the success of the plan.

This section will propose one approach to the development of a marketing plan. This is largely adapted from Kotler (1980) and is only one approach. There are other ways to develop marketing plans, but most will include the elements we give herein in some form.

A. Develop a Mission Statement

Every organization and agency should have a mission statement. Mission statements let people know why the organization exists and what it is trying to achieve. Everyone in the organization should be familiar with the mission statement and it should be posted throughout the organization. Everything that follows in the marketing plan is based directly on the mission statement of the organization. A mission statement is not a slogan; it is the philosophical basis of the organization. A mission statement may be short and to the point or it may be as lengthy as several hundred pages. The point is that a mission statement dictates all that follows in the sense that it provides a template against which all other actions are judged.

B. Determine Goals

Goals define the management philosophies within which objectives are pursued (Crowe 1983). Goals are broad and lofty statements about the desired outcome of a program. The goal of an organization that wants to increase the number of wildlife watchers might be: "To increase the number of wildlife-watching participants nationwide." The goal of a wildlife-viewing education program may be: "To inform and educate the public about raptors." Anything more specific will come in the objective-setting portion of the marketing plan. Committing goals to paper becomes more important as one gets further into the marketing plan. This is because different goals require different strategies. Different strategies and programs need to be developed based on whether the goal is related to participation, behavior, or opinions and attitudes. Often the goal is not committed to paper and different people are pursuing different goals, a situation that invites failure from the start.

C. Rationally Identify Your Business

What exactly is one's business? It is such a basic question that it often gets overlooked. In the past, however, organizations that did not fully understand the business they were in have suffered severe consequences. Levitt, in a landmark article in *Harvard Business Review* (1965), pointed out that market definitions of a business are superior to product definitions. He stated that products and technologies, as well as the services and programs offered, are transient, while basic market needs generally endure forever. For example, people will always enjoy listening to music. This desire is a market need. A product such as a record or a compact disc is the result of how that market need has been filled. When thinking strategically, focus on market needs, not on products.

When developing wildlife-viewing information, education, or outreach programs, focus on the market, not on the methods, products, or technology for conveying the information. The market is the information transfer to the customer. The product is the brochure or the slide show. Focus on the needs of the market and let the target market dictate the technology. Thirty years ago a slide show or brochure may have been the best way to disseminate information; today, because of technology changes, they may not be the best ways to transfer information or foster increased knowledge levels.

D. Identify Specific Publics

There is no such thing as a general public and there is no such thing as a general wildlife viewer. This is at the core of the science of marketing and an experience-based approach to the management of wildlife viewing. Research consistently shows that how people relate to fish, wildlife, and natural resources is affected by a variety of factors—their age, race, gender, income, level of education, and a variety of other variables. Moreover, Experience-based Management assumes that different types of wildlife viewers have different desired psychological outcomes, activities, and settings associated with participation. A list of these committed to paper is important and helpful in identifying one's place in a particular market.

E. Select Target Publics

Marketing means making choices and making choices means deciding specifically what groups will be targeted. At this juncture, choices are made as to whom to target. Once a public is targeted, it becomes a market. In most cases, it is fairly easy to choose a market, until one realizes all of the other potential markets that are being omitted. But an important premise of marketing is that a program, product, or service cannot be all things to all people. But this is why the marketing process is so powerful. All the decisions that follow in a marketing plan are based on the selection of the market. In wildlife-viewing promotion, for example, it is clear that programs for avid bird watchers will be different from programs for casual bird watchers. Different markets require different strategies. It is all right to choose more than one market to focus on, but it is important to keep in mind that each group may require different strategies. In many ways, this is the heart of Experience-based Management: What psychological outcomes or settings will be managed for?

F. Assess Current Conditions

This trend-identification portion of the marketing process allows an organization to become proactive rather than reactive. Current conditions can be assessed by identifying opportunities and threats—an organization's strengths and weaknesses. Here are some major trends as they relate to wildlife viewing nationwide between 1991 and 2001 (U. S. Department of the Interior 1997 and 2002; Aiken 1999):

• The total number of wildlife-watching participants (residential and nonresidential) decreased from 76,111,000 in 1991 to 62,868,000 in 1996. However, in 2001 the total number increased to 66,105,000 from 1996.

• The total number of nonresidential wildlife-watching participants decreased from 29,999,000 in 1991 to 23,652,000 in 1996. In 2001, the total number of nonresidential wildlife-watching participants decreased to 21,823,000.

• The total number of residential wildlife watching participants decreased from 73,904,000 in 1991 to 60,751,000 in 1996 and increased again to 62,928,000 in 2001.

• The total amount of wildlife-watching expenditures increased from $18.1 billion in 1991 to $29.2 billion in 1996. Expenditures reached nearly $40 billion in 2001.

When assessing current trends, it is important to keep in mind statistically significant differences between the years. Comparing the 2001 Survey with the two previous surveys shows a 5 percent increase from 1996 to 2001, and a 13 percent decrease from 1991 to 2001 in overall wildlife watching (U.S. Department of the Interior 2002). From 1996 to 2001 the changes in both nonresidential and residential wildlife-watching participation were statistically insignificant (U.S. Department of the Interior 2002). Total wildlife-watching expenditures showed no statistically significant difference from 1996 to 2001, but increased 41 percent from 1991 to 2001 when looking at adjusted figures (U.S. Department of the Interior 2002).

G. Decide on Specific Marketing Objectives

Once a business identifies where it is, the next step is to decide where it wants to be. Objectives are directed toward the accomplishment of goals and are specific and measurable statements of what, when, and how much will be achieved (Crowe 1983). It is important to note where the setting of objectives is placed in the marketing process: at the end of the situation assessment. This is because realistic objectives cannot be set until there is a thorough understanding of where one is presently. Many programs and initiatives fail from the start because objectives are not agreed upon and written down by those involved. Perhaps the best example of obtuse objectives occurs when it comes to "informing and educating" a market about fish and wildlife recreation. Informing and educating a target market is a laudable goal, but not a laudable objective. In the objective-setting portion of the marketing process, this needs to be refined to something more specific, such as increasing factual knowledge, increasing concern, altering opinions, changing attitudes, or altering behavior. Another example, and the purpose of this book on wildlife viewing and Experience-based Management, is that specific experiences offered by different wildlife-viewing sites are almost never identified and incorporated into the planning process. If the objectives are not defined at this point, the process will break down. For example, consider a plan with a goal of increasing the number of nonresidential wildlife watchers back to 1991 numbers. Though those involved may agree that this is an important goal, it needs to be refined further as an objective. There are multiple ways (tasks) of going about increasing the number of nonresidential wildlife watchers (the objective). One way would be to recruit people who have never taken a trip to view wildlife. Another way would be to get back the people who dropped out—many more people took trips for the primary purpose of watching wildlife in 1991 than in 2001. Another way would be to initiate retention programs for current nonresidential wildlife viewers. Still another and more specific experience-based objective would be an increase in the percentage of primitive high-involvement wildlife viewers. All of these could be objectives, but a point in every direction is no point at all. The key is to decide what exactly is going to be done and then develop an objective that communicates it. Additionally, by writing down objectives, members of an agency or organization are compelled to talk about these types of issues before the implementation phase, not after strategies have been implemented.

H. Determine Marketing Strategy

At this point, the plan has identified where the organization is, and where it wants to be. The "marketing strategy" section of the marketing plan identifies how it will get there.

1. Market segmentation

First the market should be segmented. Markets have already been chosen; this section of the marketing plan identifies the specific market segment. Who are they—exactly? As previously stated, there is no such thing as a "general public." There is also no such thing as a "general wildlife watcher." Experience-based planning assists in market segmentation because it allows the market to be segmented with the wildlife viewer in mind. For example, as noted in Chapter 5:

> ... experienced birders may regard the opportunity for a solitude experience in a remote location with no development as highly desirable and refer to it as "high quality." Conversely, these same individuals may regard opportunities designed to accommodate large numbers of recreationists at visitor centers as undesirable and refer to that experience as "low quality." Yet, depending on the motivations and expectations of the visitor, either type of recreation might be considered quality.... If both types of experiences are in demand by the public, management should attempt to provide for both.

These researchers are pointing out market segments based on what types of outcomes are desired from a particular wildlife-viewing experience. Simply put, they are different market segments. These market segments and their desired outcomes must be incorporated into planning and marketing efforts. Distinct groups of recreationists exist within the generic categories of "residential wildlife viewers" and "nonresidential wildlife viewers." Segment the market and use experience-based outcomes as the guide to each segment.

Additionally, what are the demographics of the market segment? What do they want and what do they need? What are their attitudes and opinions about the product, program, or service? Social science and market research is the key to better understanding these markets. There are numerous ways of better understanding these markets, including focus-group research and opinion and attitude surveys.

2. The marketing mix defined

Once a market has been identified and segmented, a program, product, or service is tailored to the specific market. The marketing mix—product, price, place, and promotion—is the set of controllable variables that are used to tailor the program, product, or service to the target market. These four Ps are consistent in marketing plans, although some plans tend to emphasize one over the others. This is determined by the earlier assessment phase.

Product. Product is the most important element in this mix. A product or service is what is offered to the market—from wildlife viewing to information on wildlife and habitats. It is important to recognize that an organization may have many different products. A wildlife-viewing site located minutes from a large urban center is a different product than a wildlife-viewing site located in a remote wilderness area. A wildlife-viewing site that has bathrooms is different than a site without bathrooms for those customers to whom this is an important consideration. Wildlife-viewing information accessible via the World Wide Web is a different product than information contained in a slide show. The key to success is matching the product/experience to the market. Wildlife-viewing management that matches desired experiences with those that are offered increases wildlife-viewer satisfaction, while management that does not match desired experiences with those offered decreases wildlife-viewer satisfaction.

It is also important to differentiate between a product's features and a product's benefits or outcomes. A feature is the makeup of the product or service; a benefit or outcome is what the wildlife viewer receives. Focus on the benefits and outcomes of the product, not the features. This is why an experience-based approach to planning and management for wildlife-viewing recreation is so powerful—it focuses on outcomes/benefits, not features. In the marketing world, it is often said that for most people, the drill is not as important as the hole it makes. Certainly site features such as trails, bathrooms, and brochures are important, but these need to be seen for what they are—product features. More important are the benefits the viewer receives from visiting the site. Focusing on these benefits is at the heart of Experience-based Management—experience (defined as the "bundle" of psychological outcomes desired by participants) and activity (traditional recreation activities that are linked to the attainment of these outcomes and settings). Identify the most important benefits the product can offer the market and communicate those benefits.

Price. Price, both monetary and nonmonetary, is another variable in the marketing mix. Although viewed as less important for public agencies than private businesses, price issues can have profound effects on wildlife-viewing programs and services. The monetary price of marketing is usually the direct cost to the consumer, but it needs to be remembered that consumers also have costs in time and effort. Price can be manipulated in a variety of ways; the most obvious is the actual cost. Price is an excellent way to tailor the overall product to a market to achieve an organization's objective. Using the example of a wildlife-viewing site that is crowded on weekends, but not crowded during the week, price could be used to better demand. The manipulation by airlines of demand using price is a good example. If a traveler stays over on a Saturday night, the ticket is generally less expensive, because most travel is for business and peak business travel occurs Monday through Friday. To offset this, airlines offer cheaper tickets with a Saturday-night stay.

Price and cost as they relate to wildlife viewing will become an important issue in the new millennium. Currently, many valuable programs, products, and services from public agencies are free. This seems to imply that they are of no value. This is obviously not the case. Any discussions of the future will inherently need to consider pricing and cost as an important wildlife-viewing management technique.

Place. Place refers to the physical location where the product or service is offered. Where are wildlife-viewing guides sold? Does this affect demand and sales? Are bird-watching areas located near large urban centers or are they far from an individual's residence? How about public meetings to gather input on wildlife-viewing policy—are these meetings located in areas that are easy to get to? How does the location of a public meeting influence who attends?

Identify where the product is located (or promoted or "sold") and ask if it meets the needs of the target market. Another place variable is the hours of operation. For example, are nature centers open only from 9 a.m. to 5 p.m.? If they are, what market are they catering to? Senior citizens? People who work at night? Experience-based recreation implies service to different markets. A nature center that is open when the majority of a market segment cannot utilize it is an ill-designed service. Make conscious decisions regarding these place variables and base decisions on what is best for the wildlife viewer, not the convenience of the agency. This is another example of the difference between a marketing approach and a selling approach.

Promotion. The fourth "P," promotion, is the final aspect of the marketing mix. The promotion mix includes magazines, newspapers, brochures, direct contacts, and television coverage.

Promotion options are nearly limitless and it is vital to keep in mind the target market. At this point in the marketing process, the market—who they are, the desired outcomes they want, and their opinions, attitudes, and values—have been identified. A product, program, or service has been developed and tailored that precisely fits their needs. The benefits of the product, program, or service have been identified. Because of this, the medium most likely to reach the target market can be selected effectively. Ads targeted at children are rarely seen on CNN; ads targeted at adults are rarely seen on Nickelodeon.

When developing promotional materials, keep in mind the difference between the tools of promotion and the goals of promotion. Just because an agency has developed full-color advertisements, radio ads, or a World Wide Web site does not automatically mean it has increased knowledge levels, changed attitudes, or increased wildlife-viewing participation. Real success should be measured in quantified attitude changes, total sales, increased awareness and knowledge levels, quantified increases in wildlife-viewing participation, and increased wildlife-viewing satisfaction ratings. The objective is not to develop advertisements or brochures, but to foster positive wildlife-viewing experiences and awareness, increase satisfaction levels, change attitudes, alter behavior, or increase factual knowledge. Again, it is important to separate the means and the ends of programs, products, and services.

3. Strategies and Implementation

An amazing thing happens when a marketing plan is developed. Program, product, and service strategies—the "what should be done" part—emerge naturally from the process. This is because important decisions have been made during the marketing process, such as what the goals and objectives are, who the target markets are, and their desired psychological outcomes. By proceeding through the marketing process in an orderly, step-by-step process, the organization has moved through a process of wise decision making. The organization has thought out what it is it wants to do and has committed it to writing. Developing a marketing plan takes time, but it is well worth it. And in reality, it is no different from developing a wildlife-management plan. Fish and wildlife managers would never attempt to begin to manage wildlife without a plan; neither should managers attempt to manage wildlife-viewing opportunities without a plan.

4. Field Testing and Evaluation

One reason fish and wildlife managers monitor fish and wildlife populations is to evaluate the effectiveness of different management techniques. Wildlife-viewing programs, whether public or private, should also be evaluated.

As fundamental and basic as evaluation is, it is often overlooked or not given the importance it deserves; perhaps because it is time consuming and often expensive. Evaluation, however, is one of the most important components of the marketing process, because it will answer the fundamental question, "Did the program, product, or service work?" Evaluation should be based on the goals and objectives initially set. Since the Experience-based Management approach emphasizes the psychological outcomes as the ultimate goal of recreationists, programs should be evaluated on these parameters. That is, program evaluation should focus on outcomes, not outputs. Wildlife-viewing programs should not be based on the number of trails or bathrooms built, or the number of brochures printed, but on the outcomes important to a quality wildlife-viewing experience, such as overall satisfaction ratings, as well as specific outcomes that are a part of the bundle of psychological outcomes desired by the wildlife viewer.

The Marketing Challenge

The major challenge to the fish- and wildlife-recreation and management community for the new millennium is to make their fish- and wildlife-related recreation programs as successful as the biological aspects of their fish and wildlife management programs. If the wildlife-management profession can restore once ravaged wildlife populations such as pronghorn antelope, white-tailed deer, bald eagle, wild turkey, beaver, wood ducks, and giant Canada geese, to name a few, surely it can provide, in conjunction with the private sector, satisfactory wildlife-viewing programs for a diverse American public. The key to success will be to take a marketing and experience-based approach to planning and management for wildlife viewing, while basing decisions upon a solid foundation of human dimensions research.

The challenge to that part of the fish and wildlife community interested in marketing and Experience-based Management is to demonstrate to the rest of the fish and wildlife profession the true value of these approaches and to communicate the value of these approaches focusing on outcomes of their own. Too many marketing, outreach, and planning efforts are perceived as financial black holes and only peripheral to fish and wildlife management and conservation, and as net consumers of financial and personnel resources, not net contributors. The turning point will come when programs focus on outcomes, not outputs. If $50,000 is budgeted for marketing to increase the number of wildlife viewers in a state, the outcome must be quantified data that there has been an increase in the number of wildlife viewers. If a wildlife-viewing program spends $100,000 to inform a state's residents about bald eagles, a quantified outcome could be to increase the percentage of the state's residents who strongly support bald-eagle-conservation programs. If $200,000 is spent developing wildlife-viewing sites, a quantified increase in wildlife-viewer satisfaction levels should be realized. If $25,000 is spent informing wildlife watchers, hikers, and picnickers that disturbance of desert bighorn sheep causes them to abandon water holes, a quantified outcome should be a net increase of desert bighorn using specified watering holes.

Only when marketing and planning outcomes—not output—are focused on, reached, and communicated to other fish and wildlife managers and administrators will the true value of marketing and Experience-based Management be realized.

Summary Points

• Marketing is not selling and it is more than simply promotion. A selling approach focuses on the needs of the agency or organization, while marketing focuses on the needs of the wildlife viewer. A promotional approach focuses on a single aspect of the marketing mix while a marketing approach focuses on the whole business.

• Marketing should always be practiced within the constraints of resource protection whether it is practiced by management agencies, nonprofit organizations, or for-profit private operators. We feel it is a basic consideration to marketing wildlife-viewing that marketing be done with the biological limits of the resource clearly in mind.

• Marketing is a process that begins with people (markets) and ends with programs, products, services, and strategies. Within the context of fish and wildlife management, marketing is the deliberate and orderly step-by-step process of first defining what it is—exactly—that one is trying to achieve; understanding and defining different groups of constituents (markets) through research; and then tailoring programs, products, and services to meet those needs through the manipulation of the marketing mix— product, price, place and promotion.

• The purpose of marketing in an agency or organization is both to better meet the goals and objectives of the organization and to better meet the needs of its wildlife viewers, in order to protect and conserve wildlife and its habitat and to provide wildlife viewers with quality wildlife-viewing programs, products, and services.

• The marketing process assists agencies and organizations in making the right decisions because it takes them through a series of smaller decisions and information-gathering processes that assist in reaching larger decisions. A marketing approach takes out the guesswork. It leads to the most effective decisions and the development of the most appropriate strategies, programs, products, and services.

• Market definitions of a business are superior to product definitions of a business. Products and technologies are transient, but basic market needs endure. When thinking strategically, focus on market needs, not on products.

• There is no such thing as a general public and there is no such thing as a general wildlife viewer. This is at the core of the science of marketing and an experience-based approach to wildlife viewing. Research consistently shows that how people relate to fish, wildlife, and natural resources is affected by a variety of factors, including their age, race, gender, income, and level of education. Moreover, Experience-based Management denotes that different types of wildlife viewers have different desired psychological outcomes, activities, and settings associated with participation. Wildlife managers would never attempt to manage all wildlife populations with the same wildlife-management plan. Nor should they attempt to develop wildlife-viewing programs for all wildlife viewers.

• Once a market has been identified and segmented, a program, product, or service is tailored to the specific market. Marketing mix— product, price, place, and promotion—is the set of controllable variables that are used to tailor the program, product, or service to the target market.

• As fundamental and basic as evaluation is, it is often overlooked or not given the importance it deserves. Evaluation should be based on the goals and objectives initially set. Since the Experience-based Management approach emphasizes the psychological outcomes as the ultimate goal of recreationists, programs should be evaluated on these parameters.

• The challenge to the fish and wildlife community interested in marketing and Experience-based Management is to demonstrate to the fish and wildlife profession the true value of these approaches. Too many marketing, outreach, and planning efforts are perceived as financial black holes and only peripheral to fish and wildlife management and conservation, and as net consumers of financial and personnel resources, not net contributors. Marketing and planning efforts will be more readily accepted by the fish and wildlife profession at large if they can be shown, quantitatively, how marketing and planning directly impact opinions, attitudes, behaviors, and the conservation and protection of the nation's fish and wildlife resource.

Literature Cited

Aiken, R. (1999.) *1980-1995 Participation in Fishing, Hunting, and Wildlife Watching. National and Regional Trends.* U.S. Department of the Interior, Report 96-5. Washington, DC: US Fish and Wildlife Service.

Crowe, D. (1983). *Comprehensive Planning for Wildlife Resources.* Cheyenne, WY: Wyoming Game and Fish Department.

Kellert, S. R. (1996). *The Value of Life: Biological Diversity and Human Society.* Washington, D.C: Island Press/Shearwater Books,.

Kellert, S. R., and M. O. Westervelt (1980). *Children's Attitudes, Knowledge, and Behaviors Toward Animals.*Washington, DC:U.S. Department of the Interior, US Fish and Wildlife Service.

Kotler, P. (1980). *Marketing Management.* Englewood Cliffs, N.J.: Prentice-Hall, Inc.

Levitt, T. (1965). "Marketing myopia." *Harvard Business Review.* Sept.-Oct. 26-44: 173-81.

Responsive Management (1993). *Wildlife Viewing in Maryland: Participation, Opinions and Attitudes of Adult Maryland Residents toward a Watchable Wildlife Program. Report Prepared for the Maryland Wildlife Division.*Harrisonburg, VA:Responsive Management,.

Riley, C. (1985). "Missouri's outdoor recreation skills program: Hunter education and more." *Transactions of the North American Wildlife and Natural Resource Conference* 50: 109-12.

U.S. Department of the Interior, Fish and Wildlife Service, and U.S. Department of Commerce, Bureau of the Census (1997). *1996 National Survey of Fishing, Hunting, and Wildlife-associated Recreation.* Washington, DC: U.S. Department of the Interior, US Fish and Wildlife Service.

U.S. Department of the Interior, Fish and Wildlife Service, and U.S. Department of Commerce, Bureau of the Census (2002). *2001 National Survey of Fishing, Hunting, and Wildlife-associated Recreation (Preliminary Findings).* Washington, DC: U.S. Department of the Interior, US Fish and Wildlife Service.

Westervelt, M. O., and L. G. Llewellyn (1985). *Youth and Wildlife.* Washington, DC: U.S. Department of the Interior, US Fish and Wildlife Service.

A Geomatic Approach to Wildlife Viewing-based Recreation Planning and Management

Denis J. Dean

Introduction

ONE OBVIOUS CHARACTERISTIC of wildlife viewing-based recreation is that it takes place in some particular geographic region or location. This region or location, which will be referred to as the *recreation site* (or simply *the site*), has certain physical, managerial, and social attributes that collectively define the setting component of the Experience-based Management paradigm. Furthermore, these same attributes, plus a number of additional qualities, determine the site's ability to support various species of wildlife and the activities that these animals will pursue while on the site (e.g., feeding, mating, seeking cover from predators, etc.). Collectively, the site's wildlife-support characteristics and its recreation-setting properties determine what types of wildlife-viewing activities can logically take place at the site. Thus, it is possible to view at least two of the three components of the Experience-based Management paradigm (the setting and activity components) as outgrowths of the characteristics of the recreation site.

The quantitative study of geographic locations (like recreational sites), their characteristics and their relationships to one another, is the province of the science of *geomatics*. A formal definition of geomatics is, "the acquisition, archiving, analysis, and communication of spatial data." In this context, "spatial data" are any data that can be ascribed to a particular location in space. Thus, any data that pertain to a recreation site are spatial data, because the recreation site itself is a particular location in space. Looking at the issue from a different scale, any data that relate to features within a recreation site (such as a road, trail, campground, etc.) or subregions within a recreation site (such as a developed visitor-services area, an undeveloped backcountry region, and so on) are also spatial data, but in this case, they relate to a spatial location smaller than the entire recreation site.

In recent years, with the growing availability of geographic information system (GIS) hardware and software, the advent of the global positioning system (GPS), and the decrease in cost and increase in quality of remotely sensed data, geomatics has become a very popular mechanism for addressing management issues (see Box 1 at the end of this chapter, page 344). As we have just seen, when applied to wildlife viewing-based recreation, this geomatics approach is not incompatible with the Experience-based Management paradigm. In this chapter, we will investigate how geomatics and the Experience-based Management approach can be used in tandem in the planning and management of wildlife viewing-based recreation.

Note that in the limited space available here, it is not possible to provide detailed and exhaustive tutorials describing how to perform specific geomatic tasks. Instead, the goal of this chapter is to outline some of the possibilities provided by geomatics and to show how geomatics complements the Experience-based Management paradigm. If you are unfamiliar with all the nuances of geomatics, don't get too concerned with the details regarding how certain analyses mentioned here are conducted. Instead, concentrate on the *general* process used to produce these results, and on how the results themselves can assist in developing and implementing management strategies founded in the Experience-based Management paradigm.

Geomatic Inventories of Current and Desired Conditions

Any planning or management process requires basic inventory data. In the context of the Experience-based Management paradigm, this translates into a thorough understanding of the setting component of the recreation site under consideration. Furthermore, as was mentioned previously, the site characteristics (physical, social, and managerial) described in basic inventory data not only define the setting component, they also have a strong influence on the activity component of the Experience-based Management paradigm. Finally, regardless of what management approach is adopted (Experience-based Management or some alternative model), it is not possible to formulate logical management plans without an understanding of the desired outcomes of those plans. Again, returning to the context of the Experience-based Management paradigm, it is clearly impossible to develop management plans for wildlife-viewing recreation sites without knowing what types of recreation settings and activities the plan is intended to produce. Geomatics can help managers collect, organize, and communicate both basic inventory data and information describing desired outcomes.

Geomatics-Based Physical Inventories

Perhaps the most obvious role geomatics can play in the management of sites where wildlife viewing-based recreation takes place is through the development and upkeep of inventories describing the physical conditions within those sites. Geomatic inventories can best be thought of as a series of one or more maps showing all of the salient features of the recreation site (Figure 15-1). Physical-inventory maps can show virtually any feature of the physical environment at a recreation site. Typical physical-inventory maps depict things such as topography, surface hydrological features, vegetation types, existing infrastructure (including roads, trails, human-made structures, etc.), soil types, and so on. These inventory maps can contain as much or as little detail as desired. For example, a vegetation-type map may show the entire recreation site in only a handful of broad vegetative categories (e.g., "forests," "grasslands," and "other") or it may divide the site into a myriad of individual vegetation patches, each of which is described by multiple attributes (e.g., an individual patch of forest may be described by the species of tree making up the canopy, the degree of canopy closure, the average age of the trees making up the canopy, the number of trees per acre, the understory vegetation, and so on). Figure 15-2 shows an example of some of the different levels of detail possible in geomatic inventory maps.

There are a number of advantages to this map-based inventory approach. Map-based inventories highlight the spatial relationships between objects or regions within the recreation site (Chrisman 1997). This allows managers and others to identify potential

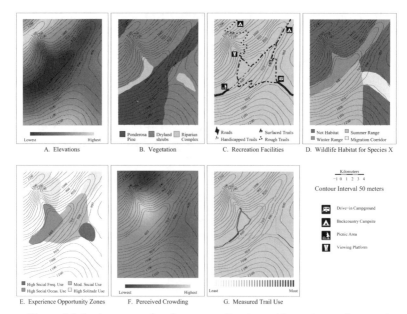

Figure 15-1. An example of a geomatics-based inventory of current conditions in a hypothetical wildlife-viewing recreation site.

 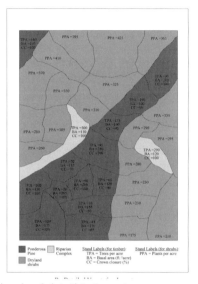

Figure 15-2. Examples of varying levels of detail in inventory maps. Left: Hypothetical vegetation-inventory map showing little detail. Right: Alternative map showing greater detail.

opportunities for wildlife viewing (e.g., an area containing underutilized recreation facilities located adjacent to an area with high concentrations of a wildlife species that can tolerate the proximity of humans), or potential problems (e.g., an area with increasing recreational use located within or adjacent to areas utilized by wildlife species sensitive to the presence of humans). Map-based inventories also can be used to keep track of change over time (Figure 15-3) (Burrough and McDonnell 1998). Perhaps most importantly, map-based inventories complement what many cognitive psychologists believe is the natural way humans conceptualize spatial information. It has long been accepted that most humans think of spatial data through the use of mental maps. This makes map-based inventories easy to understand and interpret (Kaplan and Kaplan 1982).

Maps depicting physical inventories of recreation sites are extremely common and are frequently used by both managers and recreationists. It is worth noting that at most wildlife viewing-based recreation sites, one of the first pieces of information recreationists receive from the site's managers is a map of the site. The most common drawbacks of these maps is that they are frequently out of date or not sufficiently detailed, or fail to show features that are of interest to a particular map user. Fortunately, modern geomatics techniques can address all of these shortcomings.

Figure 15-3. Inventory maps tracking changing conditions over time. This example illustrates the mapping of infestations of spotted knapweed, a common exotic-weed species causing problems in many parts of the western United States.

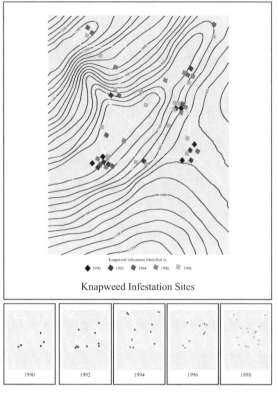

Knapweed Infestation Sites

Out-of-date maps are typically the result of the high cost of gathering new data with which to update old maps and the cost of redrafting new maps. New remote-sensing and GPS techniques provide very cost-effective means of gathering up-to-date data (see Boxes 2 and 3, at the end of this chapter, pages 345-49), and cartographic drafting functions found in GIS and desktop-mapping software make map updating virtually cost free. These same cartographic-drafting capabilities support the production of maps as detailed as the available data allow, and maps that are customized to the needs of any particular map user (Figure 15-4). In fact, it is already possible for recreationists to create their own customized maps of popular recreation sites by logging onto World Wide Web pages designed exclusively for this purpose.

Note that some physical-inventory maps are not created from survey or remotely sensed data but are derived from other data. For example, many physical inventories of wildland recreation sites include data showing wildlife-habitat areas for certain indicator, threatened, and endangered (T&E) species, or species of concern (Figure 15-1D). These maps are almost always created by applying

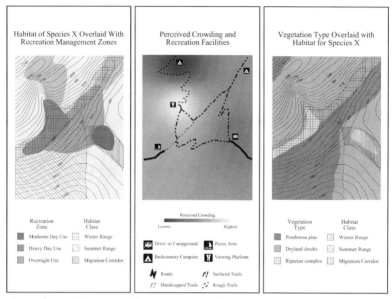

Figure 15-4. Customized maps. Each of these maps was produced using the data from Figure 15-1.

spatial data-analysis operations to other physical-inventory maps (see Box 4 at the end of this chapter, page 349). For example, if an appropriate habitat suitability index (HSI) model (Schamberger 1982) for a species of interest indicates that wintering habitat occurs within certain vegetation types and within certain ranges of elevations and aspects, a wintering-habitat inventory map can be constructed by applying the HSI criteria to topography and vegetation inventory maps, using common spatial data-analysis operations found in modern GIS software.

Geomatics-Based Social and Managerial Inventories

Map-based inventories of nonphysical features of the recreation site are also possible. These maps can describe the social and managerial attributes of the setting component of the Experience-based Management paradigm, just as maps showing inventories of physical features describe the resource attributes (Figure 15-1, parts E through G).

Social aspects of a recreation site are typically measured through survey techniques (Hammitt and Cole 1998). These surveys record recreationists' impressions of and reactions to the levels of crowding, the behaviors of other recreationists, their levels of satisfaction with their experience, and so on. There are many variations on the mechanics of how these surveys are conducted, their format, and the exact questions that are asked, but regardless of these variations, all surveys produce results that apply to the identifiable physical locations. In most cases, recreational surveys apply to the locations where the respondents recreated, but surveys designed to gather information regarding the respondents' general attitudes (as opposed to their opinions regarding a particular recreational experience) may produce data that should be ascribed to the respondents' place of residence. Regardless of the particular location to which a survey's results apply, the fact that the results apply to *any* location means that the data produced by surveys is spatial data. Spatial data can be mapped and analyzed using geomatic techniques. Thus, it is possible to develop maps showing the results of recreational surveys (Figure 15-1F), just as it is possible to develop maps showing the physical characteristics of a recreation site. Mapping of this type requires some preplanning in terms of the design of the survey and the methods that will be used to administer it, but these issues do not increase the difficulty or cost of conducting visitor surveys. Surveys designed to produce mappable results often do require more interviews than do non-mappable surveys, and this increase in sample size can produce increased costs. However, on a per-interview basis, there is no reason why a survey designed to produce mappable results need cost any more than a conventional survey.

Managerial aspects of a recreation site can also be mapped. Some managerial features of a recreation site take physical forms (e.g., trails, interpretive signs and markers, informational kiosks, etc.) and hence can be surveyed as part of a recreation site's physical inventory (Figure 15-1C). Other managerial aspects of a recreation site are less physical but still quite tangible and hence can be easily mapped. For example, recreational use levels cannot be measured using the remote-sensing or GPS techniques that can be used to map trails, interpretive signs, and so on, but use levels can be easily measured and mapped using strategically placed car and/or trail counters (Figure 15-1G).

A third set of managerial aspects of a recreational site is even less tangible, but is still crucial to Experience-based Management and is still mappable. This set includes arrangements managers make to divide large recreation areas into smaller zones. Managers typically use this approach to ensure that all management activities planned for a given zone are compatible with the zone's intended function, and to ensure that the collection of zones throughout the area encompass all the intended settings that the manager believes the overall area should provide. Managers often base their decisions on dividing large areas into smaller zones on factors such as the allowable level of human impact on natural resources, types of recreation activities to be supported, desired social settings, and so on. These zones cannot be measured in the field using conventional measurement techniques, but nevertheless they can be mapped (Figure 15-1E).

It is even possible to develop maps showing how available managerial resources such as money, time, and personnel will be consumed as they are applied to the landscape. For example, it is possible to construct a map showing the financial cost of building a scenic roadway from some starting point (such as an intersection on an existing road) to other locations on the map. It would be equally possible to conduct analyses using "costs" expressed in terms of environmental impact, recreational displacement, or any other measure desired, and these "costs" could be applied to trail building, facilities development, resource extraction, or any other activity.

These are just a few examples of how geomatics can be used to develop inventory maps showing hypothetical (e.g., the route of a proposed but currently nonexistent road) and/or non-physical (e.g., results of attitudinal surveys) features of a recreation site. Many other possibilities exist. There are relatively few conceptual and technological limits to how far the geomatics approach to developing recreation-site inventories can go; most limits result from a lack of familiarity with the possibilities and/or a lack of imagination on the part of the managers and geomaticists developing the inventories. A trained and experienced geomaticist should be familiar enough with spatial-data acquisition, management, and analysis to know what is possible, and with a little practice just about anyone can cultivate the imagination needed to develop useful ways of employing geomatics to address wildland recreation-site inventory issues.

The Role of Geomatics in Stakeholder-based Methods of Defining Desired Outcomes

While it is undeniably true that no management/planning process can succeed without a thorough understanding of existing conditions on the recreation site, it is equally true that no such process can succeed without a clear understanding of the desired outcomes of the process. In overly simplistic terms, it is clearly impossible to develop a plan to get you from some starting point to some ending point if you don't know where the starting point is located (i.e., you need good inventory data telling you what you have to start with) *and* where the ending point is located (you need a clear definition of your desired outcomes to tell you where you want to go).

Previous chapters of this book have discussed both the importance and use of stakeholder involvement in developing management plans. One of the most fundamental goals of stakeholder involvement should be the development of an agreed-upon vision of the desired future conditions of the recreation site, or where universal agreement is not possible, at least a thorough understanding of the concerns and desires of all stakeholders. This implies that stakeholders and managers must be able to develop and communicate to one another understandable visions for the future of the recreation site. Given (1) the spatial nature of data relating to recreation sites, including data describing desired future conditions; (2) the previously mentioned human tendency to conceptualize spatial data in the form of mental maps; and (3) the ability of geomatics to present spatial data in the form of physical maps that are compatible with mental maps, geomatics becomes a very attractive unifying tool for use in the stakeholder process (Dean 1994).

Stakeholder processes can use geomatics in a variety of ways. First, map-based site-inventory data of the sort discussed in the previous sections can be shared among stakeholders to ensure that all participants in the process are starting with a common understanding of current conditions. This may seem trivial, but experience has shown that it is not. Many stakeholders enter planning processes with very different views regarding current conditions. For example, it is not uncommon for some stakeholders to believe that a natural area intersected by roads has become degraded and is in need of restoration, while other stakeholders believe that this same area is highly desirable in its current condition, because of the motorized recreational opportunities offered by the roads. Understandable, accurate, and nonjudgmental inventories of areas such as this allow stakeholders to find areas of agreement (e.g., the location, size, and surface type of existing roads, vegetative conditions in surrounding forests, use levels of roads, etc.) and isolate areas of disagreement. This provides a starting place from which stakeholder discussions can begin.[1]

Another area where geomatics can assist stakeholder processes is by fostering clear communication. The heart of any stakeholder-based process is unambiguous communication between and among

---

1. In cases where there is a great deal of distrust between various stakeholder groups, some managers have successfully asked these distrustful groups to work cooperatively to actually gather basic inventory data describing contentious regions. This builds a sense of ownership of the data and (hopefully) fosters trust between stakeholder groups.

stakeholders and managers. Maps provide an nearly ideal mechanism for communicating spatial data, including data describing desired outcomes. Just as inventory maps can capture any number of physical, social, managerial, and other attributes of the site, maps can also communicate these attributes between and among stakeholders and managers.

Many newer geomatics applications are carrying this concept one step further by using visualizations and artificial reality in conjunction with maps to communicate spatial data. These visualizations and artificial realities make it possible to create realistic impressions of what a recreation site might look like in the future. These technologies produce anything from still images showing a static picture of some landscape as it might appear under some proposed conditions (Figure 15-5), through "flybys" showing what appears to be moving video of what a person traveling through or over a landscape might see under some hypothetical management regime, to interactive artificial realities that allow individuals to move through and interact with an artificial landscape (again, depicting how a region may appear in the future under certain conditions) in whatever way they desire (Harvey and Dean 1996).

Figure 15-5. An artificially created visualization of a proposed (but currently nonexistent) road. This example was created using a combination of aerial photography, image rectification, GIS and artificial neural-network techniques.

Geomatics can play other roles in stakeholder processes, but these roles are more concerned with the development and evaluation of management plans rather than the defining of desired outcomes. As such, these additional roles will be discussed in the next section.

Using Geomatics to Develop and Evaluate Management Regimes

In addition to creating inventories of current and desired conditions within recreation sites, geomatics can be used to develop possible management regimes for these areas. This is usually accomplished by developing some type of spatial model. The models used in this context are nothing more than computer programs that try to mimic, in a simplified way, some aspect of the real world. These models are called "spatial" because they operate either partially or wholly on spatial data. As such, many of the inputs and/or outputs of these models take the form of maps. Modern GIS software provides many tools that model makers can use to construct such models (see Box 4).

In general, there are two types of spatial models used to develop management plans. The first (and most common) is called a simulation model and is used to play "what if" games. Basically, models of this type are designed to predict the likely impacts of proposed activities. For example, a simulation model could be used to predict the likely impacts of a proposed road-building project on water quality in nearby streams; these impacts might take the form of erosion during road construction, changes in the quality and quantity of runoff water flowing into the streams after construction, and so on. Obviously, the outputs of a model of this type would be very interesting to a manager trying to judge the pros and cons of a road-building project (Figure 15-6).

There are a number of factors to consider when contemplating the use of spatial-simulation models. First, the scope of the model must be defined. Returning to our road-building example, is the model limited to predicting the impacts the road will have on nearby streams, or does it also consider impacts on wildlife-migration routes, recreational settings, and so on? A model that considers a wide range of impacts would have a very broad scope and would be termed *comprehensive*, while a model that considered only one or two impacts would have a narrower scope and be less comprehensive.

Obviously, in a perfect world, managers would have access to comprehensive models that could provide them with reliable estimates of all of the impacts of proposed management actions. Unfortunately, in the real world, acquiring comprehensive models

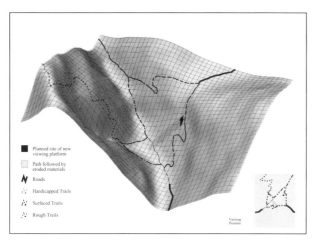

Figure 15-6. Example of simulation-model output. The simulation model shown here predicts the likely paths followed by eroded material coming off a proposed viewing-platform site.

generally requires more time and is more costly than acquiring less-comprehensive models. This is because relatively few models can simply be pulled "off the shelf" and immediately applied to a particular recreation site. If an appropriate simulation model exists, at a minimum it usually has to be *calibrated* so that it "understands" the specific conditions found at the current recreation site, and at worst it may have to be modified in some fundamental way because the functioning of the current recreation site differs from the functioning of the site where the model was developed. Returning to our previous example of estimating the impacts of a road-building project, suppose that the project in question involved building a road in a high alpine area above treeline. Road impact models do exist, and for the moment suppose that we are able to find such a model that applies to alpine areas. Unfortunately, it is unlikely that this existing model will apply to the exact soil conditions found in our area, so we will have to calibrate it to fit our soils. Alternatively, suppose we could not find a road-impact model that applied to alpine areas at all. We would then be faced with the task of either modifying an existing model that currently does not apply to alpine areas, or building a new model from scratch.

Obviously, the likelihood of finding an existing model that correctly applies to all aspects of a given situation decreases as the number of factors considered in the situation increases. Thus, it may not be too difficult to find an applicable road-impact model whose scope is limited to impacts on water quality, but it will probably be much more difficult to find a comprehensive road-impact model that considers impacts on water quality, the species of wildlife present in the area, the types of recreationists using the area, and so on. The end result is that managers often find themselves using models of limited scope, simply because they cannot afford the time and expense of developing more-comprehensive models. While this is certainly understandable, it should be recognized as a less than desirable situation. When the opportunity presents itself, more comprehensive models should be developed and used.

Other factors to consider when evaluating simulation models are the validity of the model and the understandability and meaningfulness of their outputs. Any good simulation model should have been validated to determine if its predictions accurately reflect the functioning of the real world. A wide variety of procedures are used to validate models, but at their core all of these procedures do nothing more than compare model predictions to real-world events. Ideally, a model should accurately predict real-world events under a wide range of conditions and over long periods of time. Thus, an ideal road-impact model would be able to predict road impacts in both steep and gentle terrains, under differing vegetation conditions, and so on. It would also be able to predict impacts many years into the future. In the real world, most models are seldom ideal. However, managers should understand how a model has been validated, and understand the implications of the validation results, before adopting models for use in their operations.

Finally, model outputs must be both understandable and meaningful in order to be useful. The need for understandable outputs is obvious; even if a model does a perfect job mimicking the functioning of the real world, it is useless if its outputs cannot be understood. The importance of meaningful outputs is a little less obvious, but upon reflection it too is clear. Return to our previous example of a road-impact model. Suppose the model did nothing more than predict impacts as slight, moderate, or high. While these predictions might be completely accurate and easily understood, they are not very meaningful. Suppose your model told you that your proposed road-building project had high impacts. What do you do then? Simply junk the proposal and move on to something else? Wouldn't it be more useful if your model told you where your project was impacting streams and how those impacts were occurring, and described the exact nature of these impacts? With meaningful information such as this, you would be in a much better position to know how to proceed with your road-building project.

A more ambitious alternative to simulation models are *optimization* models. As their name implies, these models are designed to find optimal management plans. In terms of scope, these models vary just like simulation models. Thus, optimization models of limited scope can be designed to find the optimal route for a new road, as can more comprehensive optimization models designed to find the best overall management regime to apply to an entire recreation site. Optimization models are more difficult to construct than simulation models, and as they become more comprehensive, they start to impinge upon highly political areas. For example, an optimization model of fairly limited scope that is trying to find the most scenic route for a new road from point *A* to point *B* is not likely to be highly political, but a more comprehensive model designed to determine if building this road in the first place is more desirable than leaving the area in question undisturbed is certainly delving into a political area. In general, past experience has shown that optimization models do not always perform well when they are forced to address highly political issues.

Both simulation and optimization models can be highly useful to recreation managers. Many simulation models of limited scope are already built into (or can easily be created using) commercially available GIS software. Other such models are available through users' groups and over the World Wide Web. More such models are becoming available every day. Furthermore, as GIS software becomes more powerful and easier to use, more and more sophisticated models become feasible. Finally, managers should keep in mind that building models can be a long-term investment. If a manager decides to invest in the construction of a large, comprehensive simulation or optimization model, that model can be used repeatedly. Thus, it may not only be useful in the short term; it may also be used repeatedly in the long term. Over time, managers can accumulate libraries of models that can be applied to a wide variety of situations. Once a library of this type is created, managers are truly able to use geomatics to its fullest advantage in the development of management plans for their recreation sites.

Monitoring Using Geomatics

As was mentioned in previous chapters of this book, it is not sufficient to simply develop a management plan, implement it, and assume the plan is producing the desired results. Plans must be continuously monitored to determine if they are behaving as intended. This is accomplished by identifying measurable physical, social, and/or managerial aspects of the recreation site that should be influenced by the plan, defining what ranges of values these measurable features will have if the plan is successful, and then repeatedly measuring these factors to determine if they are in acceptable ranges.

Identifying what particular measurable features to use as indicators of the success or failure of a given management plan, and determining what ranges of values are acceptable for these indicators, is not a geomatic issue and hence is best discussed in other chapters of this book. However, once these indicators and standards are defined, measuring them, keeping records of repeated measurements, and analyzing these measures to detect changes or trends over time, are geomatic issues

As mentioned in the previous "Geomatics Inventories" section of this chapter, any data that pertain to a particular point in space are spatial data that can be acquired, archived, and analyzed using geomatics concepts. This includes the vast majority of data likely to be gathered in a monitoring process. Physical data such as water quality, erosion levels, vegetative health and growth, and so on relate to the locations where the water-quality sample was drawn, the erosion measurements were taken, and the vegetative survey was conducted. Social data gathered through surveys of recreationists relate to the locations where the survey respondents were recreating. As mentioned previously, most managerial data can be thought of as spatial, although the exact techniques used to gather these data vary considerably. Any of these spatial data can be recorded in spatial-data formats, and analyzed to detect changes (Figure 15-7).

Long-term monitoring programs produce repeated measures of the indicators under study. At one level, keeping track of these repeated measures is nothing more than a file-management operation. However, at another level, there are complicating factors. In the long term, additional information regarding each measurement might be needed; for example, it might prove helpful to know what techniques were used to take the measurements. Consider the case of water chemistry measurements. Water-chemistry testing technology is constantly changing, and comparing a measurement taken in the past using an older technology to a newer measurement taken using modern technology might produce misleading results. Furthermore, seemingly unrelated factors that influence measurements can be significant. Returning to the water-chemistry example, consider two samples drawn from streams and shipped to a lab for testing. If one sample was kept shaded and refrigerated during shipment while the other was not, any differences detected between the two samples may be due entirely to chemical processes taking place during shipment, and may not reflect any actual differences between the two streams that produced the samples.

The only way to combat this sort of problem is to record as much information about each measurement as possible. Information of this sort is referred to as *metadata*. A certain amount of metadata is

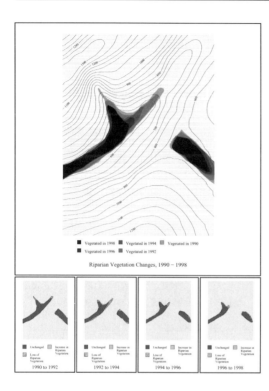

Figure 15-7. Change-detection maps showing changes in riparian habitat.

required in any monitoring program. At a minimum, when and where measurements were taken and the techniques used to take the measurements should be recorded. Depending on the specific nature of the measurements in question, additional information might be helpful as well. Preplanning and common sense are usually all that is needed to determine what metadata should be recorded, but keep in mind that it is relatively easy and inexpensive to gather metadata that ultimately turn out to be unnecessary, but omitting vital metadata can render all the data gathered during a monitoring process useless.

Conclusions

Geomatics is possibly the only field of science that provides recreation-site managers with a single, coherent method of bringing together all of the different aspects of both the physical and socioeconomic components of recreation-site management. As such, it can be a vital tool for recreation managers.

Geomatics is not a panacea. When used properly, geomatics can identify existing and desired future conditions, assist in the development of management plans designed to move a site from its current to its desired conditions, and can help monitor the progress of a plan once it is implemented. But geomatics cannot develop management plans by itself; nor can it somehow eliminate problems and controversies. However, it can help managers and stakeholders communicate more clearly and make more-informed decisions.

Finally, geomatics can only be used to its fullest advantage when recreation managers take the time and effort to obtain geomatics training themselves or work with cooperatively geomatics experts. Geomatics is more than simply making maps, and geomatics techniques cannot be learned by simply becoming familiar with a piece of GIS or desktop mapping software. When non-geomaticists use geomatics tools, the typical result is the production of a few obvious maps and little else. Given the time and effort needed to learn how to use these tools, this is usually a disappointing outcome. However, a trained geomaticist can use these tools to produce much more than a few obvious maps, and the outputs he or she can produce can be immensely beneficial to recreation managers.

Summary Points

• Geomatics is the art and science of acquiring, archiving, analyzing, and communicating spatial data. Spatial data are any data that can be attributed to a point or region in space. Geomatics makes use of geographic information systems (GISs), remote sensing, and the global-positioning system (GPS), but it is more than simply the conglomeration of these tools. Geomatics uses these tools to quantitatively study geographic features or regions, their characteristics, and their relationships to one another.
• A geomatics approach to studying wildlifeviewing-based recreation is compatible with the Experience-based Management paradigm. All aspects of the site component of the Experience-based Management approach (i.e., the physical, social, and managerial aspects) can be seen as spatial data relating to the recreation site (which is a spatial location) or smaller regions or objects within the site (which are also spatial locations), and thus can be studied geomatically. These characteristics also impact on the site's ability to support wildlife, and the types of activities recreationists can engage in (i.e., the activity component of the Experience-based Management paradigm) while at the site.
• Geomatic techniques can be used to create inventories of a recreation site's physical, social, and managerial characteristics. These inventories are easily understood and highlight the spatial relationships between and among inventoried site characteristics.
• Geomatic techniques can also be used in stakeholder processes

to foster understanding among and between stakeholders and managers. These processes can help stakeholders and managers formulate and unambiguously communicate desired future conditions for the recreation site.

• Spatial models, including simulations and optimizations, can help managers formulate, evaluate, and compare management plans. These models can vary greatly in terms of their complexity, realism, and comprehensiveness, but all such models are basically designed to answer one or both of the following questions:

1. What would be the impacts (on *A, B, C,* etc.) of implementing a proposed management action?

2. What mix of management action(s) should be applied to the recreation site, and when and where should they be applied, to maximize the site's ability to produce this desired outcome, or minimize the site's production of this undesirable outcome?

• Geomatics techniques can be used to monitor the progress of management plans by archiving measurements of indicator variables over time. When used for this purpose, metadata (information describing when and how the indicators were measured) become critically important.

Literature Cited

Burrough, P. A., and R. A. McDonnell (1998). *Principles of Geographical Information Systems.* Oxford, England: Oxford University Press.

Chrisman, N. (1997). *Exploring Geographical Information Systems.* New York: John Wiley and Sons.

Dean, D. J. (1994). Computerized tools for participatory national forest planning. *Journal of Forestry,* 92 (2), 37-40.

Hammitt, W. E., and D. N. Cole (1998). *Wildland Recreation: Ecology and Management,* 2nd ed. New York: John Wiley & Sons.

Harvey, W., and D. J. Dean. (1996). "Computer-aided visualization of proposed road networks using GIS and neural networks. In G. J. Arthaud and W. C. Hubbard (Eds.), *Proceeding of the First Southern Forestry Geographic Information Systems Conference.* Athens, GA: University of Georgia.

Kaplan, S., and R. Kaplan. (1982). *Cognition and Environment.* New York; Praeger Publishers.

Schamberger, M. (1982). *Habitat Suitability Index Models.* U.S. Department of the Interior, Fish and Wildlife Service. Washington, DC: U.S. Government Printing Office.

Box 1. Geomatics and its Tools

The tools of geomatics—geographic information systems (GISs), aircraft- and satellite-borne remote-sensing systems, global position systems (GPSs), and so on—have captured a great deal of attention in the last few years. Attitudes toward these tools range from awe ("these things are so complicated I can't even begin to understand what they are all about") to disdain ("a GIS is just another piece of software, so I can learn to use it without any formal training, just like I learned to use my word processor"). As usual, the truth is somewhere between these extremes.

On the one hand, these items *are* just tools, and while they are powerful and in some instances involve complex technologies, they are nothing more than the products of human ingenuity and creativity. They can be understood by just about anyone willing to devote time and effort to their study. On the other hand, it is a little misleading to compare them to something like a word processor. A word processor is a tool designed to assist in the task of writing. While it is undoubtedly true that most users of word processors do not receive (and probably do not need) any formal training in the use of their word-processing software, virtually all of these individuals have received extensive training, both formal and informal, in the underlying task that their word processor was designed to serve; namely, writing. It is this exhaustive training in writing that allows them to feel comfortable with their word-processing software so effortlessly.

Unfortunately, most users (and potential users) of GIS, GPS, and remote-sensing software and hardware do not have any training in the underlying task that these tools are designed to facilitate; namely, the capture, archiving, quantitative analysis, and communication of spatial data (i.e., the performing of geomatics). Thus, most people *do* find it much more difficult to use geomatics tools like GISs, GPSs and remote-sensing systems than to use writing tools like word processors. However, the problem does not lie with any inherent difficulty in the tools themselves but with differing levels of familiarity with the underlying tasks that these different tools are designed to facilitate.

Box 2. Acquiring Spatial Data I — Remote Sensing

Geomatics is a data-dependent science. This means that the first step in any geomatics project is to acquire spatial data for the area of interest. These days, virtually all of these data are digital and are processed using computers.

Whenever possible, geomaticists use existing data. This is much more cost effective than gathering new data. Fortunately, government agencies are increasingly making digital spatial data available free, and commercial firms are offering a wide variety of databases for sale. For most regions of the world, at least basic spatial data layers (such as elevation data, hydrology, roadways, etc.) are available from existing sources at minimal cost.

Unfortunately, most geomatics projects require specialized data that are not immediately available. For example, finding existing digital spatial databases describing recreation use levels, wildlife sightings, detailed vegetative inventories, and so on is unlikely. In cases such as this, new databases must be created.

Traditionally, spatial databases were created using the theodolite-and-chain techniques that have existed for hundreds of years. These techniques can produce wonderfully accurate results, but they tend to be very labor intensive and for large areas can be quite costly. As a more cost-effective alternative, spatial data can be gathered using remote-sensing and/or global-position-system (GPS) techniques.

Remote Sensing Systems: Aircraft- and satellite-based remote-sensing systems can acquire highly accurate spatial data covering large areas. Most of these systems behave like cameras, but remote-sensing "cameras" (it is more correct to call them *sensors*) are not limited to taking pictures in the wavelengths of light visible to human eyes. This allows these sensors to detect differences that are not always visible to the naked eye. For example, a healthy tree and one under stress due to the early effects of an insect attack may be indistinguishable in the visible portion of the spectrum (and thus look identical to the naked eye), but in the infrared wavelengths they appear as different as night and day. The ability of a scanner to detect various wavelengths of light is called its *spectral resolution*; a sensor that only records data for one or two wavelengths of light has low spectral resolution while some high-spectral-resolution systems record data for hundreds of wavelengths.

In addition to spectral resolution, remote-sensing systems are measured by their *spatial resolution*. This is a measure of how small an object they can "see" (the technical term is *resolve*). Some satellite remote sensing systems can't resolve any object smaller than 120 meters on a side, while others can resolve objects a mere one or two meters on a side. Many aircraft-based systems can resolve objects as small as a few centimeters on a side.

An ideal remote-sensing system would have both extremely high spectral resolution and very fine spatial resolution. Unfortunately, with current technology this isn't possible. Thus, current systems are forced to make tradeoffs between spectral and spatial resolution. Improved spectral resolution can only be achieved by degrading spatial resolution, and vice versa.

Both satellite- and aircraft-based remotely sensed data can be purchased from private and quasi-governmental agencies. A great deal of information regarding available remote-sensing data can be found on the World Wide Web. Prices, spatial resolution, spectral resolution, spatial extent (i.e., the area visible in the photo), and the degree of data pre-processing (see below) vary tremendously; it is worth doing some research to find data that will meet your needs. It is also possible to contract private firms to conduct aircraft surveys tailored to your specific situation. While this option is often cost prohibitive for small sites, it can be very viable for larger areas.

Processing Remotely Sensed Data: The raw data that come from a remote-sensing system are very similar to a conventional photograph. These images contain variations in scale (e.g., objects closer to the sensor appear larger than objects farther away) and other distortions, and lack any system of labeling or symbology to identify the objects. In most cases, these distortions and variations are removed and objects are identified prior to using the image. These tasks are termed *image processing*.

Image processing is usually a two-step process. The first step is to remove distortion and equalize scales; this is called *rectifying* the image. Once the image is rectified, the objects in the image are identified; this is called *classifying* the image. In most cases, image rectification is accomplished using either a *georectification* or *orthorectification* process. Georectification is a little easier and is generally appropriate for places where the terrain is relatively gentle; orthorectification takes a bit longer and is used in mountainous areas. In either case, a series of points (called *ground control points*, or GCPs) must be identified on the image and have their real-world coordinates (e.g., their latitude and longitude, or any other set of earth coordinates) computed. The geo- and orthorectification processes use these GCPs to develop a set of mathematical equations that convert coordinates from the image to real-world coordinates, taking out as much distortion as possible in the process.

No rectification is perfect; there will always be a certain amount of distortion left in a rectified image. This remaining distortion is usually measured by comparing the known real-world coordinates of the GCPs to the real-world coordinates of these points as they are computed from the rectified image. In a perfect rectification, these two sets of coordinates would be identical. In the real world, they will be off by some amount; the smaller the difference the better.

Once a remotely sensed image is rectified, objects seen in the image must be identified. This is the image-classification process. Classification procedures range in complexity from simple manual techniques to highly sophisticated processes based on advanced statistics and artificial intelligence. Regardless of complexity, all classification techniques share the same goal: to classify each region of a remotely sensed image into one *informational category* from a set of such categories. Thus, a blue blob on an image might be classified into the informational category "lakes," dark lines might be classified as "paved roads," and so on.

No image-classification process is perfect; all procedures will place some regions from the image into the wrong informational category. The goal in image classification is to minimize these errors. The amount of error in any classification is influenced by myriad factors, including the spatial and spectral resolution of the imagery, the classification procedure used, the level of detail in the desired informational categories (e.g., classifying all tree-covered areas into a single "forests" category is less prone to error than dividing forests into multiple cover-type categories, such as "aspen," "spruce/fir," etc.), and so on.

All image-classification processes need *ground-truthing data,* which identify the informational category of certain known sites. Ground-truthing data are typically gathered by visiting test sites identifiable on the image. Not only are these data used to perform the classification itself, but they are also used to evaluate the accuracy of the final classification results. Again, final accuracies are influenced by many factors, but it is not uncommon to see reported accuracies in the 85 percent range, meaning that 85 percent of the image has been classified into the correct informational category.

Box 3. Acquiring Spatial Data II — GPS

The Global Positioning System (GPS) was developed by the U.S. Department of Defense to assist in the navigation of military forces and for the guidance of smart munitions. Since the early 1990s, first a slightly degraded version and more recently the fully functional version of the system has been available for civilian use.

The heart of the system is 21 active satellites (as well as a few inactive satellites that are used as spares) in earth orbit. In effect, these satellites continuously transmit radio signals containing the time. GPS receivers pick up these signals, and compare their times to the current time; this allows the receiver to compute how far it is from a satellite. For example, if the satellite signal indicates the time as 10:00 A.M. and the receiver records the current time as 1/20 of a second past 10:00 A.M., it can be deduced that the radio signal took 1/20 of a second to travel from the satellite to the receiver. Since radio beams travel at the speed of light (186,000 miles per second), it is easy to compute that the receiver in our simple example is 1/20 H 186,000 = 9,300 miles from the satellite. Since the receiver can pick up signals from multiple satellites simultaneously, it can triangulate its position very easily.

GPS has come to play a huge role in geomatics. It is used to locate ground-control points for remote-sensing applications. It is used to establish benchmarks that are then used in conventional, theodolite-and-chain surveys. GPS is also used to conduct field surveys by itself. For example, it has become a common practice to place a GPS receiver in a backpack and then travel a series of roads or trails. The receiver automatically records the location of many points along these routes. By plotting these points on a map and connecting them with lines, accurate maps of the routes can be produced (this plotting and connecting of points can be accomplished automatically via computer software).

The most accurate GPS fixes are produced using a technique called *differential correction*, or DC. DC involves placing a GPS receiver (called a *basestation*) at a known location. This basestation continuously computes its GPS location and compares this computed location to its known location; the difference between these two locations is called an *offset*. By modifying locations recorded by a second GPS receiver with the offsets computed by a nearby basestation, extremely accurate corrected fixes can be computed.

GPS receivers vary quite a bit in price; in general, more expensive units are more accurate (they receive signals from more satellites) than less expensive units and can store more data. More expensive units also come with more sophisticated software designed to process the data gathered by the receiver.

Box 4. Analyzing Spatial Data Using Geographic Information Systems (GISs)

Remote sensing and GPS are the primary tools used by geomaticists to acquire spatial data; GIS is the primary tool used to archive, analyze, and communicate spatial data. A typical geomatics project uses two, and often all three, of these technologies.

GIS software is often described as being a "toolbox" that provides users with a multitude of "tools" for manipulating spatial data. This is a fairly accurate analogy. Fundamentally, it is not very different from word proccesors, spreadsheets, or other familiar software. Word processors contain assortments of tools designed to help users create and manipulate documents; spreadsheeting software contains a variety of tools designed to work with spreadsheets; and GIS software contains a large number of operators ("tools") that can be applied to spatial data sets.

Spatial data sets are the basic unit of data handled by GIS software; they are analogous to a word processor's documents or the spreadsheets used by spreadsheeting software. There are two types of data sets used in GIS, both with advantages and disadvantages. In *very* general terms, *raster* data sets are commonly used in spatial model building (because there are more operators that can be applied to raster data sets than there are for the alternative *vector* data sets) and vector coverages are used in data archiving (because vector data sets have a more developed capability for handling information describing the characteristics of geographic locations) and highly detailed map production (raster data sets can be used to produce maps, but raster maps tend to be less attractive than vector maps).

Modern GIS software will typically have hundreds or even thousands of operators that can be applied to spatial data sets. Many of these operators perform fairly mundane tasks. For example, many operators are used for data entry and editing, conversions from raster-to-vector and vice versa, and so on. Other operators are concerned solely with producing maps. There are software packages that are limited to only these types of operators (i.e., coverage entry/editing/management and cartographic functions). These software packages are sometimes mistakenly called GISs; they are more correctly called *desktop mapping* or *automated mapping* software.

In addition to basic coverage manipulation and cartographic operators, a true GIS contains data analysis operators. Individually, each of these operators is quite simple, such as the *buffer* operator (which creates a buffer of some predefined width around objects in a coverage) or the *slopes* operator (which constructs a data set showing slopes from a data set containing elevations). The trick to effectively using GIS software is knowing how to link these individual data-analysis operators in sequence to accomplish tasks for which no single operator will suffice. This requires familiarity with how the operators work, and experience—once you have found ways to perform a few multi-operator analyses, future analyses are much easier.

A Win-win Situation:
Managing to Protect Brown Bears Yields High Wildlife-viewer Satisfaction at McNeil River State Game Sanctuary

Colleen Matt and Larry Aumiller

Introduction

BEARS HAVE THE RIGHT-OF-WAY at McNeil River State Game Sanctuary. Readers might assume this implies that visitors must stay completely away from bears, that viewing is from a distance, that a spotting scope is necessary to catch glimpses of bears as they go about fishing, grazing, nursing young, playing, fighting, or mating. However, the absolute opposite is true *and* the right-of-way principle still applies. Visitors at McNeil are often close enough to smell bears as they pass by, to see the scales of the salmon they have just caught, and to hear the sounds of a female suckling her cubs. The world-renowned bear-viewing program at McNeil River in Alaska illustrates the compatibility of two mandates common to wildlife-management agencies: resource protection and recreational opportunities.

McMullin (1993) affirms that effective agencies *must* stick to what they do best: protect wildlife resources. This is as true for planning wildlife-viewing programs as it is for establishing harvest strategies. Indeed, it is often the legal mandate under which wildlife managers operate. However, this must not be interpreted to mean that recreational use and conservation are antithetical. In fact the future of wildlife conservation may rest on people's ability to have direct experiences with wildlife.

To be viable, conservation must enjoy the support of the public. For many people, understanding and valuing the natural world is not an intellectual exercise. Rather it is the outgrowth of their personal interactions with wildlife and wild lands. It is based on their immersion in the natural world and the spiritual, emotional, physical, or other feelings they derive from those experiences. In short, it is based on the benefits they personally receive from their wildlife-recreation experiences.

The Synergy of Wildlife Protection and Wildlife Viewing

Some wildlife managers have rejected wildlife-viewing development because they may wrongly assume that wildlife-viewing programs mean more facilities, trails, and roads, resulting in a loss of habitat. They may equate success of wildlife-viewing programs with numbers of visitors and want to avoid the interactions and conflicts between humans and wildlife and humans and humans associated with increased visitation. Using models such as Experience-based Management (EBM) to articulate the benefits of atypical or nontraditional wildlife-viewing programs can dispel these assumptions. EBM can help managers provide more and better opportunities for wildlife watchers while holding fast to their resource-first mission. The viewing program at McNeil is an example of how this balance can be struck.

McNeil River State Game Sanctuary was originally just what its name implies—a safe haven from human development and hunting. It was created by the Alaska State Legislature in 1967 to accomplish the following:

(1) "provide permanent protection for brown bear and other fish and wildlife populations and their habitats, so that these resources may be preserved for scientific, aesthetic, and educational purposes;

(2) manage human use and activities in a way that is compatible with (1) of this subsection and to maintain and enhance the unique bear viewing opportunities within the sanctuary; and

(3) provide opportunities that are compatible with (1) of this subsection for wildlife viewing, fisheries enhancement, and fishing, for temporary safe anchorage, and for other activities" (Alaska Statutes).

Managers of McNeil have interpreted the syntax of this legislative mandate to mean that wildlife protection dominates—that while human use of the sanctuary is important, it is secondary in purpose. The viewing program was consequently developed to affect bear population and habitat as little as possible.

The very efforts that were originally designed to impact bears as little as possible have benefited visitors more than ever imagined. From the first moment visitors step down from the floatplane upon their arrival at McNeil, it is obvious that they have entered the bears' world and that this is a place that humans do not dominate. This complete immersion in the bears' natural world has elicited the highest of praise from visitors and has set McNeil apart from other wildlife-viewing experiences. For more than thirty years, the viewing program at McNeil has been highly successful by every measurement. The number of bears using the sanctuary has more than doubled. It has been the subject of many international stories, articles, books, and films; the number of people requesting permits to visit McNeil remains high. McNeil has been used as a model for developing viewing programs involving large predators, and it may ultimately serve to enhance bears' public image and the long-term viability of brown-bear populations elsewhere in the world.

While McNeil evolved out of a desire to put bears first, the success of the visitor program holds several lessons for developing other wildlife-viewing programs. The most palpable reward at McNeil has been a uniquely personal and high-quality experience for visitors. Any wildlife program can focus on enhancing visitor experience and reap dividends similar to those found at McNeil.

Continuity and Commitment of Personnel

One of the most important reasons for the success of McNeil is found in the continuity, commitment, and cooperative work style of the staff. All too often in public lands, the only personnel who have daily contact with visitors and wildlife are seasonal workers "in the field." Yet, typically, field staff are not consulted about management issues beyond day-to-day routines. Seasonal employees run the visitor programs, yet agencies display a lack of attention to the needs of these employees. They frequently lack health or retirement benefits, or other material encouragement to sustain their commitment. Lack of acknowledgement and material encouragement may result in a high turnover of the very people who have direct experience with on-site habitat, wildlife, and visitors. At Alaska Department of Fish and Game (ADF&G), most seasonal employees are considered permanent and accrue health and retirement benefits during the period of their employment. McNeil River seasonal employees tend to return year after year. They are trusted spokespersons and are often consulted for their expertise by bear biologists and public media.

There is also a tendency for off-site managers to lack a personal connection to viewing programs and be perceived as distant from field staff. Each season ADF&G recruits administrative, research, and management staff from other programs to substitute for field staff on temporary leave. The substitutes literally step into the boots of the field staff and are trained to perform almost all of the field duties. As a result, many department personnel have the same experiences as first-time visitors while gaining professional expertise. These personnel are enriched by their experiences at McNeil. This empowerment of continuity of personnel has stimulated the creativity and responsiveness in management of McNeil. ADF&G

has traditionally had very little turnover of both off-site and field staff and the longevity of program managers has made growth and change of management philosophy possible. ADF&G's decentralized decision-making culture contributes to this as well.

The field staff developed the unique relationships between bears and humans at the sanctuary. Although bear hunting was eliminated prior to the creation of the sanctuary, there was no direct management of either bears or visitors prior to 1973. Alaskans and other visitors were generally quite fearful of brown bears and unguided visitors often carried firearms. Managers eventually became concerned about incidents involving armed visitors who threatened to remove or displace bears. In 1967 and again in 1970 photographers killed aggressive bears who had become agitated by their stealthy approach. Visitors were camping near prime bear-fishing sites at the river. Several bears had become food conditioned due to visitors' careless treatment of food and garbage. The number of bears using the sanctuary was at an all-time low.

The department responded to the growing number of conflicts at the sanctuary by instigating a permit system in 1973, and providing field staff in 1975. The prevailing belief in 1975 was that bears were unpredictable carnivores that would, given a chance, approach humans with predatory intent. Field staff were originally charged with protecting bears from people and people from bears, and often used heavy-handed deterrents like firecrackers and rubber bullets to aversively condition bears to stay away from humans and their belongings.

However, with repeated seasons at McNeil, the field staff gained confidence in their abilities to "read" bear behavior. They recognized that the behavior of wild bears is predictable and malleable. They also began to understand the type of visitors who came to McNeil. People willing to make the effort necessary to get to McNeil were different from people casually viewing wildlife by the roadside. They were people who would willingly restrict their own behavior and movements to create the conditions necessary to safely encourage close proximity of bears. As human behavior became more predictable to the bears, the bears became habituated and learned to see humans as neither threat nor attractant. Visitors were thrilled by the tolerance shown by habituated bears as they continued to engage in their daily activities with close human presence. Staff found that these large carnivores could learn to respect human boundaries with the use of rather subtle deterrents. Young bears are now discouraged from wandering into the campground by field staff merely yelling, banging a pot, or clapping their hands.

Clear Objectives

The field staff used their experiences to identify the actions that brought about the least impact to bears and the greatest satisfaction to visitors. These actions were guided by clear objectives. While the objectives for McNeil were the result of an evolutionary process in management philosophy, a structured planning model like EBM can help managers express and establish clear management objectives early in the development of a viewing program.

The success of the McNeil bear-viewing program is largely due to the achievement of three objectives: (1) to avoid adversely impacting bears by limiting the number of visitors and controlling their behavior; (2) to encourage safe, close contact with bears through their habituation; and (3) to enhance viewer experiences through close but unobtrusive contact with bears once objectives (1) and (2) have been met. "Unobtrusive contact" means placing visitors near areas of interest to bears (e.g., next to available salmon), while causing minimal displacement of natural activity.

The first objective—limiting visitor numbers and behavior— is necessary to achieve the second objective, habituation. Habituation is defined as the reduction in the frequency or strength of response following repeated exposure to an inconsequential stimulus (Jope 1985; Gilbert 1989). The stimulus at McNeil is proximity to people in a nonthreatening interaction. Brown-bear habituation would not be possible without limiting the numbers of visitors and guiding visitors' interactions with bears .

Bears are less stressed and habituate more easily to smaller groups of people. Visitor numbers are controlled via permits issued by lottery. Permits are limited to ten per day at bear-viewing areas. Accumulated experience has shown that a group of seven to ten people is optimal to achieve the goal of minimally impacting bears.

Achieving Habituation through Visitor Cooperation

McNeil demonstrates that managers of wildlife-viewing programs need to consider the characteristics of both the wildlife and the visitors involved. Highly motivated visitors are eager to participate and present managers with different challenges and opportunities than do casual viewers. They can be educated and recruited to behave in ways that protect wildlife and enhance their own experiences. They are willing to forgo physical comforts to retain natural surroundings.

The McNeil visitor program may have developed in an entirely different direction had it been accessible to less-avid wildlife viewers.

Habituated bears behave neutrally toward humans and allow visitors close views. (Photo by Colleen Matt)

Because McNeil is located in the roadless wilderness of Alaska, it takes money and effort for visitors to come to here. People typically arrive by commercial floatplane service from Homer, Alaska. Prior to the establishment of the permit system and throughout the 1980s, visitors were often wildlife photographers by profession or avocation and had considerable outdoor experience. Advertising was strictly by word of mouth though there was growing media attention through televised nature documentary films. Even today, the primary market (using the survey item described Chapter 5), classify themselves as people who are highly interested in wildlife and enjoy wildlife photography, sketching, or painting (Table 16-1).

The proximity to bears at McNeil is often much closer than visitors expect or have previously experienced. These visitors are willing to suspend their fear of bears and follow the few rigid rules set by the staff. Visitors' cooperation and trust in the staff has been essential to the development of the program. Inappropriate behavior prompted by fear or carelessness could have created a serious incident and halted the progress of the visitor program. It could have led to tighter restrictions on visitors, possibly with physical barriers between humans and bears. Instead a safety record that includes no human or bear injuries for more than thirty years has been established. Managers now have considerable confidence in the ways things have been allowed to evolve at McNeil. A serious incident now would be seen as an anomaly.

Visitors to McNeil are guided and instructed to perform actions that encourage habituation and safety (Aumiller and Matt 1994).

Table 16-1. Wildlife viewing types (adapted from Bright 1998).

Wildlife viewing type	Percent
Type 1: People who are highly interested in wildlife. They take several trips during the year specifically for wildlife viewing, and they enjoy opportunities to study and teach about wildlife.	24.5
Type 2: People who are also highly interested in wildlife. However, they particularly value opportunities to photograph, paint, or sketch wildlife.	48.7
Type 3: People with a general interest in wildlife. They take occasional trips to see wildlife, but mostly they enjoy wildlife viewing when engaged in other activities such as fishing.	24.8
Type 4: People with a slight level of interest in wildlife. They rarely take trips specifically to see wildlife, but enjoy wildlife viewing when it is associated with activities such as fishing.	2.0

• *Predictable and consistent interactions.* Humans are instructed to be as predictable and consistent as possible. The main camp location has remained in the same area, two miles from the viewing site, for over thirty years and camping is not allowed elsewhere in the sanctuary. Visitors follow the same trails, use the same viewing sites, and are limited to viewing hours between 10 A.M. and 8 P.M. The camp, trails, and viewing areas are detectable to bears by both scent and sight. Hence, bears make their own choice about their proximity to humans; wary bears avoid areas frequented by humans and their avoidance is proportional to their level of wariness. As a result, people are most likely to interact with bears that are most comfortable with humans. Groups of visitors are managed so that bears perceive little variation in human behavior. Staff find that inconsistent behavior in interactions causes bears to avoid humans.
• *Nonapproach.* In all areas of the sanctuary, with the exception of the campground and personal boundaries, staff try to allow bears to choose their proximity to humans. Personal boundaries, in this context, refer to the distances at which a guide takes action to discourage a bear from closer approach. When bears make the choice to come near to humans they generally show little or no signs of stress. However, when humans approach bears they may induce high stress levels in the bears.
• *Proximity to food sources.* The process of habituation is hastened when bears have to be near humans in order to gain access to food. However, for safety reasons, it is equally important that humans do not block access to or put themselves directly in the midst of the desired food sources. The viewing pad at McNeil River Falls is

an example of this principle. The viewing site is near the river yet it does not impede bears' approach to or use of the falls since there are alternate routes.

• *Calm demeanor.* Staff are careful to avoid loud noises and fast or exaggerated body movements when bears are close to people. Staff find that slow movements and low-level talking cause less stress in nearby bears than loud noises and quick, jerky movements. A highly habituated bear might pass a group of visitors within 2 m if that group is relatively calm and quiet. However, the same bear might be alarmed by a visitor's loud cry and wild gesticulation or movement and swerve to avoid the group.

• *Adaptation to bears' stress level.* In all interactions with bears, staff tailor their responses to the level of stress observed in nearby bears. For example, during the daily trek through the sanctuary, guided groups occasionally approach bears in order to pass them to get to the viewing sites. As they do so, staff watch each bear closely for signs of stress or an escalating pattern of stressed behavior. If stress is apparent, the staff stop or change course to avoid the bear.

These actions allow habituated bears to routinely pass or sit within 8 m of visitors without showing signs of stress. Other partially habituated bears may choose to stay further away. For example, some bears fish on the opposite bank of our viewing site. In recent years, field staff have counted over ninetyfive adult and subadult habituated and partially habituated bears. Wary bears characteristically flee from human encounters and are seen from a distance if at all. The interaction between humans and bears is dynamic; some wary or partially habituated bears may become more tolerant of humans and eventually become highly habituated.

Close and Extended Interactions between Wildlife and People

Closeness is one of the keys to the success of viewing programs for highly involved and creative viewers. These types of viewers also value extended viewing opportunities. The drive-through model of wildlife viewing found in many national parks is not adequate to serve viewers interested in photographing, painting, studying, and teaching about wildlife. Close and extended access to wildlife is primary. Access to and interaction with staff is also important.

The quality of visitors' experiences at McNeil is enhanced by the wilderness setting, their sense of guidance and safety, and abundant personal contact with the staff (Table 16-2). ADF&G has been deliberate about maintaining the wilderness character of the setting and minimizing services at McNeil. Each day one or two sanctuary personnel take a group of ten visitors to watch bears. Daily trips generally last five to nine hours and the groups usually spend the viewing period at a single site. Permits allow visitors to view bears up to four days, with three days as the average number of viewing days.

Visitors to McNeil are not always supervised in their movements and actions. However, they are educated upon arrival at the sanctuary about proper behavior near bears and consequently they tend to behave in ways that contribute to bear habituation even when unaccompanied by sanctuary staff. Because the ratio of staff to visitors is high, visitors have many opportunities to interact and speak to experts about bear natural history and behavior. The staff remain on call when they are not guiding excursions to viewing sites. The lack of a formal natural-history interpretive program, intrusive signs, or trails at the sanctuary creates an atmosphere of adventure and discovery. Visitors feel part of a unique experience.

Ongoing Evaluation and Refinement

Refinement of goals and methods are two reasons why evaluation can improve recreational programs. Evaluation can also be a tool to justify the structure and cost of a viewing program. If, for example, a viewing program limits the number of users at a site, agencies may be challenged to defend the program to lawmakers and to the general public. The public and commercial services may expect government-supported recreational programs to accommodate everyone, even though such action often drives down the quality of recreationists' experiences. Several such challenges have been directed at McNeil through the years, and evaluation efforts have helped support and defend the services offered there.

Program Structure Evaluation. In 1997 managers at the department wanted to make changes to the McNeil viewing program so that it could "pay for itself," while maintaining the values that have made the program successful. They needed an instrument to measure applicants' desires and perceptions. A survey (Bright 1998) addressed applicant perceptions of current and proposed permit systems, and applicant experiences at McNeil and other bear-viewing sites.

Before making substantive changes, managers wanted deeper discussions with users of the sanctuary. They formed a stakeholder group composed of citizens with a direct interest in the operation of the sanctuary. After discussing the issues and Bright's survey results, the stakeholders made several recommendations for small changes, but otherwise endorsed the existing visitor-permit system. It is interesting to note that both the surveyed applicants and the

Table 16-2. Importance of experience attributes for a trip to McNeil River State Game Sanctuary (adapted from Bright 1998).

	A	B	C	D	E	Mean
Having the bears behave as if people were not around	0.2	1.1	3.9	26.5	68.4	4.62
Not competing with others for good viewing locations	0.5	0.9	6.2	27.9	64.5	4.55
Watching bears interact (e.g., playing, fighting, courting, nursing)	0.2	0.6	5.4	33.9	60.0	4.53
Learning more about bears and their behavior	0.4	1.5	11.2	38.1	48.9	4.34
Watching bears catch fish	0.2	2.2	15.6	38.4	43.7	4.23
Sharing the McNeil camp and viewing areas with only a small number of other people	2.6	4.0	15.0	33.7	44.7	4.14
Being able to spend three or four days at the viewing areas	2.7	8.4	18.6	29.3	41.0	3.98
Personal conversations with the guides about bears and bear behavior	1.7	6.6	20.0	37.0	34.6	3.96
Seeing a large number of bears at the viewing areas	1.0	4.6	28.9	33.1	32.4	3.91
The reassurance and security provided by the guides	2.7	9.1	20.7	30.8	36.6	3.90
Enjoying the scenery	1.6	7.9	22.5	36.6	31.4	3.88
Enjoying the primitive setting of the McNeil camp and viewing areas	4.1	8.5	19.2	33.8	34.4	3.86
Having bears come within just a few yards of people at the viewing areas	6.7	8.8	23.1	26.5	34.9	3.74
Identifying and learning about individual bears	3.0	8.1	29.7	31.1	28.0	3.73
Viewing other wildlife in the area	2.4	10.1	30.9	33.9	22.7	3.64
Watching bears away from the viewing pad	5.4	11.7	32.5	31.3	19.1	3.47
Spending several days camping near bears	12.2	19.0	29.9	23.9	15.1	3.11
The potential for encountering bears on trails or beaches	14.3	14.2	33.4	23.8	14.3	3.10
Developing friendships with the visitors and guides	14.4	25.6	33.6	18.2	8.3	2.80

Numbers are percent of respondents. A = Not at all important. B = Slightly important. C = Moderately important. D = Very important. E = Extremely important

Newly arrived visitors are briefed about rules for their behavior around bears while visiting McNeil River. (Photo by Colleen Matt)

stakeholders rejected suggestions to reduce the four-day permit length. The stakeholders went further and recommended that managers establish six-day permits for highly motivated viewers (Slemons 1999)

Economic Valuation. Economic valuation gives managers a tool for assessing and prioritizing the highest use of potential wildlife-viewing sites. The value of wildlife watching can go far beyond direct on-site use. In most existing and potential viewing areas, multiple uses such as tourism and fisheries compete with high-involvement wildlife viewing. Economic valuation of wildlife watching may be necessary to justify management for viewing as the highest and best use of an area. The methods used by Swanson et al. (1992) should be applied to evaluate and mitigate various conflicting uses.

In addition to comparing the value of wildlife watching to other resource values, economic valuation can help set fees for services. Two studies have looked at willingness to pay (Clayton and Mendelsohn 1993; Bright 1998). McNeil permit and application fees have been adjusted based on the results of these studies.

Evaluation of visitor perceptions. Articulation of the benefits of wildlife-viewing programs is often important for their long-term survival. Bright's (1998) survey of applicant experiences showed that visitors to the sanctuary are overwhelmingly satisfied with their trip to view bears. Respondents were asked to indicate how important specific experiences were for them when they visited McNeil River (Table 16-2). The following attributes derived from the sanctuary's goal to limit and control visitors' behavior: lack of competition for viewing sites; the opportunity to learn more about bears; being with only a small number of people; three- or four-day viewing permits; opportunity for individual conversations with guides; security provided by guides; primitive camping conditions; opportunity to learn about individual bears; and opportunity to develop friendships with guides and other visitors. The following attributes derive from the viewing program goal to habituate bears to the proximity of humans: opportunities to view bears that are unconcerned by the proximity of people; watching bears interact; watching bears catch fish; the opportunity to see a large number of bears; close encounters with bears; watching and encountering bears in all areas of the sanctuary; and camping around bears. These attributes, along with viewers' comments about the high quality of their experiences, illustrate the compatibility of resource protection with recreational service.

Willingness to Try Something Different

Good management is creative. If managers are creative, almost any species of wildlife can be observed with the same enthusiasm and dedication demonstrated by the viewers at McNeil River. As noted in Chapter 5, application of EBM depends upon the creativity of managers and their ability to articulate a full range of wildlife-viewing opportunities and their benefits. Innovative programs might offer extended viewing opportunities, close proximity to wildlife, minimal crowding, and personal attention from knowledgeable guides. Agencies must be willing to adopt methods that create the conditions in which these benefits occur. These methods could include high staff/visitor ratios, limitations to the numbers and behavior of visitors, and low staff turnover.

Today's wildlife managers are challenged to provide opportunities for a growing segment of viewers seeking the same sorts of creative and intimate wildlife-viewing experiences as can be viewed nightly on television documentaries. This challenge is intensified by a diminishing inventory of wildlife and habitat. It may seem impossible to balance increasing viewer demands with managers' mandate to protect the supply from further encroachment of human contact. Yet viewing programs at such places as McNeil meet the supply-and-demand challenges while fulfilling the agency's mission to protect the concentration of bears.

Summary Points

- The McNeil River State Game Sanctuary viewing program was developed to protect wildlife and provide high-quality experiences for visitors.

• Long-term, highly committed, expert field personnel maintain the quality of the McNeil River viewing experience. Alaska Department of Fish and Game supports seasonal field personnel by providing employment benefits and seeking their input in managerial decisions.

• Clear management objectives provide better protection for wildlife, safer interactions between wildlife and visitors, and enhanced viewing experiences.

• Closeness to wildlife is important for highly involved and creative wildlife viewers. At McNeil River visitors willingly perform actions that encourage habituation of bears that leads to close contact.

• Periodic review and evaluation has helped maintain the success of the McNeil River viewing program.

Literature Cited

Alaska Department of Fish and Game (1996). *McNeil River State Game Refuge and State Game Sanctuary Management Plan*. Anchorage, AK.

5 Alaska Statutes. §§ 16.20.092 (1998).

Aumiller, L., and C. Matt (1994). "Management of McNeil River State Game Sanctuary for viewing brown bears." In J. J. Claar and P. Schullery (Eds.), *International Conference of Bear Research and Management* 9(1), 51-61.

Bright, A. D. (1998). *Evaluation of the Bear Viewing Program at McNeil River State Game Sanctuary: Applicant Experiences*. Pullman, WA: Washington State University.

Clayton, C., and R. Mendelsohn (1993). "The value of watchable wildlife: a case study of McNeil River." *Journal of Environmental Management*, 39, 101-6.

Gilbert, B. K. (1989). "Behavioral plasticity and bear-human conflicts." In M. Bromley (Ed.), *Bear-People Conflicts: Proceedings of a Symposium on Management Strategies*. Yellowknife, NWT, Canada: Northwest Territories Renewable Resources.

Jope, K. L. M. (1985). "Implications of grizzly bear habituation to hikers." *Wildlife Society Bulletin*, 13, 32-37.

McMullin, S. L. (1993). "Characteristics and strategies of effective state fish and wildlife agencies." *Transactions of the 58th North American Wildlife and Natural Resources Conference*, 58, 206-10.

Slemons, J. (1999). *Final Report: McNeil Task Force*. Anchorage, AK: Alaska Department of Fish and Game.

Swanson, C. S., J. C. Bergstrom, and J. N. Trent (1992). "The total value of the McNeil River State Game Sanctuary in valuing wildlife resources in Alaska." In G. L. Person et al. (Eds.), *Valuing Wildlife Resources in Alaska*. Boulder, CO: Westview Press.

About the Authors

Larry Aumiller is Sanctuary Manager at the Alaska Department of Fish and Game

Steven J. Bissell is a consultant at Responsive Management, Harrisonburg, Virginia

Alan Bright is Assistant Professor in the Department of Natural Resource Recreation and Tourism at Colorado State University, Fort Collins

Lisa C. Chase is a Natural Resources Specialist with University of Vermont Extension in Brattleboro, Vermont

Denis J. Dean is Associate Professor of Geospatial Science in the Remote Sensing/GIS Program, Department of Forest Sciences, Colorado State University, Fort Collins

Daniel J. Decker is Professor in the Department of Natural Resources, and an Associate Dean in the College of Agriculture and Life Sciences, Cornell University, Ithaca, New York

Maureen P. Donnelly is Associate Professor in the Department of Natural Resource Recreation and Tourism, Human Dimensions in Natural Resources Unit, Colorado State University, Fort Collins

Mark Damian Duda is a consultant at Responsive Management, Harrisonburg, Virginia

Peter J. Fix at the time of writing was Research Associate in the Department of Natural Resource Recreation and Tourism, Human Dimensions in Natural Resources Unit, Colorado State University, Fort Collins. He is now Assistant Professor in the Department of Resources Management, University of Alaska, Fairbanks

David C. Fulton is Assistant Unit Leader in the Minnesota Cooperative Fish and Wildlife Research Unit, USGS-BRD, University of Minnesota, St. Paul

R. Bruce Gill at the time of writing was Mammal Research Leader with the Colorado Division of Wildlife

Sandra Jonker is a Research Assistant in the Department of Natural Resource Recreation and Tourism, Human Dimensions in Natural Resources Unit, Colorado State University, Fort Collins

T. Bruce Lauber is Research Associate in the Department of Natural Resources, Cornell University, Ithaca, New York

John B. Loomis is Professor in the Department of Agricultural and Resource Economics, Colorado State University, Fort Collins

Michael J. Manfredo is Chair of the Department of Natural Resource Recreation and Tourism, Human Dimensions in Natural Resources Unit, Colorado State University, Fort Collins

Colleen Matt is Lands & Public Services Coordinator at the Alaska Department of Fish and Game

Cynthia L. Pierce is Professor in the Department of Resource Recreation and Tourism at the University of Idaho

Bo Shelby is Professor in the Department of Forest Resources, College of Forestry, Oregon State University, Corvallis

Jerry J. Vaske is Professor in the Department of Natural Resource Recreation and Tourism, Human Dimensions in Natural Resources Unit, Colorado State University, Fort Collins

Doug Whittaker is a researcher/consultant with Three Rivers Research, Anchorage, Alaska

Daniel J. Witter is Policy Coordination Chief at the Missouri Department of Conservation, Jefferson City

Index

Economic-impact analysis, 256-59
Ecotourism: defined, 61-62, 198-200; importance of wildlife to, 201-2; local involvement and, 210-12; marketing and, 313; participation in, 200-201; partnerships and, 212-14; wildlife-based, 202-14
Education and communication, 138, 177-80, 183-85, 187-88, 277-303, 311, 317; effectiveness of, 292-93; *See also* messages
Elasticity. *See* price elasticity of demand
Endangered Species Act, 35
Equality in allocation, 111
Equity in allocation, 111
Evaluation: of education campaigns, 299-303. *See also* planning for recreation
Expectancy theory, 72
Expenditures for outdoor recreation programs, 18, 29
Experience-based management (EBM), 6-7, 21, 44-46, 62-64, 70-84: economic analysis and, 254; ecotourism and, 197, 204; geomatics and, 325-26; in decision-making, 172-73; in education, 278; in planning frameworks, 99-101, 105-119; in public involvement, 132-39; marketing and, 309, 317, 319, 322; wildlife protection and, 352

F

Fees: license, 36-38; user, 189-90
Fire. *See* habitat manipulation
Fishing: participation. *See* participation in wildlife-related recreation
Flow theory, 59-60
Funding. *See* wildlife viewing funding

G

Galapagos Islands, 208
Game species management, 35
Geographic information systems (GIS), 326, 330-31, 336, 339, 342
Geomatics, 326-43; and defining outcomes, 333-36; and development of management plans, 336-39; and monitoring, 339-41; defined, 326; physical inventories, 327-31; social and managerial inventories, 331-33
Goals in planning for recreation. *See* planning for recreation
Grazing. *See* habitat manipulation
Gross Domestic Product, 28-29
Guides, 206-10

H

Habitat manipulation, 177, 225-29, 232
Habituation, 176, 239-41, 354-58
Human dimensions of fish and wildlife management, 21
Human needs in outdoor recreation, 2, 9, 32-33, 44, 59